The Living State

The Living State

EDITED BY
R. K. Mishra
Professor & Head
Department of Biophysics
All India Institute of Medical Sciences
New Delhi, India

A HALSTED PRESS BOOK

JOHN WILEY & SONS
New York   Chichester   Brisbane   Toronto   Singapore

Copyright © 1984, WILEY EASTERN LIMITED
New Delhi

Published in the Western Hemisphere by
Halsted Press, A Division of
John Wiley & Sons, Inc., New York

Library of Congress Cataloging in Publication Data

Main entry under title:

The Living State.

"A Halsted Press Book".

1. Life (Biology)—Addresses, essays, lectures.
2. Biology—Philosophy—Addresses, essays, lectures.
I. Mishra, R.K., Professor.
QH501.L57 1984   577   84-555

ISBN 0-470-20027-8

Printed in India at Prabhat Press, Meerut.

Preface

The phenomena exhibited by the living systems in general which qualify them to be living are so puzzling that one is apt to consider them beyond existing laws of physics. We may then need new physics and mathematics to understand them. Or we may consider these sciences in the last quarter of the century to be adequate for the task. Rather, one must examine them as possible consequences of subtilities in the structure and energy flows in these systems. In short, they could be epiphenomena of an undescribed state of matter. Thus, one may pass beyond confines of physiology of the organs, tissues, cells, organelles, molecules to the properties resident in new principles of structuration of matter. Gilbert Ling used the word Living State in 1966 but he confined himself to only one major principle, namely, the association and induction as the basis of several biophysiological phenomena. But the matters have not stayed at that level and while now biological phenomena like long distance correlation, solitonic waves, and superconductivity have been suspected in living systems, powerful tools of both theoretical analysis and experimental verification have come in the hands of investigators in biology. For example, a whole range of phenomenology can be explored in thermodynamics of far from equilibrium situation, wherein formation of structures, chemical oscillations and morphogenesis all seem to be possible. Another view is furnished by cooperative interactions described in the field of 'synergetics'. Physics of condensed matter, topology and imperfection sensitivity of perfect structures are all illuminating sources of knowledge in the field. We are now much ahead of the theory of finite automata and self organization for understanding living organisms and it is possible now to conceive of a new "state" of matter: the Living State; a very special case and the most important one of self organization in the Universe so far as mankind is concerned. It was therefore very fortunate that the International Seminar on the Living State could be organised in New Delhi in December 1981 to which most of the leaders of the field contributed personally as shall be seen in this volume, which comprises of fuller accounts of the presentations made of papers in the Seminar. While George Wald inaugurated the Seminar with an impassioned appeal for uniqueness of life and its value the key note was furnished by Ilya Prigogine, in which concepts of instability and symmetry breaking are recalled. Prigogine later deals with time, life and special circumstances in the far from equilibrium. This is followed by statements on the actual state of matter by a modern version of association-induction hypothesis by Gilbert Ling. Mishra later reviews the consequence of the choice between atoms and bonds for the fundamental role played by them in basic metastability and functionality of the living organisms. Later some other aspects of this problem are examined. Clegg draws attention to a new phase of Living State in which living systems are maintained at minimum amount of water. Misra and Prigogine describe the actual function of time, and the essential constructive role of irreversibility in dynamical systems in a very fundamental manner. It is probable that the Living State has many properties that might seem to arise by complexity of the systems enjoined by a multiunit character. Perhaps this is a very necessary aspect for functions of these systems. Babloyantz attends to this problem. Other pioneers represented are Haken for a lucid view of 'Synergetics' and Thom with his catastrophe theory for biological morphogenesis. A significant summary of cooperativity and long range correlations is provided by its founder H Fröhlich. Thus Bose-Einstein condensation and role of polar modes have been examined. These ideas are followed by contribution on the same theme by Bhaumik *et al.*, while an experimental approach for the same objective is outlined by Pohl *et al.* Solitons, i.e. non dissipative waves might have strong relevance in biology for explaining repetitive phenomena like nerve impulse or muscle contraction. A lucid account of the Davydov solitons which form an alternate theory of muscle contraction is furnished by Davydov himself. Morowitz gives a somewhat deterministic account of evolution as opposed to the Prigoginian probabilistic view. Hans Kuhn gives a detailed account of a model of emergence of living matter involving a model of primordial translation machinery for information. He emphasises the formation of an adaptable device possessing the self replication and translational machinery and prebiotic earth where energy rich compounds would pose no problem. Das refers to the uncanny role of reverse transcriptase in evolution. Sundaram and Viswanadhan proposed a model for quantitative three dimensional features of proteins utilizing the concepts of a supersource and self information which permit one to compare various proteins. Goodwin examines morphogenesis by a 'field' approach to biological form and phase transformations. S. Kauffman confines to the concept of self organization in the dynamical order relevant for evolution of regulatory processes for the gene.

Nanjundiah and Lokeshwar examine the postulate of scale-invariance for patterning of embryo and the theory of regulation in the cells and conclude that there must be a system of intercellular communication for this purpose. Stuart Newman examines a dynamical model and suggests that fibrino-actin concentration provides a spatial prepattern for skeletal differentiation of the limb with the help of cyclic AMP and of developmentally regulated DNA-binding protein. Parikh and Pratap analyse a neuronal network model for acquisition and storage of memory and its decay.

After these papers dealing with basic theory of structure, thermodynamics and phase changes attention is focussed on some problems that are relevant from the point of view of a complete theory for the Living State. Thus E. Neumann considers the theory of metastable membrane states in relation to electrical signals and Bhatia describes the visuoperceptual space and the psychophysics of vision. A fundamental problem in control and regulation of cell function is posed by polymorphism and microheterogeneity of DNA which is examined quantum chemically by Pullman *et al*. Further doubt on determinism as the only mechanism of the Living State seem to be engrained in the work of Saran and Dhingra in quantum chemical investigation of intermediates in the photoreaction of bacteriorhodopsin, and of Phadke *et al* in the anomalous pK values of histidine in proteins.

As already stated the phenomenology associated in the word "consciousness" cannot be ignored in any description of the Living State. Quite naturally this is emphasized by George Wald. Sudarshan deals with superposition, coherence and 'choice' in relation to the quantum theory and a complete cosmology. Sinha deals with universe on information theoretic relationship between initial state of the universe and the information content of the Living State. Lastly Wojciechowski in a detailed article reveals the influence of knowledge in shaping and influencing human behaviour individually and in groups.

Grateful thanks are due to Prof. Abdus Salam and L. Bartoci of International Center for Theoretical Physics, Trieste for giving the heart warming initial financial support which catalysed the holding of this Seminar. Particular thanks are also due to Prof. M.G.K. Menon, Secretary, Department of Science and Technology for his constant encouragement, advise and support. Prof. Eigen was specially active in supporting our search for funds for this Seminar. And additional encouragement was provided by A. Szent Gyorgi, Britton Chance, David Green, Linus Pauling, and late Eric Jantsch.

We acknowledge our gratitude for financial support to the Silver Jubilee Fund of the All India Institute of Medical Sciences, New Delhi, Indian Council of Medical Research, Council of Scientific and Industrial Research, Indian National Science Academy and National Academy of Sciences, Dept. of Defence, the ISCON, The British Council, the Govts. of Austria, Belgium, Canada, France, Deutsche Forschungsgemeinschaft of West Germany, the Govt. of USSR, The USSR Academy of Sciences,

National Science Foundation of USA, the COSTED, and most importantly the Department of Science and Technology. The ITDC, Ministry of Tourism and the Ministry of Information and Broadcasting, Govt. of India, gave considerable help for which we thank them.

R.K. MISHRA

Contents

Preface	v
The Living State—Inaugural Remarks by *George Wald*	1
THE KEYNOTE FOR SELF ORGANIZATION AND THE LIVING STATE	3–10
The Living State by *I. Prigogine*	5
THE LIVING STATE: SOME CONCEPTUAL ASPECTS	11–142
Time, Life and Entropy by *I. Prigogine*	13
The Living State According to the Association-Induction Hypothesis by *Gilbert N. Ling*	21
The Living State by *R.K. Mishra*	35
Metastable States and the Soft Mode in Liquid Crystalline Matter by *R.K. Mishra* and *G.C. Shukla*	64
An Aspect of Self-Organization and Symmetry Breaking in Biological Systems by *G.C. Shukla, K. Bhaumik* and *R.K. Mishra*	70
Cryptobiosis: A Phase of the Living State by *James S. Clegg*	78
Time, Probability and Dynamics by *B. Misra* and *I. Prigogine*	89
Dissipative Structures in Multiple Unit Systems by *A. Babloyantz*	112
Irreversible Thermodynamics of Living State by *R.P. Rastogi*	122
SELF ORGANIZATION: SOME ASPECTS OF BIOLOGICAL PHENOMENOLOGY	143–432
Some Aspects of Synergetics by *H. Haken*	145
Towards a Theory of Biological Morphogenesis by *R. Thom*	151
Emergence of Self-organizing Systems at the Molecular and Cellular Levels Viewed in a Darwinian Framework by *U.N. Singh*	165

Coherent Excitations in Biological Systems by *H. Fröhlich* — 175

Co-operativity and Long-ranged Correlations in Biosystems by *Debajyoti Bhaumik, Kamales Bhaumik* and *Binayak Dutta-Roy* — 184

Variations of the Micro-Dielectrophoresis of Cells During Their Life Cycle by *Herbert A. Pohl, Tim Braden* and *Douglas G. Pohl* — 194

Solitons in Biology and Muscle Contraction by *A.S. Davydov* — 219

Molecular and Global Perspectives on Life by *Harold J. Morowitz* — 225

A Model of the Origin of Life by *Hans Kuhn* — 235

DNA Synthesis by Reverse Transcriptase: Role in Evolution by *M.R. Das* — 254

Protein Information and the Living State by *K. Sundaram* and *V.N. Viswanadhan* — 262

Biological Fields, Morphogenesis and Phase Transitions by *B.C. Goodwin* and *J.L. Rius* — 278

The Self-Organization of Dynamical Order in the Evolution of Metazoan Gene Regulation by *Stuart A. Kauffman* — 291

Biological Patterns and the Problem of Regulation by *Vidyanand Nanjundiah* and *Balakrishna L. Lokeshwar* — 308

Embryonic Development as a Mode of Activity of the Living State: The Case of the Vertebrate Limb by *Stuart A. Newman* — 323

An Evolutionary Model of a Neural Network by *Jitendra C. Parikh* and *Ram Verma Pratap* — 336

Metastable Membrane States and the Generation of Electric Signals in Biology by *Eberhard Neumann* — 340

Biological Implications of Liquid Membrane Phenomena by *R.C. Srivastava, R.P.S. Jakhar, S.B. Bhise, P.R. Marwadi* and *S.S. Mathur* — 342

Ascent of Sap and Translocation by Means of Electrical Double Layers by *M. Amin* — 345

The Visuoperceptual Space by *Balraj Bhatia* — 356

Quantum-Mechanical Exploration of the Polymorphism and Micro-heterogeneity of DNA by *Bernard Pullman, Alberte Pullman* and *Richard Lavery* — 369

PCILO Investigation on the Structure of Intermediate Species of the Photoreaction Cycle of Bacteriorhodopsin by *Anil Saran* and *M.M. Dhingra* — 385

Role of Secondary Interactions in the Anomalous pK Values of Histidine in Proteins by *Ratna S. Phadke*, *R.V. Hosur* and *Girjesh Govil* — 396

Life and Mind in the Universe by *George Wald* — 402

Superpositions, Coherence and Choice: Towards a Physics of Life by *E.C.G. Sudarshan* — 404

The Universe and the Living State: An Information Theoretic Relationship by *K.P. Sinha* — 411

Knowledge and the Evolution of the Living State by *Jerzy A. Wojciechowski* — 417

Bibliography — 433
Author Index — 435
Subject Index — 445

THE LIVING STATE
Inaugural Remarks

George Wald

Emeritus Professor of Biology
Harvard University
Cambridge, Mass, USA

Professor Mishra has prepared for us an enormous feast, not only of science but of things that frequently are divided off from science and called the humanities, disregarding the reality that science is ineluctably one of the greatest of the humanities. Physics has been going through another revolution. We are prospering very well in biology, but there is much left out of the present biological scene. We have had enormous advances and victories in coming down to what seemed to be our ultimate particles of life. There is much that needs to be put together. I like to think that we scientists see our history in an enormous perspective—something like 15 billion years of the universe, 6 billion years of the solar system, 4.7 billion years of the earth, 3 billion years of life on the earth, something like 3 million years of something like human life, some 10,000 years of civilization and then, quite lately, something happened. Mere trivial 200 years ago came the industrial revolution and humanity began to live on what is almost of exponential curve and we have been living exponentially. That exponential curve includes, among its various coordinates, information. We have been living lately in an information explosion and it is as uncomfortable to live with as many other characters of this living exponentially. So, I hope this meeting will make history, a history of trying to put what we know of life and perhaps what we hope of life together.

Though physicists, particularly particle physicists have, now for a long time being living with the problem of consciousness day in and day out, but it is rather curious that biologists tended to avoid that problem, in fact that biologists for one reason or another tend to avoid whatever they cannot measure. I think that this is perhaps inspired by some envy of what biologists imagine physicists to be like. Quite mistakenly, but nevertheless that is the conception they have had of physicists. Perhaps part of the breaking down just now is that so much of biology is beginning to be contributed by physicists. It is not only consciousness that makes biologists a little uncomparable, it is that, it is very curious. I have been interested in the subject of death for certain time, but one day I went to our library, our biology department library at Harvard, and I ran down shelves of general texts looking at the index for the word 'death'. I never found it.

So let me remind you—and I do this with some hesitation, indeed I wondered until a moment ago whether to remind you of it at all. Here we are starting with glorious meetings in which we hope to be treated to all kinds of illuminations but we do so at a point at which that industrial revolution, that began a trivial 200 years ago and for a time promised humanity endless leisure and abundance, has brought us as a species to the brink of self-extinction. I have been living for some time now, two lives, that I think of simplistically as inner and outer. I look upon most of what this meeting will be discussing as part of that inner life, but most of my time has been occupied lately, with that I think of as the outer life, what I think of as survival politics, because that part of reality too, survival, is surely part of biology and it is not only human survival, that is at peril, but much of the rest of life is threatened now as never before in the history of this planet. One can take some comfort in realization, because I think we can realize, that there is a universe-life; where life occurs, given enough time, if conditions exist that make it possible. A great many places. The smallest estimate that I would give any credence to is 10^{18}, so a billion billion— those are American billions. A billion billion places in the already observed universe as a lowest estimate. So one could take some comfort in the realization that if we do anything as stupid as to wipe out our species, as could happen now at any time—all the hardware is already prepared that would probably end by doing that. If we should do anything so phenomenally stupid, a self-made catastrophe, one could take some comfort in the fact that the universe will not miss us. It has a lot, lots else of life, and I am sure, many other places where such contemplative science, art and technology-making animals as we, exist. So sometimes I am asked, "Why do you care" and all I can answer is, I have developed a sentimental attachment to life on the earth. This is our home, these are our people, we are now the custodians of life on earth and it's a big responsibility and those are our children. So it is with this background that we advance our science.

The Keynote for
Self Organization and
The Living State

The Living State

I. Prigogine

Center for Studies in Statistical Mechanics
The University of Texas
Austin, Texas-78712, USA

and

Faculté des Sciences
Université Libre de Bruxélles
1050 Bruxélles, Belgium

I would like first to thank the organizers, and specially Professor Menon and Professor Mishra, for the kind invitation to participate in this conference. This gives me the opportunity to meet old friends and to visit once more this great country whose scientific and cultural traditions I deeply admire.

I believe that the timing of this conference (The Living State of Matter) is indeed excellent. This title brings together two ideas: the concept of life and the concept of matter. We know of course since about thirty years the extraordinary role played by biomolecules in the functioning of life. However, till recently, there seemed to exist a striking contrast between the laws of physics and the evolving world of life. The great French biologist, Jacques Monod, has called living systems "strange objects". Life would be compatible with the laws of physics but in some sense does not follow from these laws.

I believe that today the situation has changed drastically. One may state that the interest in most fields of science is shifting towards the temporal, the multiple and the complex [1]. As often stated, the two great revolutions in physics of this century are quantum mechanics and relativity. Both started "simply" as corrections to classical mechanics, made necessary once the role of the universal constants, c (the velocity of light) and h (Planck's constant) was discovered. Today both have taken an unexpected turn: quantum mechanics deals in its most interesting part with the description of unstable particles and their mutual transformations. Similarly, relativity started as a geometrical theory, but today this theory is mainly associated with the thermal history of the early universe. Thermodynamics is also going through a similar transformation to which I shall come back later.

I am also very happy that the conference takes place in India. The basic problem of Indian philosophy deals with the reconciliation of the pluralism of the forms of reality with the intrinsic unity of nature. I shall not deal in more detail with this aspect which has been beautifully described by Professor Kothari in his introductory lecture. I believe that we come today closer to this dialectical view of nature with its emphasis on both unity and diversity.

Here we are far from the classical western dualism as expressed by Déscartes or Kant. The feature which, more than any other, has led to a more unified view of the physical world, is the rediscovery of time. Time is in itself a dialectical concept. Classical physics emphasized the static aspect of time. Today the complementary aspect of time, dealing with irreversibility, appears more and more as being essential. Let me first present two examples which illustrate the role of time as associated to irreversible processes.

You may all have heard of chemical clocks (for more details see [1]). Briefly speaking, we are dealing with two substances which may transform one into the other. Suppose one type of molecule would be red, the other blue. We perform the experiment in some open vessel to prevent the system from reaching equilibrium. The classical vision of a chemical reaction is chaos, molecules collide at random and perform random motions, somewhat as do particles in dust. In such circumstances, we would have to expect that the number of red and blue particles would fluctuate: we would see occasionally some flashes of blue, but no well defined colour pattern. This is however not true: recent experimental investigations have shown that in a wide range of conditions, such a system may behave as a chemical clock.

We see the system, as a whole, become alternatively blue or red. I believe that this is one of the most striking discoveries made in this century. First of all, the existence of this coherent behaviour shows that the distribution of molecules may present long range correlations. As well known, intermolecular forces extend only over short distances, of the order of molecular diameters. Here, on the contrary, the molecules may "communicate" over macroscopic distances and over macroscopic times. Moveover, irreversible processes play an essential role in the building up of the chemical clock. Whenever the flow of matter maintaining the reaction is suppressed, the system evolves towards equilibrium and the coherent patterns are suppressed. Let me emphasize two very important aspects. First, this example shows that matter far from equilibrium may acquire new properties (such as long range "communications") which, till now, have been generally associated with living systems. Secondly, we see that one of the most important concepts of physics, "coherence", wave properties as expressed through the appearance of a chemical clock or chemical waves, may also emerge through nonlinear interactions in far from equilibrium conditions. This is a most unexpected development of the idea of coherence. In classical 19th century physics, wave properties were associated mainly to light. The basic discovery of quantum theory was the wave particle duality implicit in the existence of Planck's constant. This duality is universal as matter has both wave and particle properties. Now we are in presence of a new extension of the concept of coherence to chemical or biochemical systems. However, this new extension is highly specific. The new wave properties appear only if specific conditions, including the distance from equilibrium are satisfied.

Let me briefly deal with a second most important discovery, the cosmic black body radiation, which fills all the accessible universe [2]. We have been used to consider fossils in the realm of biology or ruins left by past history, but the existence of a fossil form of energy which originated in an early stage of the universe is certainly a great surprise. We are probably only in the beginning of new developments, whose outcome is impossible to predict. Anyway, a reconceptualisation of physics is going on at present. These ideas of nonlinearity, of fluctuations, of instability diffuse into wider and wider fields including biology, ecology, and even the description of human behaviour. Everywhere we discover structures with characteristic lengths and times. Obviously, biology can be seen in a new light which transcends the classical opposition between molecular biology with its emphasis on specific molecules and holistic views centering on the behaviour of biological systems as a whole. We may now go from one level to another. The possibility, which is based on the microscopic behaviour such as catalytic properties of molecules or enzymes, is certainly a most significant new development.

In the perspectives which I have just described, one of the central problems common to many fields is the emergence of structure. How may structures originate from thermal chaos? This problem has interested me over practically all my scientific career. I remember vividly the great impression Schrödinger's well known book (*What is Life?*) produced on me. However, Schrödinger had little to say about the origin of biological order. The mechanism which I proposed in 1945 was that order may be generated through dissipation, through non-equilibrium conditions. In this way the functioning of the system (as expressed by the equations of hydrodynamics or of chemical kinetics) becomes closely connected to the space-time structure. We have today, more than 35 years since this idea was suggested, ample evidence for the transformation of irreversible processes, associated with energy or matter flow, into "structure".

The simplest examples refer to hydrodynamics [3]. A well known case is the Bénard instability: a liquid layer is heated from below where, beyond a critical difference of temperature, convection patterns appear.

Obviously, convection patterns are highly organized states corresponding to millions of millions of meolcules which remain correlated for macroscopic times. Such problems as well as other hydrodynamical instabilities

(Taylor instability, turbulence, etc.) are widely studied today. A first characteristic feature is the ease with which a flow of energy can be converted into structure.

You may take a bottle containing oil with metal particles in suspension. We have only to rotate it to produce regular ring patterns, as a consequence of the Taylor instability. The flow of energy associated to rotation has produced beautiful patterns, in apparent contradiction with the usual interpretation of the second law of thermodynamics according to which the normal evolution of a system is towards randomness. Moreover, if we increase the distance from equilibrium, for example increasing the heat flow in the Bénard experiment or the rotational energy in the Taylor experiment, we observe a succession of patterns of increasing complexity. The variety of these patterns is staggering. Some of these are time-independent, others present both space and time regularities; some are basically periodic, others appear as chaotic. An unexpected point is that, when we repeat the experiment, different patterns may appear in spite of all the precautions we may take to control the experimental conditions. There is therefore a basically random element in the occurrence of these non-equilibrium dissipative structures. The closest analogy in classical chemistry would be the problem of nucleation. Here also the outcome of experiment may depend on minute circumstances such as the presence of impurities, the nature of the vessel, and so on. Still the appearance of macroscopic stochastic features is most unexpected in a field which is ultimately described by the equations of hydrodynamics which were always considered as a prototype of deterministic equations. Again, as with the problem of coherence we mentioned above, there is an interesting analogy with quantum mechanics. We have become used to the idea that probabilities may be essential for the description of the microscopic world. Now probabilities appear even at the macroscopic level.

Next, we would like to mention the sensitivity of this type of experiments to outside fields such as gravitation or electromagnetic fields. The whole Bénard instability can be viewed as due to the interaction of the gravitational field and of a nonequilibrium process, the flow of heat. In a recent paper, my coworker Kondiputi has calculated the sensitivity of the gravitational field [4]. He has concluded that gravitational field, even a million times weaker than the one on earth, would still give rise to a Bénard instability. This may lead to an interesting experiment for spatial research; anyway, we see that instabilities lead to an increased sensitivity of matter to outside conditions. We may consider instabilities as non-equilibrium phase transitions. Phase transitions have of course been studied since long but here the variety of situations exceeds by far the one observed in equilibrium processes. No wonder that this has led to a renewal of non-linear mathematics.

Till now, I have concentrated on hydrodynamics but similar phenomena appear also in chemistry, especially in non-equilibrium autocatalytic processes [1, 5]. The variety of possibilities is even much greater in chemistry as there exists nothing in chemistry like the general equations of hydrodynamics and, therefore, the number of possibilities for the appearance of instabilities is vastly larger. Of special significance is the fact that, in chemistry, we may have intrinsic characteristic lengths expressed which are constructed in terms of diffusion constants and kinetic constants. This leads to symmetry breaking processes very much as in liquid/solid phase transitions which give rise to spatial structures. In the lectures which will be presented during this seminar, there will be ample opportunities to discuss specific instabilities as they arise in chemical reactions, in biological systems or even in insect societies.

Let me only mention that successive instabilities may induce quite different effects; in hydrodynamical systems, they lead ultimately to chaos. This chaos is due to the appearance of a large number of characteristic times and lengths, and should not be confused with thermal chaos. In biological systems, and specially in the problem of morphogenesis, reaction-diffusion equations may, on the contrary, give rise to increasing differentiation and organization. Interesting work, along these lines, has been made by Kaufman, Ortoleva and members of our group in Brussels [6], following the original suggestion by Turing in 1952. I hope that there will be time during this symposium to stress the importance of non-equilibrium structures in cellular problems, and specially in the problem of cancer and of diseases, involving the perturbation of rythmic activities so essential in biological systems.

It is not astonishing that these developments as well as the ones in elementary particle physics and cosmology lead to an intense search for reconceptualization on all levels. The concepts of 'law', of 'order', can no longer be considered as given, and the mechanism of emergence of laws and of order out of disorder, of chaos, has to be investigated. In my lecture at this conference, I shall discuss more in detail the emergence of irreversibility from the reversible laws of classical or quantum mechanics through a specific type of symmetry-breaking processes [1, 7]. Here I want simply to mention that we were looking for general, all-embracing schemes which could be expressed in terms of eternal laws, and we have everywhere found time, events, evolutionary patterns. We were also looking for symmetry, and here also we came to a surprise; on all levels we discover symmetry-breaking processes. The features of self-organization, of amplification of fluctuation through evolution appear to become central to biology, to physics and to chemistry. In a sense, the laws of biology, instead of being an exception to the general laws of nature, seem to illustrate some aspects of these laws in a specially striking way.

We are in a period in which it is easy to predict doom. But from the point of view we have considered, there are also some positive aspects in the intellectual evolution of this century. In the 17th century, at the dawn of Western science, the founding fathers such as Déscartes, Newton and Leibniz, were not only what we now call scientists: they were first 'natural philosophers', deeply embedded in the culture of their time.

The development of science in this century has also brought science and culture closure together. It is essential to keep open all possible channels between science and culture. Science has now grown beyond the cultural background of the Western world of the 17th century. It has a more pluralistic character. It has become more respectful of other forms through which men continue their everlasting dialogue with nature.

By loosing its somewhat imperialistic character related to the specific circumstances in which modern science has originated, the very essence of scientific creativity becomes more apparent and more acceptable in the frame of the cultural diversity of our time.

I don't want to deny the risks involved in the link which exists between modern science and weapons of destruction. Still I believe that the past should not be overidealized; without nuclear weapons, millions of people died during the last war. Whatever the difficulties of the moment, science remains an essential element, perhaps our basic hope to insure for the large humanity of tomorrow a standard of life compatible with our idea of human dignity.

References

1. Prigogine, I. *From Being to Becoming*, Freeman, 1980.
2. Weinberg, S. *The First Three Minutes. A Modern View of the Origine of the Universe*, Basic Books, 1977.
3. Swinney, H.L. and Gollub, J.P. (eds.) *Hydrodynamic Instabilities and the Transition to Turbulence*, Springer, 1981. Topics in Applied Physics, Vol. 45.
4. Kondiputi, D.K. and Prigogine, I. "Sensitivity of the Non-equilibrium Structures", *Physica*, **107A** (1981), 1–24.
5. Nicolis, G. and Prigogine, I. *Self-organization Processes in Non-equilibrium Systems*, Wiley-Interscience, 1977.
6. Chadam, J. and Ortoleva, P. "On Growth and Form", *Proceedings of the Conference on Mathematical Biology*, Southern Illinois University at Carnbondale, 1980.
7. Misra, B. and Prigogine, I. "Time, Probability and Dynamics", *Workshop on Long-Time Prediction in Nonlinear Conservative Dynamical Systems*, organized by L. Reichl, Center for Studies in Statistical Mechanics, University of Texas, Austin.

THE LIVING STATE
Some Conceptual Aspects

Time, Life and Entropy

I. Prigogine

Center for Studies in Statistical Mechanics
The University of Texas
Austin, Texas-78712, USA
and
Faculté des Sciences
Université Libre de Bruxélles
1050 Bruxélles, Belgium

In my preceding lecture at this conference, I have considered the constructive role of irreversible processes. Many examples of self-organization in far-from-equilibrium systems can be found in the interesting communications of Lefever, Babloyantz and others at this conference. Therefore, I will not go into more details about self-organization and its relation to biological systems, but will treat a different theme. Living systems have been called strange systems by Jacques Monod. Indeed they are quite different already through the fact that they are autonomous; they actively interact with their environment. Obviously one of the ultimate aims of the reconceptualization of physics which is going on at present is to understand how life, including man, could be generated in the universe. From this point of view the emphasis on irreversibility, on randomness, is certainly an important feature we have to take into account. But there is an additional feature. Life is not the mere passive result of cosmological evolution; it introduces a supplementary feedback. In other words, life is a result of irreversible processes, but can also induce in turn new irreversible processes. There is the old observation: life comes only from life. But more generally we may say irreversibility generates irreversibility. This is the problem I would like to discuss here.

First, we have to be a little clearer about what we mean by irreversibility. We know today that there are many arrows of time: the cosmological arrow related to the expansion of the universe, a microscopic level related to the so-called violation of the T invariance. This list could be continued. Every time we have classes of asymmetric events, we may speak about an arrow. However, this is not what we mean by thermodynamic irreversibility. Here the concept of entropy plays an essential role. Irreversible processes can be embedded in a monotonously increasing entropy at least as long as we consider closed systems. From the macroscopic point of view, the meaning of the second law is quite simple; it introduces a kind of selection principle supplementing the information given to us by the other laws of thermodynamics such as conservation of energy or mass. From the point of view of conservation of energy, heat and work play the same role. However, as everybody knows, it is easy to transform work into heat but not so simple to transform heat into work. The so-called perpetual motions of the second kind, in which we would use the thermal energy contained in the ocean to drive a ship, are explicitly excluded by the second law of thermodynamics. We may therefore say that the second law limits our action on matter. More so, it limits the type of processes which we can find in nature. Only processes leading to a positive entropy production are permitted.

There is not much disagreement about these macroscopic ideas. However, the basic question is then: "Can we transpose the concept of entropy into the microscopic world; can we give a meaning to irreversible processes on the level of dynamics, be it classical or quantum?"

Even today after a hundred years we are immediately led to Boltzmann's work when we ask this question. As is well known, Boltzmann has shown that entropy can be defined in terms of evolution of a population of molecules. It may be noticed that Boltzmann felt that his work was in some sense parallel to Darwin's field, but now in physics. The driving force behind biological evolution is that the natural selection in Darwin's theory cannot be defined for one individual either but only for a large population. It is therefore a statistical concept. Following this line of thought and using a model of dilute gases, Boltzmann discovered his fundamental formula;

$$S = k \log P$$

which relates entropy to the probability. The increase of entropy is in this way described in terms of a probabilistic process which expresses that the probability tends to its maximum, time going on. This is a fundamental result; Boltzmann has definitively linked entropy and probability.

However, many questions remained open. Over the century since Boltzmann's ideas were developed, a huge literature has grown, and it is out of the question even to summarize it here. More information can be found elsewhere. However, I want to emphasize two aspects:

First, the failure of Boltzmann to be able to define an arrow of time. Boltzmann thought at first that he was able to prove that the arrow of time was determined by the evolution of dynamical system towards states of higher probability, but as a result of objections from Poincaré, Zermolo and Loschmidt, he changed completely his mind. He gave up his attempt to prove there existed an objective arrow of time and introduced instead a subjective point of view which, in a sense, reduced the law of entropy increase to a tautology. The arrow of ime would only be a convention which we (or perhaps all living beings) introduce into a world in which there is no objective distinction between past and future. This is difficult to accept in our time, in which not only physics but also history seems to indicate the importance of unidirectional change. As Popper has written: "Boltzmann's idea is untenable, at least for a realist. It brands unidirectional change as an illusion. This makes the catastrophe of Hiroshima an illusion. Thus it makes our world an illusion and all our attempts to find more about our world."

The second point has to do with the limitation of Boltzmann's work to dilute gases. It it true that if we compress the molecules of a gas into a small part of a container, time going on, we expect to find them uniformly distributed. This is in agreement with the idea of progressive disorder as postulated by Boltzmann. However, the situation is not always so simple. We could perform a computer experiment in which a few hundred molecules interact in a two-dimensional system through the usual attractive and repulsive potentials. We may place at the initial time the molecules at random positions with random velocities. In suitable circumstances we see that crystallization is going on; the particles become ordered. We may even see the appearance of nuclei of crystals, as well as dislocations in the final lattice. From the dynamical point of view, the system evolves to order, and still the second law requires that the evolution be to disorder. What may be the solution to this riddle?

"We begin now to have some answer to these questions as the result of an effort which is going on since some years both in Brussels and Austin. In a recent paper, "Time, Probability and Dynamics", Misra and the author concluded in the following terms:

The preceding considerations thus lead us to the view point that irreversibility expressed in the second law results from a special form of symmetry breaking at the dynamical level that causes the dynamical group to be 'realized' as a dissipative semigroup associated with a probabilistic process admitting an H-function. The physical origin of the symmetry breaking in question is a limitation on physically observable states. Such a limitation comes, in the first place, from (strong) instability of dynamical motion as a consequence of which the concept of phase space trajectories ceases to be physically meaningful and the physically realizable state of the system need to be described in terms of (Gibbs) distribution functions But the existence of symmetry breaking under consideration is the expression of a further limitation: not all distributions but only a suitable proper subset of them can correspond to physically realizable states. We have presented arguments indicating that this second limitation is a consequence of the fact that certain types of "future directed" correlations cannot exist in physical systems so that only those distributions which do not contain such 'future directed' correlations can represent physically realizable states.

Thus the second law, which implies at the microscopic level a limitation on the possibilities of "manipulation" of matter (e.g., the impossibility of perpetual machines of the second kind), implies a limit to our manipulation also at the *microscopic level*. To put it differently, the second law makes explicit on the macroscopic level a basic structure referring to the microscopic level. It expresses an essential new element foreign to the laws of dynamics but, of course, compatible with them. The analogy with quantum statistics may perhaps clarify what we want to say. The limitation to symmetrical (or antisymmetrical) wave functions is of course not a consequence of the Schrödinger equation. However, once a restriction on the symmetry of wave functions is formulated it is propagated by the laws of quantum mechanics.

This is admittedly a rather abstract conclusion. Let me make a few comments. Classical or quantum dynamics can for a suitable class of systems be transformed into a probabilistic process. Such a transformation involves always a symmetry-breaking. A probabilistic process has a single direction. It leads to the most probable state, either in the future or in the past. On the contrary, the starting dynamical process was invariant in respect to time reversal. It is at this very point where the arrow of time appears. We have to understand why only one of the probabilistic processes is realized in nature. We shall give an example in the next paragraph. In addition, the transformation from the dynamical description to the probabilistic description involves a change of representation. In the dynamical description the basic quantity is the density function as described by Gibbs' ensembles (or the density matrix in quantum mechanics). In the probabilistic process it is a new distribution function satisfying a Markov type of equation and which refers to new entities. This is the reason why the dynamical process may well lead to order in terms of the units considered, while the probabilistic process leads to increased disorder.

Order and disorder are not necessarily mutually exclusive concepts. They may refer to different descriptions. Let us try to illustrate these general conclusions in terms of a more physical model. Many dynamical systems can be expressed in terms of "collisions" and "correlations". Consider a cloud of particles which is directed on a target (a heavy motionless particle). The situation is described in Fig. 1. In the far distant past, there were no correlations between particles. Now, scattering has effects. It disperses the particles (it makes the velocity distribution more symmetrical) and, in addition, produces correlations between the scattered particles and the scatterer. The appearance of correlations can be made explicit by performing a velocity inversion (i.e., placing a spherical mirror). Figure 2 represents this situation (the wavy lines represent the correlations). Therefore, the role of scattering is the following: in the direct process, it makes the velocity distribution more symmetrical and creates correlations; in the inverse process, the velocity distribution becomes less symmetrical and correlations disappear. Therefore a basic distinction between the direct and the inverse processes is introduced through the consideration of correlations.

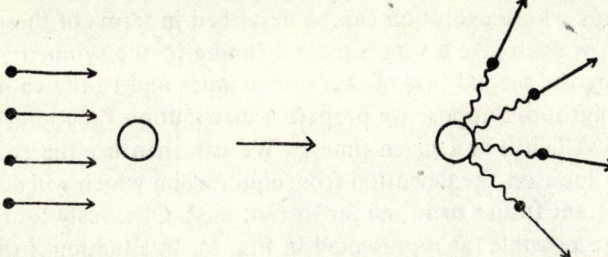

Fig. 1. Scattering of particles. After the collision scatterings remain correlated with the scatterer (wavy lines).

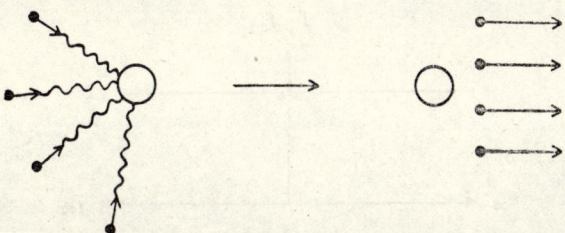

Fig. 2. Velocity inversion after a collision. The correlations are destroyed after the impact on the scatterer.

We may apply our conclusions to many-body systems. Here also we may consider two types of situations: in one, uncorrelated particles come in, are scattered, and correlated particles are produced (Fig. 3). In the opposite situation, correlated particles come in, the correlations are destroyed through collisions, and uncorrelated particles come out (Fig. 4).

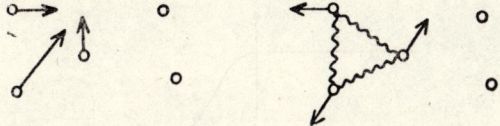

Fig. 3. Creation of *post*-collisional correlations.

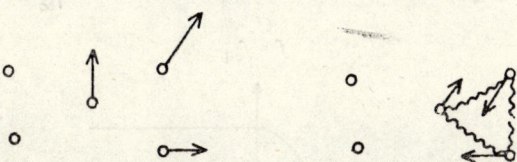

Fig. 4. Destruction of *pre*-collisional correlations.

The two situations differ through the temporal order between collisions and correlations. In the first case, we have "post-collisional" correlations, in the other, "pre-collision" correlations. (A clear distinction between these two cases is not always possible. To make this distinction easy, it is convenient to consider large, "infinite" systems.)

For systems whose evolution can be described in terms of these dynamic correlations, we can give a very simple meaning to the symmetry breaking introduced by the second law of thermodynamics and to the choice of the relevant semigroup. Suppose we prepare a distribution function (say of the one particle velocity) at a given time t_0. We can then use the equations of dynamics to look on the deviation from equilibrium which will occur either in the far distant future or in the far distant past. Obviously four types of situations are possible (as represented in Fig. 5). In situation A the velocity distribution would reach equilibrium neither for $t \to +\infty$ nor for $t \to -\infty$. On the contrary, in situation B equilibrium is reached in both directions of

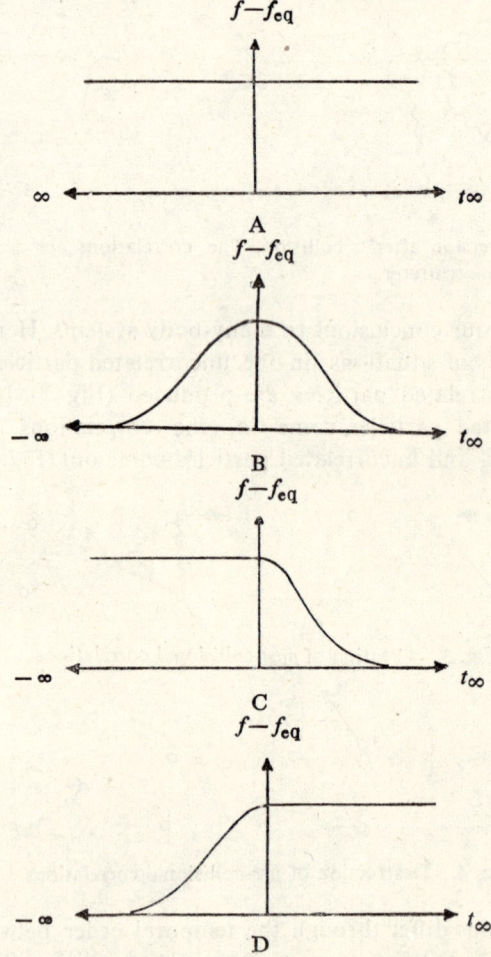

Fig. 5. Evolution of the deviation from equilibrium. According to the initial conditions, we may have four different cases (see text).

time. In situation C equilibrium is reached for $t \to +\infty$ but not for $t \to -\infty$; and finally, in situation D it is reached for $-\infty$ and not for $+\infty$. The type of situation which can be realized depends on the initial conditions. Now it can be shown, and that is the essential point, that cases A and D can only occur if at the initial time we had long-range persistent correlations between the particles which prevent the system reaching equilibrium through successive collisions. In B and C we may also have correlations, but here the correlations are *post*-collisional correlations, which do not prevent the system from going to equilibrium. In A and D we may have particles coming from infinity and being correlated before they have collided. In B and C we have correlations only after collisions (Fig. 5).

We may now formulate a microscopic selection law which is the very basis of the second law of thermodynamics. No persistent long-range *pre*-collisional correlations can be prepared or may be found in physical systems. Correlations are always the result of previous dynamical interactions. It should be noticed that this selection principle breaks time symmetry and permits us to choose precisely the right semigroup with the arrow of time which we see around us. Once this selection principle is formulated, it is a question of technical matters which I cannot describe here, to go from the initial dynamical process to a probabilistic process and to reconcile the dynamical evolution with an ever-increasing probability in isolated systems.

Let us now come back to the relation between time and life. First, let me emphasize that the second law of thermodynamics expresses perhaps the most basic symmetry breaking in the physical world which makes then possible all other symmetry breakings, including that of matter and antimatter. In principle, we could imagine the world populated by two types of beings some living towards the future and others towards the past. Some amusing consequences of such a situation have been discussed in the beautiful book by Martin Gardiner, *The Ambidextrous Universe*. We could recognize such physical or biological systems because for us their evolution would be the opposite of the one to which we are accustomed. Microscopic analysis would show that these systems transform correlations into collisions, while the systems as we know them proceed in the opposite direction.

The second law is a statement of the unity of the physical world. Only a single arrow of time exists. We may now in this perspective discuss the relation between time and life. Certainly life is one of the most striking manifestations of this universal arrow of time. From this point of view, it may be considered as a consequence of existence of irreversible processes, but the point I want to emphasize is that in turn it transmits its intrinsic time symmetry-broken situation to objects in the physical world which, without its intervention, would have a time symmetrical behavior. If we look back on situation B in Figure 5, we shall interpret this situation as giving rise to an approach to equilibrium in our future. Beings which would live in the opposite direction would, on the contrary, interpret this situation as giving rise to approach to equilibrium in their future, which is our past.

In a sease, we transform a basically symmetrical situation into a time dissymmetrical situation using our own temporal dissymmetry. As I mentioned already, it is usual to say that life gives rise to life. But here we see life transmitting irreversibility: duration giving rise to duration.

There is at present a lot of speculation on cosmology. Certainly the concepts which we have described are consistent with what is generally called the standard cosmology. In its earliest stages the universe was causally disconnected. Only uncorrelated elements could appear from the horizon of each observer. Time going on, their interaction introduces supplementary correlations. The overall evolution of the physical world begins to sound in this way as a progressive extension of correlation through matter coming into contact either directly or through intermediary fields. But the large-scale history of our universe is still largely unexplained. Certainly two essential features have to be taken into account: gravitation and entropy. We are still far from any synthetic view encompassing these two concepts, and therefore many possibilities are still open.* From my previous studies of complex systems on a much more modest scale, I have kept the strong impression than it is difficult to even imagine or to enumerate all the possibilities open to nonlinear systems far from equilibrium. This seems to me even more true if we consider the universe as a whole with the striking nonlinearities as described by Einstein's equation and with the enormous deviations from equilibrium which must have prevailed at the earliest stages. Therefore I want to finish with an optimistic note: there will be no end to history.

References

1. Misra, B. and Prigogine, I. "Time, Probability and Dynamics", Proceedings of the Workshop on Long-Time Prediction in Nonlinear Conservative Dynamical Systems," edited by C. W. Horton, Jr., L. E. Reichl and V. Szebehely, to be published by Wiley-Interscience, New York.
2. Prigogine, I. *From Being to Becoming*, W. H. Freeman and Co., San Francisco, 1980.
3. Prigogine, I. and George, Cl. PNAS 1982 (to appear).
4. Popper, Karl, *Unended Quest*, La Salle, Illinois, Open Court Publishing Co., 1976.
5. Gardner, Martin. *The Ambidextrous Universe*, Charles Scribner's Sons, New York, 1979.
6. DeWette, F. W., Allen, R. E. Hughes, D. S. and Rahman, A. Physics Letters, **29A** (1969), 548–549.

*As we have seen, the introduction of the second law of thermodynamics is closely related to the discussion of possible initial conditions. This question acquires an even greater urgency in general relativity. We hope to report on this question soon.

The Living State According to the Association-Induction Hypothesis

Gilbert N. Ling

Department of Molecular Biology,
Pennsylvania Hospital, Eighth and Spruce Streets
Philadelphia, Penna 19107, USA

One time biologists believed that there is a "living substance" and called it protoplasm. Thomas Huxley (1853) eloquently described protoplasm as the physical basis of life. However, in the ensuing 130 years this concept was all but abandoned. A set of concepts under the title, the membrane pump theory, became widely accepted. In this view the substance of life was believed not to be one, but a complex of matter, owing their defined structure and functions often to membranes and membrane pumps. This view of life is sometimes compared to a flame representing a state of "dynamic equilibrium". An apparently quiescent and unchanging shape and composition belie unresting influxes and effluxes of both matter and energy across the enclosing cell membrane boundary and other membrane boundaries. Stopping these reactions and pumping activities, life ceases.

According to a third view, called the association-induction hypothesis (Ling, 1962, 1969, 1977a) being alive signifies not merely the presence of the right kind of substances at the right places but that these components must interact in a specific way so that together they maintain a high energy state, called the living state. The living state is basically a static state; its maintenance, per se, does not call for continual energy-consuming activities.

A Physical State

First, what is a state? A state describes the pattern of interaction among the individual atoms, molecules and ions constituting the systems. Ice is in a solid state while water is in a liquid state. The difference in state between ice and liquid water lies in the different relations on both the space and time coordinates.

Liquid water and solid ice are both made of a single kind of molecule, water. The living state refers to a far more complex system but it also without exception, contains water which exceeds in quantity all other components. Water comprises also the major component of the immediate environment of living cells. In terms of number, the second largest component in the living cell is K^+; in the external environment, it is Na^+. In terms of bulk, the second largest component in the living cell is protein. Recognizing the composition of living matter one can then define the living state as that in which the assembly of water, proteins, and K^+, plus other essential but quantitatively minor components including ATP, are closely associated and energetically linked to one another in such a way that together they are poised at a high energy state.

A High-Energy State
What is high energy? Consider the scenes at a sea shore. At the right time, when the sun and moon become properly aligned so that their gravitational forces reinforce each other, one observes high tide (Ling, 1981). Water reaches higher levels. If now one quickly built a dam with a turbine before the sun and moon go out of alignment, water captured at the high level may be utilized to generate electricity. The high energy state of high tide involves not just ocean water but the sun and moon, etc. as a system.

Another model could be if several nails are tied end-to-end with pieces of string and placed among iron filings on a plate (Ling, 1969). There is no significant interaction among these components. The system is at random and thus a high entropy state. If now a strong magnet is placed near the end of one of the terminal nails, magnetization of the first nail will induce magnetization of the second nail and this can repeat itself until a whole chain of nails become magnetized.

In the process, iron filings in the surrounding area will be picked up by the nails and assume more ordered arrangement dictated by magnetic fields locally produced and now propagated through the iron filing pieces also. Here again the magnet-nails-iron filings are closely associated through the progagated magnetic induction and together they exist in a high energy-low entropy state. Therefore whether it is the alignment of the sun and moon, or the approach of the horse-shoe magnet, each creates a disturbance of the individual elements of the system: ocean water in one case, nails and iron filings in the other, so that energy-wise it is no longer at its usual comfortable equilibrium position. This disturbance from equilibrium represents a high energy state.

The High-Energy Living State: Association and Induction
Note that in the magnet-nail model, the high energy state is one in which the major components are in close association with one another through the operation of magnetic induction effect. In the high-energy living state of protoplasm according to the association-induction hypothesis, electric induction effect also serves a similar role to hold together the major components

of the living protoplasm together, including interaction (i) between water and protein, (ii) between water and water, and (iii) between proteins and ions. The equivalent of the polarizing magnet is ATP, one of a class of critically important substances operative usually at low concentration and are called *cardinal adsorbents*.

So far the discussion has centred around theory. It attempts to point out that the basic elements of the high-energy living state: close association of protein-K^+-water through propagated inductive effect generated by the cardinal adsorbent, ATP, which interact with a key specific protein site called cardinal site. Next we shall examine some recent experimental findings.

Experimental Evidence of Association

According to the membrane theory, the bulk of cell water and cell K^+ are free. In the last half century, there have been repeated attempts to put to test these basic assumptions. There were a number of impressive experiments which at the time seemed to indicate that these assumptions of free K^+ and free H_2O are correct. With time, however, many of these findings and conclusions became disproved and others became highly equivocal. Complete analyses of these history-making findings of the past and their present status are to be presented in a forthcoming book I am in the prosses of finishing. (For less complete analyses see Ling et at., 1973, Ling and Negendank, 1980.). However, during the last four years, primarily due to the young German scientist, Ludwig Edelmann, it is now established that the bulk of intracellular K^+ in frog muscle cells, and by inference in all living cells, is in an adsorbed state. I shall review briefly these critical findings before presenting the reasoning that these same findings indirectly also established that the bulk of cell water must be also adsorbed not singly but in multilayers.

Experimental Evidence for the Adsorbed State
Cells of the Bulk of K^+ in Muscle

According to the AI Hypothesis, K^+ is selectively adsorbed on the β- and γ-carboxyl groups belonging to the aspartic and glutamic acid residues of intracellular proteins (Ling, 1952, 1962). More than 60% of the β- and γ-carboxyl groups belong to myosin and myosin is located exclusively in the dark, or A bands of voluntary muscle cells (Ling, 1977b). Hence,

PREDICTION 1: K^+, or Cs^+ or Tl^+ which can stoichiometrically and reversibly replace K^+, must be concentrated in the A bands. Cationic electron microscopic stain, like uranium, also binds to β- and γ-carboxyl groups (Hodge and Schmidt, 1960).

PREDICTION 2: K^+ or Cs^+ or Ti^+ in living frog muscle should be found on all cytological structures that normally looks dark in an EM plate of muscle stained with uranium. As illustrated in Fig. 1A uranium stained dark area includes the A band, especially its two edges, and the Z-lines. So far these predictions have been unanimously confirmed by three different laboratories, one in the U.S. (Ling, 1977b), one in the West Germany (Edelmann, 1977, 1978, 1980a), and one in Hungary (Tsombitas and Tigyi-Sebes, 1979), using three different techniques: transmission electron microscopy, two variations of autoradiography, and dispersive X-ray microprobe analysis. Since all these works has been already published and reviewed (Edelmann, 1981a, Tigyi et al, 1981) I shall present only the most dramatic, the transmission electron microscopy work of Ludwig Edelmann (1977).

Electron microscopes see electron density, not color. Light atoms like ^{1}H with atomic mass of 1, cannot be visualized. Conventional procedure of EM uses as stains heavy atoms like uranium (^{238}U), tungsten (^{148}W), osmium (^{190}Os), and lead (^{207}Pb). Potassium (^{39}K) is too light to be seen clearly (vide infra). However cesium (^{133}Cs) and thallium (^{205}Tl) are both heavy atoms. They also can replace K$^+$ in frog muscle stoichiometically and reversibly without serious impairment of normal physiology (Ling and Ochsenfeld, 1966; Ling 1978b). Using an elegant new freeze-drying technique he developed, Edelmann (1978b) demonstrated, in muscle cells without chemical fixation or staining, Cs$^+$ and Tl$^+$ distribution in frozen-dried, dry-cut muscle section, (Fig. 1B, C) virtually indistinguishable from similar sections pre-

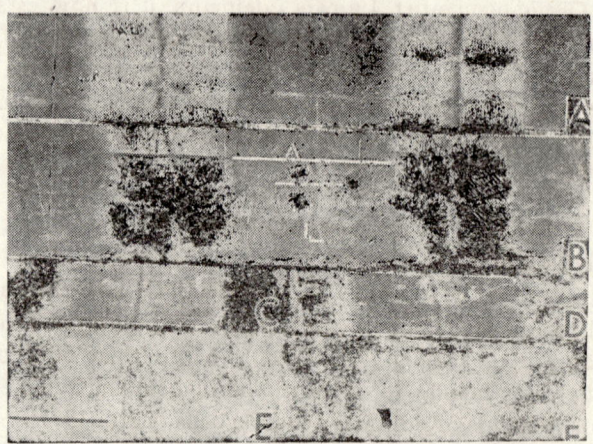

Fig. 1. Electron micrographs of frog sartorious muscle. (A) Muscle fixed in glutaraldehyde only and stained with uranium by conventional procedure. (B) EM of section of freeze-dried Cs$^+$-loaded muscle, without chemical fixation or staining. (C) Tl$^+$-loaded muscle without chemical fixation or staining. (D) Same as C after exposure of section to moist air, which causes the hitherto even distribution of thallium to form granular deposits in the A band. (E) Section of central portion of B after leaching in distilled water. (F) Normal "K-loaded" muscle. A from Edelmann, unpublished, B to F from Edelmann, by permission of *Physiol. Chem. Physics.*

pared by glutaraldehyde fixation and uranium (only) staining (Fig. 1A). Leaching Cs^+-loaded muscle sections in water removes the dark images (Fig. 1E). In later work Edelmann (1981) showed that Rb^+ (^{85}Rb)-loaded muscle can be visualized about as well Cs^+-loaded muscle. Under favourable conditions, and by comparing with wet-cut section, even K^+ in normal K^+-loaded muscle can be weakly but definitely seen also.

The question may be raised, "Can K^+ be localized in regions with fixed anions but only as free-floating counterions?" The answer is, no. Experimental evidence exists showing that the accumulation of K^+ as well as other alkali metal and Tl^+ ions are ion-specific, highly dependent on short-range attributes which can only be "felt" if these counter cations are in close association with the fixed anions. Furthermore, this short-range attribute-dependent specificity persists in muscle cells without a functional cell membrane (Ling, 1977b; 1977c). In other words K^+ and its surrogates, Cs^+, Tl^+, and Rb^+ are adsorbed one cation to one fixed anion in the A bands and Z-line as predicted.

As mentioned earlier, Edelmann (1980a, 1981) and Ling (1977b) had confirmed these findings with autoradiograph. X-ray microprobe of Edelmann (1978a) also confirmed the same prediction, which in turn was completely confirmed by Trombitas and Tigyi-Sebes (1979) from Pecs in Hungary.

More recently, Edelmann (1980b, 1981b) produced yet a fourth and even more spectacular confirmation of the theory, using a technique called LAMMA (laser microprobe mass analysis). In this, focused small laser beams punch minute holes on thin EM sections of frog muscle. The vaporized A band of the muscle is analyzed for its atomic contents by a mass-spectrometer. The even more spectacular aspect of this new method is that he is not loading fresh-living muscle with Cs^+ but exposing in vitro cut EM sections to a solution and analyzes the uptake of Cs^+ and other alkali-metal ions.

Figure 2a shows the mass-spectrometer analyzed peak heights for four elements in a thin gelatine film containing 50 mM K, 50 mM Na, 50 mM Li and 10 mM Cs. A thin section of frozen-dried, imbedded normal frog muscle exposed for 5 min. in a solution containing 50 mM Na^+, 50 mM K^+ and 10 mM Cs^+ is exposed to the laser beam and its vaporized A-band shows a spectrum as in Fig. 2b. Comparing with Fig. 2a, a modest selective accumulation of K^+ and Cs^+ over Na^+ is shown. If Cs^+ is withheld, the K^+/Na^+ preference is made more prominent (Fig. 2d). Figure 2c shows as even more exciting aspect. Here inclusion of 50 mM Li in the incubation solution caused a marked increase in the Cs^+ preference, over both K^+ and Na^+. This section is 0.5 μm thin cut from a single muscle fiber 60 to 100 μm in diameter. There is no question of a membrane pump, yet selective K^+ and Cs^+ adsorption over Na^+ is demonstrated offering definite proof of the adsorbed state of K^+ in living cells and that the selective preference of K^+ over Na^+ is an adsorption phenomenon and not the result of a membrane pump.

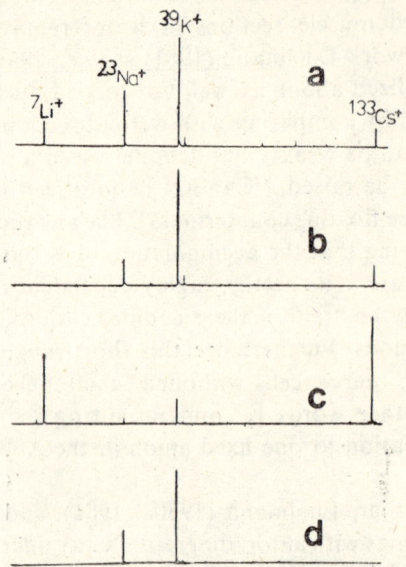

Fig. 2. LAMMA-spectra (a) from a gelatine standard containing 50 mM LiCl, 50 mM NaCl, 50 mM KCl, 10 mM CsCl from A-band regions of muscle sections exposed to a solution containing (b) 50 mM NaCl, 50 mM KCl, 10 mM CsCl; (c) 50 mM NaCl, 50 mM KCl, 10 mM CsCl in addition to 50 mM LiCl; (d) only 100 mM NaCl, 10 mM KCl (Edelmann, 1980; by permission of *Physiol. Chem. Physics*).

Experimental Evidence for the Adsorbed State of Cell Water

In most living cells K^+ is by far the major cation, matching in concentration Na^+ in a Ringer solution (i.e., 100 mM). The proof of the adsorbed and hence osmotically inactive state of K^+ leaves an osmotic imbalance that must be redressed. To do so one must make clear that one understands what is osmotic activity. Osmotic activity is an expression of how much water activity is reduced. In conventional system of aqueous solution this loss of water activity follows the dissolution of solutes and in this case the osmotic activity is quantitatively related to the number of solute molecules or ions. Thus we must find some other entity that can exert an osmotic effect equal to that of 100 mM Na^+. There is, however, no solute in the cell beside K^+ at this concentration level. This difficulty which cannot be overcome in the membrane theory is, however, already overcome by another part of the AI Hypothesis, the polarized multilayer theory of cell water. In this theory water activity is lowered due to multilayer polarization by the matrix protein present throughout the cell (Ling, 1965, 1972).

According to this theory, one condition for the multiple polarization of water is the existence of a matrix of chains alternately carrying negatively charged (N) and positively charged (P) fixed sites at distance roughly equal to that of the diameter of a water molecule, constituting what was called an NP-NP-NP system. A modification of this is an NO-NO-NO system where only fixed negative sites are present but each is separated from its nearest neighbour by a distance equal to diameters of two water molecules. Protein chains satisfy NP-NP-NP system, only if the chains are extended and not when it forms intracellular H-bonds. Thus one expects most native globulin protein to have little long-range polarizing effect, but proteins which for one reason or another exist in extended conformation will polarize multilayers of water. One unambiguous way to demonstrate multilayers of polarized water is, according to the AI Hypothesis, the ability to exclude solutes of large molecular size and complexity including hydrated Na^+, sugars, and free amino acids. Table 1 from Ling et al (1980) shows data confirming the expectation; globular protein as a rule does not exclude Na^+; gelatine, which due to its large proline and hydroxyproline content, cannot form helical folds, does. So do polyvinyl-pyrrolidone (PVP) and poly(ethylene oxide) (PEO), both NO-NO-NO systems.

Direct osmotic activity measurements fully confirm the expectation that proteins like gelatine and polymers like PVP and PEO at millimolar concentration, reduce water activity as indicated by vapour pressure depression equal to that of NaCl solution hundreds of times more concentrated. Globular proteins at equal concentration has little effect. These findings make it plausible that it is not free K^+, but rather the matrix protein, which acts on the intracellular water activity so that osmotic equilibrium is maintained.

The Critical Role of ATP in the Maintenance of the Living State
Proof of the adsorbed state of K^+ in living cells indirectly also provided evidence for the theory of polarized multilayer state of the bulk of cell water. These adsorbed states of K^+ and water are maintained conditionally much as the state of "ordered" nails and iron filings requires the magnet. The condition of the maintained adsorbed state of K^+ and water is the adsorption of the cardinal adsorbent ATP. Here the interaction is electric induction rather than magnetic induction. The interaction is primarily between nearest neighbouring sites. In terms of statistical mechanism, the interaction is cooperative.

When the cooperative interaction is of a kind, which in terms of the AI hypothesis autocooperative, adsorption of one K^+ increases the affinity for K^+ at the two nearest neighbouring sites. This type of adsorption tends to exhibit all or none behavior (Fig. 3a), where the all ith-solute adsorption switches to all jth-solute adsorption in consequence of a change in the ratio for the external ith and jth solute concentration. This type of autocooperative transition is most relevant to oxygen transport of red blood cells where effective and full loading and unloading of oxygen occurs with minor change in the oxygen tension in tissues and in the alveolar air. Figure 3b

Table 1. ρ-Values of Na⁺ in Water Containing Native Proteins (A), Gelatin (B), PVP (C, E), and Poly (Ethylene Oxide) (D)

Group	Polymer		Concentrations of medium (M)		Number of assays	Water content (%) (mean±SE)	ρ-Value (mean±SE)
(A)	Albumin (bovine serum)		1.5	a	4	81.9±0.063	0.973±0.005
	Albumin (egg)		1.5	a	4	82.1±0.058	1.000±0.006
	Chondroitin sulfate		1.5	a	4	84.2±0.061	1.009±0.003
	α-Chymotrypsinogen		1.5	a	4	82.7±0.089	1.004±0.009
	Fibrinogen		1.5	a	4	82.8±0.12	1.004±0.004
	γ-Glubolin (bovine)		1.5	a	4	82.0±0.16	1.006±0.005
	γ-Glubolin (human)		1.5	a	4	83.5±0.16	1.006±0.005
	Hemoglobin		1.5	a	4	73.7±0.073	0.923±0.006
	β-Lactoglobulin		1.5	a	4	82.6±0.029	0.991±0.005
	Lysozyme		1.5	a	4	82.0±0.085	1.009±0.005
	Pepsin		1.5	a	4	83.4±0.11	1.031±0.006
	Protamine		1.5	a	4	83.9±0.10	0.990±0.020
	Ribonuclease		1.5	a	4	79.9±0.19	0.984±0.006
(B)	Gelatin		1.5	a	37	57.0±1.1	0.537±0.013
(C)	PVP		1.5	a	8	61.0±0.30	0.239±0.005
(D)	Poly(ethylene oxide)		0.75	a	5	81.1±0.34	0.475±0.009
			0.5	a	5	89.2±0.06	0.623±0.011
			0.1	a	5	91.1±0.162	0.754±0.015
(E)	PVP	Q	0.2	b	4	89.9±0.06	0.955±0.004
		S*	0.2	b	4	87.2±0.05	0.865±0.004
		Q	0.5	b	3	83.3±0.09	0.768±0.012
		S	0.5	b	3	81.8±0.07	0.685±0.007
		Q	1.0	b	3	67.0±0.26	0.448±0.012
		S	1.0	b	3	66.6±0.006	0.294±0.008
		Q	1.5	b	3	56.3±0.87	0.313±0.025
		S	1.5	b	3	55.0±1.00	0.220±0.021

Temperature was 25±1°C and test tubes were agitated, except in the experiments of E, which were carried out at 0±1°C and in which some test tubes, marked Q, were quiescent and unstirred. S represents sacs shaken in test tubes at 30 excursions/min (each excursion spans 1 inch) axcept the first set (S*) for which agitation was achieved by to-and-fro movement of silicone-rubber coated lead shot within the sacs. The symbols a and b indicate that the media contained initially 1.5 M Na_2SO_4 and 0.5 M Na-citrate respectively. In D, poly(ethylene oxide) (mol. wt. 600,000) was dissolved at a 10%(w/w) solution, and the viscous solution was vigorously stirred before being introduced into dialysis tubing. In E, the quiescent samples contained more water. This higher water content accounts for only a minor part of the difference, as shown by comparison of the 6th and 7th sets of data; even with a larger water content, the ρ-value is lower in the stirred samples (6th). Na was labelled with ^{25}Na and assayed with a γ-counter.

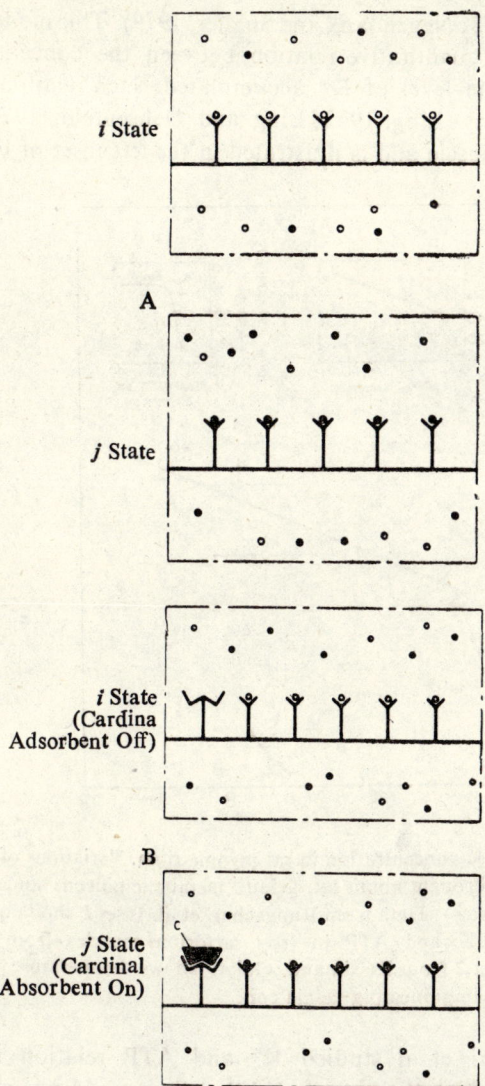

Fig. 3. Cooperative shifts between *i* and *j* states due to a change in the relative concentration of the *i* and *j* solutes in the environment B. Cooperative shifts between *i* and *j* states due to adsorption-desorption of cardinal adsorbent C in an environment with unchanging *i* and *j* concentrations (Ling, 1977, by permission of *Mol. Cell.-Biochem.*).

illustrates how in the presence of constant ratio of *i*th and *j*th solute in the environment, *i*th to *j*th state transition occurs as a result of the installation and removal of the cardinal adsorbent ATP. Thus ATP adsorption leads to autocooperative adsorption of K^+ in an all-or-none manner as we have shown for frog muscle and many other living tissues (Ling and Bohr, 1971; Gulati, 1973, Jones 1973; Negendank and Shaller, 1979). This model predicts that there should be a quantitative relation between the concentration of ATP in a cell and the level of K$^+$ accumulated. Such relation has been demonstrated repeatedly (Ling, 1962; Ling and Ochsenfeld, 1978; Gulati et al, 1971) for frog muscle and is illustrated in the left inset of Fig. 4.

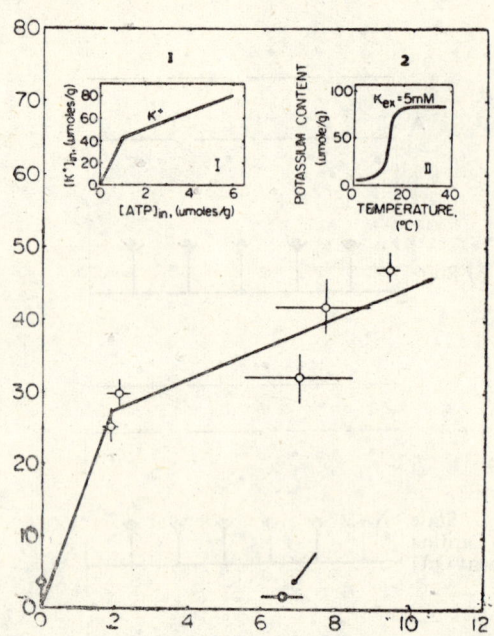

Fig. 4. Plot of ATP vs. K concentration in rat myometrium. Variations of ATP concentration were brought about by various metabolic poisons and by cooling (marked with arrow). Data from Rangachari et al. Inset *I* shows quantitative relation between K+ and ATP in frog sartorious muscles from Ling and Ochsenfeld. Inset 2 from Reisin and Gulati shows temperature transitions from K+ to Na+ in guinea pig taenia coil.

In 1972, Rangachari et al studied K$^+$ and ATP relation in uterine smooth muscle from which they concluded that they could not observe the ATP vs. K$^+$ relation reported for frog muscle. I plotted in graphic from their tabulated experimental data (Ling, 1974), (Fig. 4). It would seem that by and large the data agree with the frog muscle data well except for one point marked with an arrow where profound K$^+$ loss was not accompanied by an ATP concentration fall, it turned out in this case alone, the drop of K$^+$ content was produced by cooling the temperature to 0°C. That mammalian smooth muscle and other tissues respond to cooling by a loss of K$^+$ and gain of Na$^+$ has already been known (Reisin and Gulati, 1972), and is illustrated in the right inset of Figure 4 and illustrates a cooperative temperature transition, not involving ATP concentration change. In conclusion, Rangachari et al's data confirm the AI Hypothesis on two points: ATP dependency of K^+ at one temperature; ATP independent K^+ loss at below transition temperatures.

Several Interesting Experimental Findings that the Concept of the Living State as Suggested in the AI Hypothesis Offers Some Insights

The Freezing Preservation of Living Tissues
With the aid of glycerol and DMSO it is now widely practiced to preserve living cells such as spermatozoa, red blood cells, tissue culture cell lines over a long period of time. Indeed to these frozen cells, time has lost its bite. While I have no direct evidence, it would be fully expected that this state of preservation can be maintained indefinitely at or near absolute zero. At these low temperatures all chemical reactions come to a halt. In the definition of life that we can construe from the membrane theory, the cells must be dead, since dynamic equilibration cannot be sustained at the temperature. Thawing and reviving of the frozen cell would then amount to the creation of new life—a very awkward position, to say the least. On the other hand, the frozen state is fully compatible with the living state which is primarily a metastable equilibrium state. Solid ice is in a solid state at $0°C$ or $0°K$.

The Remarkable Preservation of the Ability of Selective K^+ Accumulation of the EM Section of Frozen-dried Muscle
Using very rapid freezing, Edelmann caught the muscle before it can be triggered into the contracted lower energy (1977a). Since water is essential for the cooperative transition it is not surprising that the frozen dried cell can survive in its normal living state as shown by the maintenance of the fragile ability of selective K^+ adsorption over Na^+. That it should also survive the infiltration of the low viscosity Spurr medium is truly remarkable suggesting that the medium containing such items a nonenyl succinic anhydride, vinyl cyclohexene dioxide, all oxygen-containing molecules, may neatly fit in sterically and electronically space originally occupied by water in the state of polarized multilayers and their polymerization further stabilizes the assembly in the living state, as evidenced by the continued ability to selectively adsorb K^+ over Na^+ and to exhibit a Li^+ (cardinal adsorbent) control of selectivity of Cs^+ over Na^+.

The Ability of Increasing External K^+ to Counter the Fall of ATP Concentration in Maintaining the Living State
Figure 5 reproduces a remarkable finding of Riggs, Walker and Christensen (1958). The ability of Ehrlich ascites cells to accumulate methione depends on metabolism. Poison like cyanide and 2-4, dinitrophenols (2,4D) stops

metabolism causing a loss of this ability of accumulating methione. Yet these authors showed that the action of cyanide and 2,4D can be countered by increasing external K+ concentration to a level so that the intracellular K+ level is maintained at its normal level. According to the AI Hypothesis, the adsorption of methione like that of K+ is due to ATP dependent specific adsorption. In this case, the decrease of ATP is compensated by the increase of external K+ because then the K+-protein-water-ATP system is cooperatively maintained at its normal living state.

Very recently it was suggested that increase of dietary K+ and reduction

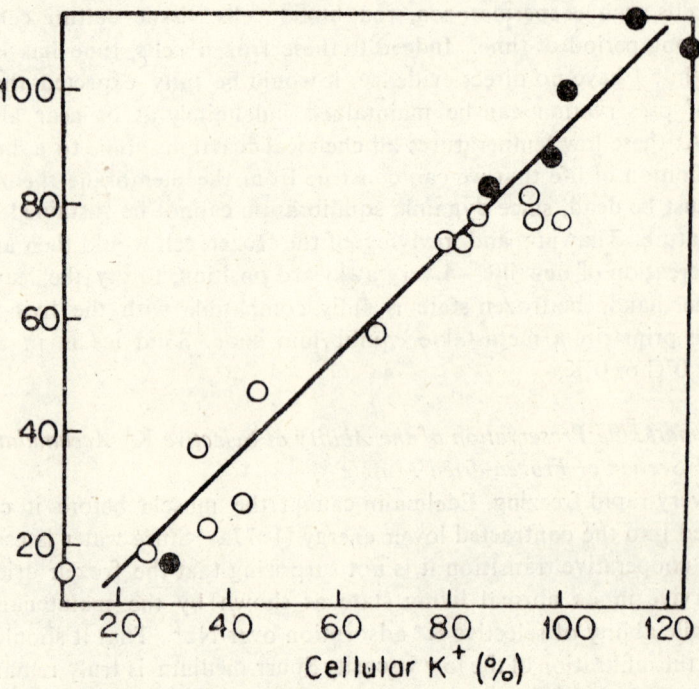

Fig. 5. Association between the degree to which K+ of the Ehrlich cell is displaced by Na+, and the residual ability to concentrate glycine, after treatment with cyanide or 2,4-dinitrophenol. The cells were first incubated 30 min. in the usual medium, to which either 5 mM NaCN(o) or 0.15 to 0.5 mM dinitrophenol(o) had been added. The uptake of glycine in m moles/kg cell water for each millimolar unit of concentration in the medium during an hour of incubation at 37° in medium containing 2 mM glycine was then determined, a subnormal or supranormal external K+ concentration having been selected to obtain the indicated cellular K+ content. The cell K+ is expressed as a percentage of the level found in cells incubated in the normal medium in the absence of any inhibitor. The results were taken to be much the same whether the inhibitor continued to be present or not during the uptake phase of the experiment, those for dinitrophenol having been selected previously, for detailed documentation (from Riggs and coworkers, 1958).

of dietary Na^+ may help to correct debilitating diseases including heart failure by the same theoretical concept of the living state which I think the experimental data of Riggs et al, experimentally have demonstrated (Ling, 1981b; see also Cope, 1977).

The foregoing work was supported by NIH Grants 2-R01-CA16301-03 and 2-R01-GM11422-13, and by Office of Naval Research Contract N00014-79-C-0126.

References

1. Cope, F. W., *Physiol. Chem. Phys.*, **9**, 547.
2. Edelmann, L., *Physiol. Chem. Phys.*, **9**, 313 (1977).
3. Edelmann, L., *Microsc. Acta Supp.*, **2**, 166 (1978a).
4. Edelmann, L., *J. Microscopy.* **112**, 243 (1978b).
5. Edelmann, L., *Histochemistry*, **67**, 233 (1980a).
6. Edelmann, L., *Physiol. Chem. Phys.*, **12**, 509 (1980b).
7. Edelmann, L., in *Intern. Cell Biol*, 1980–1981, p. 941, ed. by H. G. Schweiger, Springer-Verlag, Berlin (1981a).
8. Edelmann, L., *Fregnius Z. Anal. Chem.*, **308**, 218 (1981b)
9. Gulati, J., *Ann. N. Y. Acad. Sci.*, **204**, 337 (1973).
10. Gulati, J., Ochsenfeld, M. M., and Ling, G. N., *Biophys. J.*, **11**, 973 (1971).
11. Hodge, A. J. and Schmidt, F. O., *Proc. Nat. Acad. Sci.*, **46**, 186 (1960).
12. Huxley, T. H., *Brit. and Foreign Med. Chir. Rev.*, **20**, (1853).
13. Jones, A. W. W., *Ann. N. Y. Acad. Sci.*, **204**, 379 (1973).
14. Ling, G. N., The Role of Phosphate in the Maintenance of the Resting Potential and Selective Ionic Accumulation in Frog Muscle Cells in *Phosphorus Metabolism* (Vol. II), W. D. McElroy and B. Glass, eds., The John Hopkins University Press, Baltimore, MD. pp. 748–795 (1952).
15. Ling, G. N., *A Physical Theory of the Living State: The Association-Induction Hypothesis*, Blaisdell Waltham, Mass. (1966).
16. Ling, G. N., *Ann. N. Y. Acad. Sci.*, **125**, 401–417 (1965).
17. Ling, G. N., *Int'l Review of Cytology*, **26**, 1–61 (1969).
18. Ling, G. N., Hydration of Macromolecules in Structure and Transport Process in *Water and Aqueous Solutions*, ed. A. Horne, Wiley-Interscience, New York, pp. 201–213 (1972).
19. Ling, G. N., *Physiol. Chem. Phys.*, **6**, 285–286 (1974).
20. Ling, G. N., *Molecular and Cellular Biochemistry*, **15**, 159 (1977a).
21. Ling, G. N., *Physiol. Chem. Phys.*, **9**, 319 (1977b).

22. Ling, G. N., A Theoretical Foundation Provided by the Association-Induction Hypothesis for Possible Beneficial Effects of a Low Na, High K Diet and Other Similar Regimens in the Treatment of Patients Suffering from Debilitating Illnesses (Symp. Int. on Metabolic Treatment of Heart Conditions, Mexico City, 1980), *Gazeta de la Faculted de Mecina*; *Aggressology* (in Press).
23. Ling, G. N., *Physiol. Chem. Phys.,* **13**, 29 (1981).
24. Ling, G. N., and Bohr, G. R., *Physiol. Chem. Phys.*, **3**, 431 (1971).
25. Ling, G. N., Miller, C. and Ochsenfeld, M. M., *Ann. N. Y. Acad. Sci.*, **264**, 6 (1973).
26. Ling, G. N., and Negendank, W., *Persps. in Biol. and Medicine*, **23**, 215 (1980).
27. Ling, G. N. and Ochsenfeld, M. M., *J. Gen. Physiol.*, **49**, 819 (1966).
28. Ling, G. N., Ochsenfeld, M. M., Walton, C. and Bersinger, T. J., *Physiol. Chem. Phys.*, **12**, 3 (1980).
29. Negendank, W. and Shaller, C., *J. Cell Physiol.*, **98**, 95 (1979).
30. Rangachari, P. K., Paton, D. M., and Daniel, E. E., *Bioch. Biophy. Acta*, **274**, 462 (1972).
31. Reisin, I. L. and Gulati, J., *Science* **176**, 1137 (1972).
32. Riggs, T. R., Walker, L. M. and Christensen, H. N., *J. Biol. Chem.*, **233**, 1479 (1958).
33. Tigyi, J., Kallay, N., Tigyi-Sebes, A., and Trombitas, K., in *Intern. Cell Biol.*, 1980, 1981, p. 925. ed. by H. G. Schweiger, Springer-Verlag, Berlin 1981.
34. Trombitas, K., and Tigyi-Sebes, A., *Acta Bioch. Biophys. Acta Sci. Hung.*, 14, 271 (1979).

The Living State

R. K. Mishra

Department of Biophysics
All India Institute of Medical Sciences
New Delhi-110029, India

Are the properties of living systems which distinguish them from the non-living, functions of the state in which their chosen atoms, ions and molecules find themselves? Certainly the states of matter like solid, liquid or gas, of the same elements or compounds express themselves by distinct properties which are ascribable to their physical state. Is the same true for the living systems? Or, in other words, what are the distinctive properties and behaviour which could emerge from mere fact of accociation of molecules in a particular manner as seen in the living system? We describe herewith the essential units and forces that should lead to the functions we associate with the living state.

Essential Atoms

It is a well-known fact that there is sufficient terrestrial abundance of a large number of elements. Many of them, or perhaps all of them, can form soluble salts in the Earth which is awash with water and there is therefore no reason why these proportions should not be reflected to a considerable extent, if not entirely, in the composition of an organism. But as a matter of fact there is very wide difference. For instance, silicon at 16.08% is 135 times more plentiful than carbon, 0.119% in the earth's crust, yet silicon cannot be called essential for all forms of life. Not only that, some elements like cobalt are tenaciously held and occur in very small amounts, yet life can not be sustained without them: such are the *essential* elements. Attempts have been made to unravel the basis of their biological essentiality. Quite clearly same critical functions must be related to them. Is there some relationship between their electronic structure and biological essentiality? Several workers have sought to explain or predict the essentiality of elements by their positions in the periodic table, just as it was done for chemical or physical properties in the early days of periodic tables. Pirschle [1]

Table 1. Elements which are Indispensable to Plant Nutrition

Series	0	I	II	III	IV	V	VI	VII	VIII	0
1st period	H									He
2nd period	He	Li	Be	B	C	N	O	F		Ne
3rd period	Ne	Na	Mg	Al	Si	P	S	Cl		Ar
4th period	Ar	K	Ca	Sc	Ti	V	Cr	Mn	Fe Co Ni	Kr
5th period	Kr	Rb	Sr	Y	Zr	Nb	Mo	Ma	Ru Rh Pd	X
		Cu	Zn	Ga	Ge	As	Se	Br		
		Ag	Cd	In	Sn	Sb	Te	J		
6th period	X	Cs	Ba	La	Ce usw	Ta	W	Re	Os Ir Pt	Rn
		Au	Hg	Tl	Pb	Bi	Po	—		
7th period	Rn	—	Ra	Ac	Th	Pa	U			

FREY - WYSSLING LINE

Table II. Designations of Element Groups

	s1	s2	p1	p2	p3	p4	p5	p6	d1	d2	d3	d4	d5*	d6	d7	d8	d9	d10	f1	f2	f3	f4	f5	f6	f7	f8	f9	f10	f11	f12	f13	f14
1	1 H 1s1	2 He 1s2																														
2	3 Li 2s1	4 Be 2s2	5 -B- 2p1	6 C 2p2	7 N 2p3	8 O 2p4	9 -F- 2p5	10 Ne 2p6																								
3	11 Na 3s1	12 Mg 3s2	13 -Al- 3p1	14 Si 3p2	15 P 3p3	16 S 3p4	17 Cl- 3p5	18 Ar 3p6																								
4	19 K 4s1	20 Ca 4s2	31 Ga 5p1	32 Ge 5p2	33 -As- 5p3	34 -Se- 5p4	35 -Br- 5p5	36 Kr- 5p6	21 Sc 3d1	22 Ti 3d2	23 -V- 3d3	24 -Cr- 3d4	25 Mn 3d5	26 Fe 3d6	27 Co 3d7	28 -Ni- 3d8	29 Cu 3d9	30 Zn 3d10														
5	37 Rb- 5s1	38 -Sr- 5s2	49 In 5p1	50 -Sn- 5p2	51 -Sb- 5p3	52 Te 5p4	53 -I- 5p5	54 Xe- 5p6	39 Y 4d1	40 Zr 4d2	41 Nb 4d3	42 -Mo- 4d4	43 Tc 4d5	44 Ru 4d6	45 Rh 4d7	46 Pd 4d8	47 -Ag- 4d9	48 -Cd- 4d10														
6	55 -Cs- 6s1	56 -Ba- 6s2	81 Tl 6p1	82 -Pb- 6p2	83 -Bi- 6p3	84 -Po- 6p4	85 At 6p5	86 Rn 6p6	71 Lu 5d1	72 Hf 5d2	73 Ta 5d3	74 W 5d4	75 Re 5d5	76 Os 5d6	77 -Ir- 5d7	78 Pt 5d8	79 -Au- 5d9	80 -Hg- 5d10	57 La 4f1	58 Ce 4f2	59 Pr 4f3	60 Nd 4f4	61 Pm 4f5	62 Sm 4f6	63 -Eu- 4f7	64 Gd 4f8	65 Tb 4f9	66 Dy 4f10	67 Ho 4f11	68 -Er- 4f12	69 Tm 4f13	70 Yb 4f14
7	87 Fr 7s1	88 -Ra- 7s2							103 Lw 6d1										89 Ac 5f1	90 Th 5f2	91 Pa 5f3	92 Uu 5f4	93 Np 5f5	94 Pu 5f6	95 Am 5f7	96 Cm 5f8	97 Bk 5f9	98 Cf 5f10	99 Es 5f11	100 Fm 5f12	101 Md 5f13	102 No 5f14

☐ ESSENTIAL ELEMENT

-│- ELEMENT CAPABLE OF AFFECTING LIVING SYSTEMS PATHOLOGICALLY OR THERAPEUTICALLY

and Lendle [2] thus reported on similarity amongst the effects of elements homologous in the Periodic Table. Thatcher [3] proposed a similar extended classification. Frey-Wyssling [4] described the 'Frey Wyssling Line' on the Periodic Table (Table I) which indicates elements essential or active in plant nutrition. But this is soon disproved if one writes the table taking the shapes and features of orbitals into account as done in Table II. Steinberg [5] had claimed on the basis of a similar table a correlation between filling of electron "shells" in atoms and the biological essentiality. However, these have too many exceptions and his predictions about essentiality have not been fulfilled. Wald [6] explained the abundance of H, O, N, C on the ground that these are the smallest atoms in the periodic system that achieve stable electronic configuration by adding respectively 1, 2, 3, and 4 electrons. Even if this be true, still it is not understood why an element should be essential for the living organism. Undoubtedly there should be and often is, a trend in *physical* properties of elements with reference to their positions in the periodic table but no correlation of a single factor can be found between position and *chemical reactivity* [8], let alone biologic essentiality since many complicated functions may contribute to the appearance of an observed biological activity. Indeed, perhaps even in equilibrium-state biological structures and functions require a fine balance of several such properties of constitution of an atom such that it becomes essential for the given biological function [9]. All one can say is that excepting possibly iodine of 5th period the largest number of essential elements are found upto the 3rd period (Table II). The list ends at 4th period (K and Ca) which could well be in 3rd period by atomic number and the list may be *considered to end in the 3rd period*. In fact all but titanium and scandium are biologically active, i.e. physiologically or pharmacologically or toxicologically. Silicon is found in diatoms; one cannot say how essesntial it is in them. In case of iodine, it is by itself a toxic element, the toxicity of which is prevented by thyroxine [12].

Functions to be expected from such a selection of atoms

(i) *Interaction with water* due to hydrogenic (1s) orbitals leading to water binding, translocation of water and maintenance of volume, shape, turgidity. This forms the main element of a thermalising bath and facilitates transfer of mass and momentum.

(ii) *Functions due to properties of Carbon*: (a) Capability due to its tetravalence to bear *both hydrophilic and hydrophobic* groups and thereby to separate local zones of different solvability or polarity. Because molecular conformation is a result of the balance of attractive ond repulsive forces, domains of specific conformations and activity are created.

(b) *Amenability to modification of functional groups and of resultant conformation by addition or subtraction of electrons or elements of water*, resulting in modifications of molecules. Tetrahedral carbon, with single bond sub-situation or double bonds can give rise to fibres, helices, folds, kinks, random coils and globulated masses which may be stabilised further by hydrogen bonds, solvophilic and solvophobic forces. These last ones act on arrays of carbon structures with substituents which differ in their interaction to solvent. To a lesser extent phosphorous and sulphur based groups may do such a function.

(iii) *Functions due to multiple bonding and hybridisability. For this capability carbon and other elements* are used, see (c) below, and primary or secondary pumps for energy or for information are generated.

It would be seen than *s and p outer orbitals* of elements are required in bulk by entering into essential structures of molecular edifies; amongst *d* orbitals d^5 and d^8 *are required in traces.*

Consequences of functions that can be performed

Above elemental properties lead to structures that bring about or permit the following sonsequences:

(a) If the primordial earth had water to begin with and its crust had soluble salts like chlorides or oxides there would be an abundance of water and its subspecies. The hydrogenic orbital species belonging to H^+, OH^- and others, Na^+, K^+ with smaller amounts of Ca^{++}, Mg^{++}, lead to turgidity, solution phase reactions, ionogenic potentials, fluidity and transfer of mass and momentum.

(b) 25 elements Ca, Mg contribute to complexation and bridging.

(c) With structural edifice of hydrogenated carbon mutability and metastability of conformation, association, micellisation, formation of sheets, pleats, helices, balls, phase-compartmentalisation into hydrophilic and hydrophobic phases, contraction and relaxation follows.

(d) N, O lead to electron and proton based interaction.

(e) S, P and transition metals are concerned with energy pumps and coupling of energetic processes. They also act as electron and charge pools. Energy transduction involves C, H, N, O, S, P and transition element based compounds.

With this background it is interesting to note that virtually all enzyme functions, in equilibrium state, are within the confines of these potentialities (Table III).

Table III. Classification of Enzymes

1. *Oxidoreductase*: Acting on CH—OH group of donors; acting on the aldehyde or keto-groups; acting on the CH—CH group of donors; acting on the CH—NH group of donors.
2. *Transferases*: Transferring C_1-groups; acyltransferases; glycosyltransferases; enzymes transferring N-containing groups.
3. *Hydrolases*: Cleaving ester linkages; cleaving glycosides; cleaving peptide linkages;
4. *Lyases*: C—C lyases; C—O lyases; C—N lyases.
5. *Isomerases*: Racemases and epimerases; cis-trans isomerases; transmolecular oxidoreductases; intramolecular transferase.
6. *Ligases*: Forming C—O bonds; forming C—N bonds; forming C—C bonds.

Note the predominent role of elements of water (including electrons) as the most important single factor in the above.

The Dynamism of the Living State

Having considered the above mentioned facts one can virtually ascribe a basis to phenomena of "Living State" as manifested by isolated systems in a laboratory or a test tube. *But lacking would be the dynamism, self-organisation, self-reflection, self-reference, permissive or conditioning action, information and energy transfer, number-dependent rates and concerted action of a body as a whole in any responsive activity, programmed development, modelling of organs and disease and resistance to disease, in far from equilibrium situation.*

If we take this view then *we need the following*:

(a) a *"connected" system*;

(b) a *"fluctuating" system*; a system which can oscillate between the two or more stable states in some law-abiding manner;

(c) a state which has *an internal energy pump to help self-organisation*;

(d) a state which is *open*.

If these would be available, then given the primordial environment on the surface of the planet the selection of the compounds and the state are obligatory, or we can say the reverse, given the selected atoms, the Living State would result. Indeed asymmetries in time and space and non linearities would ensue.

We describe below a prototype of such a state and characteristic thereof which would make all these possible and may indeed be the precise nature of the Living State.

Prototype of the Living State

A state that does occur in nature, and what is more, in carbon-hydrogen bonded systems as seen in living systems, qualifies for the required state. This would be the *lyotropic liquid crystalline* state with modifications as included in Table IV below. It can be shown that it possesses the necessary connectivity, cooperativity, metastability. The state is also in the correct energy range for transformations observable in the living systems. Indeed it may be superior to chalcogenide glasses, as a model of the Living State.

The significance of this state with reference to lipids, proteins, nucleic acid and higher systems in general have been reviewed [12a, 12b]. Table IV gives the classification and Table V lists lipids, protein, polypeptides, nucleic acids, viruses, mixed systems in which it has been documented [12b, 12c].

Table IV [12b]. Different types of lyotropic liquid crystalline phases

Class	Description	Common notation in literature
	Structural arrangement displaying Bragg spacing ratio $1 : \frac{1}{2} : \frac{1}{3}$ with one-dimensional (one-dimensional periodicity)	
L_{-1}	Lamellar packing with coherent double layers of molecules and ions separated by water Neat phase type. In case of polypeptides extended β structures, hydrogen bounded to each other, are held in stratified arrangement with solvent around.	Neat phase
L_{-2}	Lamellar packing with coherent single layers of molecules and ions separated by water. Single-layered lamellar type.	
L_{-3}	Lamellar packing with coherent double layers of molecules and ions separated by water. Mucous woven type.	
	Lyotropic liquid crystals with particle structure displaying Bragg spacing ratio $1 : \frac{1}{2} : \frac{1}{3} : \frac{1}{4}$	
P_1	Rod-like particles with organic core surrounded by water. Rods with predominantly quadratic cross-section in tetragonal arrangement. Normal two dimensional tetragonal type. In case of polypeptides similar form is called the ω-form. The helices constitute "solid" rodlets and those that have been observed are 4_{13} helices packed in a square lattice.	White phase
P_2	Rod-like particles with water core in organic environment. Rods with rectangular cross-section in an orthorhombic array. Normal two-dimensional rectangular type.	Rectangular phare
	Lyotropic liquid crystals displaying Bragg spacing ratio $1 : \frac{1}{3} : \frac{1}{4} : \frac{1}{7}$ particle structure with molecules arranged in two-dimensional hexagonal symmetry.	

Table IV [12b]. Different types of lyotropic liquid crystalline phases (*Contd.*)

Class	Description	Common notation in literature
P_{H-1}	Rod-like particles with organic core in aqueous environment. Cylindrical to hexagonal cross-section in hexagonal array. Middle phase types; normal two-dimensional hexagonal type.	(1) Middle phase-I (2) Hexagonal phase-I
P_{H-2}	Rod-like particles with acqueous core in organic environment. Cylindrical to hexagonal cross-section in hexagonal array. Reversed two-dimensional type.	Hexagonal phase-II
P_{H-3}	Rod-like particles with complex structure in aqueous environment. Complex two-dimensional hexagonal type. In case of polypeptides the complex phase is formed by more than one, for example, three rodlets in the place of each one. Solvent is reduced to a constant low amount.	Complex hexagonal phase
	Lyotropic liquid crystals displaying cubic symmetry. Isotropic lyotropic liquid crystal with spherical to dodecahedral particles arrrnged in face-centred cubic lattice.	
C_{f-1}	Particles with organic core in aqueous environment. Normal face-centred cubic type.	Cubic phase C_{1-1}
C_{f-2}	Particles with water core in organic environment. Reversed face-centred cubic type.	Cubic phase C_{1-2}
C_{f-3}	Particles with complex structure. Complex free-centred cubic type.	
	Isotropic lyotropic liquid crystals with spherical particles packed in body-centred cubic lattice.	
C_{b-1}	Particles with organic core in aqueous environment. Normal body-centred cubic type.	
C_{b-2}	Particles with complex structure. Complex body-centred cubic type.	
	Cholesteric lyotropic liquid crystals. X-ray diffractogram shows only one fairly diffuse reflection in the small angle region, the sharpness of which increases and the spacing decreases as concentration is raised.	
CH	Planar sheet of parallel rods, arranged hexagonally in the sheet, each sheet deriving from the next by a translation perpendicular to the plane and a very small systematic rotation in the plane (Mishra).	

Table V [12b]

Molecules and structures from living systems claimed to exhibit liquid crystalline characteristics

Lipids

    Lecithin
    Sphingomyelin
    Cephalin
    Phrenosin
    Kerasin
    Nervone
    Cerebron
    Various phospholipids and monoglycerides
    Cholesterol esters: oleate, benzoate, propionate, stearate, palmitate, butyrate, ammonium stearate and palmitate
    Lipids from erythrocytes
    Lipids from beefheart mitochondria
    Lipid-water model systems
    Lecithin, cholesterol, bile salt, water as ternary system

Proteins and Polypeptides

    "Muscle globulin", myosin
    Histones
    Sickle-cell hemoglobin
    Hemoglobin
    Trypsin
    Poly-γ-benzyl-L-glutamate
    Poly-γ-methyl L-glutamate
    Poly-γ-ethyl-D-glutamate
    Poly-β-benzyl-L-asparate
    Poly-α-L-glutamic acid
    Poly-α-carbobenzoxy-L-lysine
    Poly-α-sodium-L-glutamate
    Poly-L-Lysine hydrochloride

Proteins and Lipids

Nucleic acids

    DNA
    t-RNA

Viruses

    Tobacco mosaic virus
    Cucumber virus

Various mixed systems

    Myelin forms
    Brain extract
    "Neurokeratogenic colloid"
    Nerves
    Nerve myelin
    Smooth muscle fibres

Table V [12b] (Contd.)

    Muscles, tendon, nerves, viscera, bone
    Adrenal extract
    Ovaries
    Structural aspects of colored beetles
    Cuticle
    Living sperms-Sepia officinalis
    Red cell interior and red cell membrane
    Atherosclerotic plaques

General comment on liquid crystallinity in living systems and theroies related to it
    General implications
    "Solubilising potential" of cholesterol esters, and lecithin as "common carrier" in red blood cells
    Vision
    "Mind"-brain interrelationship
    Activity and growth of merismatic cells
    Theory of anisotropy of tobacco maic virus suspensions
    Relation of lipids to membrane function

Reference listed in [12b]; [14].

At this stage it is worth pointing out that the liquid crystalline state may occur in at least two different forms: the thermotropic and the lyotropic. In the former the phase transition are induced by variation of temperature and in the latter by changing relative proportion of medium in which the substance is suspended. The cholesteric phase may be generated in both. Of these the thermotropic phase is to be excluded from consideration since the temperature variation in the body is irrelevant for phase changes in the molecules concerned in the constitution of the body. Indeed, thermotropic deposits, if anything, signify pathology. We, therefore, focus our attention to the lyotropic phase in which the relative proportion of fluids determines phase changes. The essential requirement is possession of a "loose structure" as defined below.

The Mechanisms of the Living State

Having made this overview, and seen that the state discussed does occur in living systems at least in some components and possibly in all, it became necessary to consider the mechanisms whereby number-dependent cooperative transformations would be possible at the body temperature. However, such a concept faces a formidable barrier, namely, there exist organs and macroscopically visible tissues of very appreciable solidity and structural integrity and activity. Modifications in these due to proximity of unbonded molecules alone or due to weak forces are not easily conceivable. The system does not appear to be a fluid with a structure that can support chemical waves or bring about gross modifications of sufficient durability.

However, cooperative behaviour of weak modifications can bring this about. Thus excitations, fluctuations induced by energy flows may be effective. The consequences are particularly rich if the system has non-linear couplings. Then these factors can drive structures into different macroscopic states. *In other words, energy flows produce material structuration as its consequence, rather than the structure arising out of geometric consideration per se.* The following formulation bridges the apparent gap:

(A) *The "loose structure" of N elements and the solvent effect*

We invoke a "loose" structure:
(a) which can be transformed at the "classical level" into structures of macroscopic assymetry in space or time by input of mass or momentum; and
(b) which can be driven into structures of altered symmetries and stabilities, by cooperative interaction of elemantary excitations at quantum or microscopic level.

We define for our purpose *"loose structure" as a structure wherein the energy barriers between conformations and elastic recoil are of the same order as the interaction energies* [14a]. Thus coherent conformational change is nearly as easy as expansion or contraction of an aggregate of molecules. This is a realistic situation because the interaction energy between two methylenic groups 5 A apart, are of the order 0.1 kcal/mole and in a 17 methylenic chain of the order of a few kcal/mole while barriers within various conformations are also of the orders of a few kcal/mole. Indeed in spite of apparent solidity, density and viscosity, macroscopic structures can be driven by the cooperativity of weak forces and energy released in the system. They may exhibit behaviour like Navier-Stokes fluid to which the formalism of Prigogine can be readily applied.

Stability of structure in the "loose structure" model
We present three aspects of this problem : (I) classical level; (II) at quantum level, and (III) the need of an internal pump for self-organisation.

The stability of structures is readily appreciated by very simple arguments. A system of molecules with residual charges, as are ubiquitous in living organisms, may be approximated by a system of dipoles with a net dipole. The other component would be ions.

In a dipole oscillator model, oscillators R apart, of normal frequency Ω, coupling strength $g(R)$, yield an eigenmode $(\Omega) = (\omega^2 - g(R))^{1/2}$ leading to a soft mode $(\Omega_-)^2$ at a critical distance R_0 when $g(R_0) = \omega^2$. This instability is prevented by non-linear effects, the lowest being quartic in the Lagrangian with a cusp at R_0. This R_0 is quite large for macromolecules. Would this soft mode be induced by physical presence of other couples ? It happens to be true. If there are N-interacting dipole oscillators, the equation of motion is

$$\ddot{X}_l = \sum_{j \ne l}^{N} a_{ij} X_j - a_{ll} X_l \qquad (1)$$

where $a_{ij} \sim (R_{ij})^{-3}$, R_{ij} distance between ith and jth oscillator, $x_l(i)$ = displacement at time t, $a_{ll} = \omega_i^2$. The Eigenfrequencies would be prescribed by the roots of

$$X^N - P_1 X^{N-1} - P_2 X^{N-2} \dots - P_N = 0 \qquad (2)$$

where $X = \Omega^2$, $P_1 = \Sigma a_{il}$, $P_K = (-1)^{K-1} A_K$ for $K > 1$.

Here, A_K ($K < N$) is the sum of all principal minors of the order K of the matrix A whose i-jth element is a_{ij}; obviously for $K = N$, A_K is the determinant of matrix A.

When all particles are in a rarefied ensemble off diagonal (a_{ij}) are smaller than diagonal a_{il}. It is reasonable to assume that all $a_{ij} = a_{ji}$ ($i \ne j$) and a_{ij} are real and positive, so that when all the particles are far away from each other, then the Cartesian roots of the equation (2) above are real and positive. If however A_n changes its sign (irrespective of N being even or odd) then at least one of the roots of X becomes negative, so that one of the eigenfrequencies is imaginary and we get catastrophic mode instability.

The condition of the mode instability ($A_2 < 0$) can occur if $a_{12} a_{21} > a_{12} a_{22}$ in two body system; far three body instability ($A_3 < 0$) occurs when $a_{12} a_{21} + S > a_{11} a_{22}$, where

$$S = \frac{1}{a_{33}} (a_{11} a_{23} a_{32} + a_{22} a_{13} a_{31} + a_{12} a_{23} a_{31} + a_{13} a_{32} a_{21}), \text{ a positive quantity.}$$

a_{ij}'s are inversely proportional to distance; so this inequality can be achieved at a greater separation between the first and second particles in the 3 body system.

For $N > 3$ situation the condition is $A_N < 0$; for $N = 3$

$$a_{12} a_{21} + S_1 + S_2 > a_{11} a_{22}$$

where $S_2 > 0$, and S_1 is a function of d_{ij}, where $d_{ij} = a_{ii} a_{jj} - a_{ij} a_{ji}$ (for $ij \ne 1, 2$). The value of d_{ij} is positive only when the particles 1 and 2 are far apart. Then instability can be achieved at a greater distance than isolated 3 body. If d_{ij} is negative the instability is not favoured. In short, if we add a particle to a two body system they will separate for stability, but if there are many more added, some of them will come so close that they make a negative contribution and reduce the value of R_0.

Anticipating a mode instability we, a priori, assume a non-linear quartic interaction and set up a three body Lagrangian

$$L = \tfrac{1}{2}(\dot{X}^2 + \dot{Y}^2 + \dot{Z}^2) - \tfrac{1}{2}\omega^2(X^2 + Y^2 + Z^2) + g_1(R_{12}) XY$$
$$+ g_2(R_{13}) XZ + g_3(R_{23}) YZ - \tfrac{1}{4}\epsilon(X^4 + Y^4 + Z^4) \qquad (3)$$

where ϵ is the strength of anharmonic interaction. From this, linearising the equations of motion around the stable point of origin, we get normal modes. If we put the oscillators as equidistant at vertices of an equilateral triangle the locus of the stable point ($\ddot{X}_0 = \ddot{Y}_i = \ddot{Z}_0 = 0$) undergoes a bifurcation at $R = R_{0l}$ for $g < \omega^2$ it goes in a linear branch $X_0 = Y_0 = Z_0 = 0$; for $2g > \omega^2$ it goes in a parabolic branch where $X_0 = Y_0 = Z_0 = [(2g-\omega^2)/\epsilon]^{1/2}$. In a two body system the bifurcation occurs at $g > \omega^2$ while here it occurs at $g > \omega^2/2$. So there is an increase of R_0 from a two body to three body case. The non-additive component of the van der Waals potential $V_{(NA)}$ is calculated from $V_{(123)} = V_{(12)} + V_{(23)} + V_{(31)} + V_{(NA)}$; the three body potential is sum of two three body potentials and the V_{NA}. We get in the extreme quantum limit, the potential

$$V_{(123)} = -\hbar(\Omega_1 + \Omega_2 + \Omega_3 - 3\omega)$$

$$= -\frac{\hbar}{4}\frac{(g_1^2 + g_2^2 + g_3^2)}{\omega^3} - \frac{5\hbar}{64}\frac{(g_1^2 + g_2^2 + g_3^2)^2}{\omega^7} + 0(g^6)$$

For small ω leading term in V_{NA}

$$V_{NA} \cong -\frac{5\hbar}{32}\frac{(g_1^2 g_2^2 + g_1^2 g_3^2 + g_2^2 g_3^2)}{\omega^7}\frac{1}{R_{12}^3 R_{23}^3 R_{31}^3}$$

which is similar to Axilrod and Teller [15] and Sinanoglu's [16] relation

$$V_{NA} = -\frac{5}{32}\frac{(g_1^2 g_2^2 + g_1^2 g_3^2 + g_2^2 g_3^2)}{\omega^7}(1 + 3\cos\theta_1 \cos\theta_2 \cos\theta_3)$$

θ_1 is the angle whose vertex is at the dipole i.

Thus both quantitative and qualitative agreement with literature can be established by properties of the "loose structure". The net consequence is that addition of extra molecules in an assemblage as a "loose structure" enjoins a new spectrum of intermolecular distance governed by instabilities. Pumps of mass and momentum like heart, lungs, muscle could bring this about to govern autorhythmicity of structures.

I. *Energy of a dissimilar molecule*

Interactions due to vibrational energy of constituent molecules explain the role of "*solvent*" i.e. a dissimilar molecule which is not forming a covalent bond [19]. The coupling constant D (below) determines lowering of bimolecular interactions. In a few body ($n = 2, 3, 4, \ldots$) interaction of one dimensional harmonic oscillator interacting with off-diagonal coupling in displacement of charge we get the Hamiltonian

$$\mathcal{H} = \sum_{i=1}^{n}(p_i^2 + v^2 X_i^2) + \sum_{ij} DX_i X_j$$

Let us assume that we are dealing with a case where off-diagonal coupling D and force constant K (or $v^2 = k$ when $m = 1$) are unequal. Let us assume that molecules are too far apart for electronic coupling. Then it can be shown, as done by Longuet-Higgins [20] that ΔE (internal energy), ΔS

(entropy) and ΔF (free energy) are all attractive, i.e. $-$ve. The Δ refers to change as a result of off-diagonal coupling.

In the case of like atoms 2, 3, 4 etc., the relevant changes are

$$\Delta E_2 \sim -\tfrac{1}{4}\frac{D^2}{v^3} + 0(D^3)$$

$$\Delta E_3 \sim -\tfrac{3}{4}\frac{D^2}{v^3} + 0(D^3)$$

$$\Delta E_4 \sim -\tfrac{3}{2}\frac{D^2}{v^3} + 0(D^3)$$

and this shows

$$\Delta E_4 < \Delta E_3 < \Delta E_2$$

Thus addition of like atoms increases stability. We expect this to be true to a point determined by the elastic limit of the assembly.

If one has unlike atoms, i.e. "solvent as per our definition, so that $v_0^2 \neq v^2$,

$$X^2(X+\alpha) - D^2(2X+\alpha) + 2D^3 = 0$$

$X = v^2 - \omega^2$, $\alpha = v_0^2 - v^2$. ω^2 is normal frequency of the system. ΔE_3^{sol} refers to change in energy due to unlike body. If $\alpha \approx 2D$, we have

$$\Delta E_3^{sol} \approx -\tfrac{1}{6} - \frac{D^2}{v^3} - \frac{\alpha^2}{36v^3} + 0(D^3)$$

From above equation we have

$$\Delta E_3^{sol} > \Delta E_3$$

Now we choose a solvent which gives $\alpha^2 D$ and D^3 as negligible; then it can be shown that

$$\Delta E_3^{sol} \approx \frac{\alpha}{2v} - \tfrac{1}{4}\left(\frac{\alpha^2}{v^3} + \frac{2D^2}{v^3}\right) + 0(D^3)$$

Thus ΔE_3^{sol} can be both greater or less than E_3 depending upon the solvent i.e.

$$v_0^2 \gtrless \left(\frac{v^2 - D}{3}\right)$$

Thus the interaction leads to greater or decreased stability depending upon D between off-diagonal elements. We thus see, using simple arguments, *that stability and symmetry of association and conformational state can be altered* by the fact of gathering of similar or dissimilar molecules in a "loose" structure.

A similar conclusion may be suggested [17] by a study using Bremmerman's optimisation of unconstrained global minimum [18] wherein an influence of length or size of the molecule also becomes apparent (see Table VI) in constraining other molecule. Energy barriers are such that they can be reached by sources of thermal radiation within the body. Number

Table VI. Summary of the Results: Number of Allowed Orientations

Number of possible orientations for one molecule = 324

N-Pentane

Without "solvent"		With "solvent"	
2 molecules	3 molecules	2 molecules	3 molecules
272	178	261	169

N-Nonane

Without "solvent"		With "solvent"	
2 molecules	3 molecules	2 molecules	3 molecules
226	138	211	102

of allowed conformations in a "loose structure" in a couple are markedly reduced by the addition of a third molecule.

"Solvent" refers merely to a third molecule which is simply present without carrying any chemical reaction in the three body aggregate. It is to be distinguished from a true macroscopic solvent.

In order to discuss the implications one may ask: do molecules exchange places in an aggregate in living system? If they do, the stability of structures may be affected in a reproducible manner. This appears to be probale due to simple macroscopic as well as quantum consideration.

(a) *Macroscopically*: the molecules are added or removed by pumps like cardiovascular, respiratory, or intracellular contractile elements which bring about transport of mass and momentum in molecular assemblies all through the body. Thus conditions for ordering in space and tlme are generated.

(b) At the quantum chemical level the specificity and recognition (of molecules, substrates, hormones, drugs etc.) involve specific molecular shape and electronic structure. For understanding effects of a very large number of components one may direct the attention to elementary excitations. A way of considering them, in a uniform formalism, permitting at the same time derivation of macroscopic properties like susceptibility, is pointed out by considering the boson-like states. It appears boson and pseudoboson formalism is adequate for deriving London charge fluctuation forces (*vdW*) and susceptibility and coherence of elementary excitation may *drive* structural variations not permitted in unexcited state.

II. *Bosons in Living State—a discussion*

(A) *London Excitations*: Fröhlich in his work on long range coherence in living systems has considered boson ordering in open systems due to a pump when there is a thermalising bath. He has considered simple harmonic motion due the proton oscillating in hydrogen bonds as the bosons that get so ordered. If this would be true it would require nnrealistically high density of protons or their oscillations. However, this is not required. It is easily seen that London excitations as quasi-"excitons" are quite suitable objects. In the traditional quantum electrodynamic derivation of $v\,dW$, the molecules are assumed to be localised clusters of electrons. If we were to consider localised uncertainty $v\,dW$ generated virtual excitations (London) as virtual excitons or in other words if we consider the molecules as localised clusters of bosons the expression of potential remains formally the same.

Thus we may write the Hamiltonian for the interacting cluster of boson, in natural unit, as

$$\mathcal{H} = \sum_i \Omega_i b_i^\dagger b_i + \sum_q \tfrac{1}{2}\omega_q(a_q^\dagger a_q + a_q a_q^\dagger)$$
$$+ \sum_{i,k,q}[g^+(q)a_q^\dagger + g^-(q)a_q]b_k^\dagger b_i$$

where b's represent boson and a's photon operators and $g^\pm$ coupling at vertex. The zero of the energy scale is shifted to zero points enery of bosons, to maintain correspondence with electronic formulation. The unperturbed eigenstates and energy eigenvalues are given by

$$\psi_0 = |N_l N_k \ldots \mu_q, \mu_r \ldots\rangle$$

and

$$E_0 = \sum \Omega_i N_i + \sum \omega_q(nq + \tfrac{1}{2})$$

where N and n are boson and photon occupancy number respectively. The second order diagrams contribute self energy because the nature of process is fluctuation and not scattering.

Renormalising to second order, one gets

$$\langle \Omega_i \rangle = \Omega_- - \tfrac{1}{2}\sum_{kq}(N_k+1)\left(\frac{g^+(q)g^-(q)}{\Omega_k - \Omega_l + \omega_q} - \frac{g^-(q)g^+(q)}{\Omega_k - \Omega_l - \omega_q}\right)$$

and

$$\langle \omega_q \rangle = \omega_q - \tfrac{1}{2}\sum_{lk} N_l(N_k+1)\left(\frac{g^+(q)g^-(q)}{\Omega_k - \Omega_l + \omega_q} + \frac{g^-(q)g^+(q)}{\Omega_k - \Omega_l + \omega_q}\right)$$

The lowest order contribution is from the fourth order diagram and the interaction potential can be shown to be

$$V = -\tfrac{1}{2}\sum_{\pm}\sum_{qr}\sum_{ij}\sum_{\alpha\beta} N_i(1+N_j)N_\alpha(1+N_\beta)$$
$$\times\left[\frac{n_q + \tfrac{1}{2}}{\pm\omega_g \pm \omega_r}\left(\frac{S_r g^\pm(q)g^\pm(r)}{\Omega_j - \Omega_l \pm \omega_q} + \frac{S_r g^\pm(q)g^\pm(r)}{\Omega_j - \Omega_l \pm \omega_q}\right)\right.$$
$$\times\left(\frac{g^\mp(q)g^\mp(r)}{\Omega_\alpha - \Omega_\beta \mp \omega_q} + \frac{g^\mp(q)g^\mp(r)}{\Omega_\alpha - \Omega_\beta \pm \omega_q}\right) - \frac{S_q S_r g^\pm(q)g^\pm(r)}{(\Omega_j - \Omega_l \pm \omega_q)(\Omega_j - \Omega_l \mp \omega_r)}$$
$$\left.\times\left(\frac{g^\mp(q)g^\mp(r)}{\Omega_\alpha - \Omega_\beta + \Omega_j - \Omega_l} + \frac{g^\mp(q)g^\mp(r)}{\Omega_\alpha - \Omega_\beta - \Omega_j + \Omega_l}\right)\right]$$

S_r and S_q are signature factors, i.e., $S_r g^+(r) = +g^+(r)$, $S_r g^-(r) = -g^-(r)$ and in the same way for $S_q g^\pm(q)$.

The summation $\sum_{\pm}$ tells that there is no net emission or absorption of photons, hence care should be exercised for use of signatures. We see that upto the fourth order the total energy is linear in $(n_q + \frac{1}{2})$ and one can write

$$\langle \omega_q \rangle = \omega_q - \chi(\omega_q, +q, -q) = \sum_{\pm} \sum_r \frac{S_r}{\pm \omega_q \pm \omega_r} \chi(\omega_q, +q, \pm r) \chi(-\omega_q, -q, \mp r)$$

in which

$$\chi(\rho, \pm q, \pm r) \equiv \sum_{i,k} N_i(1 + N_k) \left(\frac{g^{\pm}(q)g^{\pm}(r)}{\Omega_k - \Omega_i + \rho} + \frac{g^{\pm}(q)g^{\pm}(r)}{\Omega_k - \Omega_i - \rho} \right)$$

Thus one derives the general susceptibility relation of McLachlan, i.e. the Fourier transfer of the response function or in other words the correlation function of the fluctuation of this system.

For the statistics of the quasi-bosons and quasi-photons we start with diagonalised Hamiltonian with quasi-boson and quasi-photon creation and annihilation operators, and it can be shown [14a] that quasiphotons are an independent system of bosons, while quasi-bosons are not. Since the process is fluctuating any change in the occupancy N_i is associated with a concomitant change in $N_j (i \neq j)$ and we obtain a Bose-Einstein distribution, with the usual chemical potential μ from

$$N_i = 1/(e^{(E_i - \mu)/KT} - 1)$$

with the usual symbolisation and the identity

$$\coth \left(\frac{E_k - E_i}{2KT} \right) = \frac{N_i(N_k + 1) + N_k(N_i + 1)}{N_i(N_k + 1) - N_k(N_i + 1)}$$

and the interaction potential

$$V = (8\pi i)^{-1} \int_c d\rho \, \coth \left(\frac{\rho}{2kT} \right) \times \sum \frac{\chi(\rho, \pm q, \pm r)}{\omega_q \pm \rho} \frac{\chi(-\rho, \mp q, \mp r)}{\omega_r \pm \rho}$$

By contour distortion we arrive at

$$V = (4\pi i)^{-1} \int_{-i\infty}^{+i\infty} d\rho \, \coth \left(\frac{\rho}{2kT} \right) \times \sum_{qr} \frac{\chi_1(\rho, \pm q, \pm r)}{\omega_q \pm \rho} \frac{\chi_2(-\rho, \mp q, \mp r)}{\omega_r \pm \rho}$$

Thus one can express the identity with McLachlan relationship of usual susceptibilities. This has very wide implications. It leads to proper formulae of a variety of potentials of interaction as follows:

(i) For small isotropic molecules at moderate distance where l = length of molecule, R = intermolecular distance any λ wavelength of photon, the London interaction energy is

$$W = -\frac{3h}{2\pi^2 R^6} \int_0^\infty d\xi \, \alpha(i\xi) \beta(i\xi)$$

where $\alpha(i\xi)$ and $\beta(i\xi)$ are the polarisabilities of two molecules at imaginary frequency $i\xi$.

(ii) For small anisotropic molecules at moderate distance one gets

$$W = -\frac{h}{4\pi^2}\int_0^\infty d\xi \alpha_{ij}(i\xi)T_{jk}\beta_{kl}(i\xi)T_{li}$$

subscripts being tensor indices and dipole-dipole interaction tensor is given by

$$T_{jk} = \frac{\delta_{jk}}{R^3} - \frac{3R_jR_k}{R^5}$$

(iii) If molecules are dissolved in a uniform isotropic dielectric, $\alpha(i\xi)$ and $\beta(i\xi)$ are excess polarisabilities and one uses the frequency dependent interacting tensor

$$T_{jk} = \epsilon(i\xi)^{-1}\left[\frac{\delta_{jk}}{R^3} - \frac{3R_jR_k}{R^5}\right]$$

where ξ is the dielectric constant of the medium.

(iv) For polyatomic molecules at moderate distance in vacuum, one uses α and β as the mutual susceptibility and one gets

$$W = -\frac{h}{4\pi^2}\int dr_1 \int dr_2 \int dr_1' \int dr_2' \int_0^\infty d\xi \frac{\alpha(r, r_1, i\xi)\beta(r, r_1, i\xi)}{|r_1 - r_1'||r_2 - r_2'|}$$

Primed quantities refer to a molecule, and unprimed to the other.

(v) Attraction of an isotropic molecule near the surface of dielectric is given by

$$W = -\frac{h}{8\pi^2 R^3}\int_0^\infty d\xi \alpha(i\xi)\left[\frac{\epsilon(i\xi) - 1}{\epsilon(i\xi) + 1}\right]$$

(vi) Two molecules reasonably far apart on the surface on interacting will have an effective potential energy

$$W = -\frac{3h}{2\pi^2 R^6}\int_0^\infty d\xi \alpha(i\xi)\beta(i\xi)\left[\frac{2(\epsilon^2 + 5)}{3(\epsilon + 1)^2}\right]$$

One can also connect this to Casimir-Polder interaction where R^{-6} dependence is changed to R^{-7}.

In sum, the coupling in an aggregate of atoms or molecules can determine number and size dependent instabilities, spectrum of distance, and in a "loose structure" provides energy sufficient to derive a variety of conformations.

(B) *Structural consequence of condensation or coherence of elementary excitations. Role of elastic restituting forces in the molecule*: For a substratum or a system as mutable as this, Bose-Einstin like "condensation" under the condition of a pump and a thermalising bath becomes a distinct possibility as discussed earlier [21]. So far as obtaining experimental proof is concerned, we may consider the following argument (from 22) to establish a route for experimental verification. Since quanta are pumped in at a critical pumping rate S_c, population and "condensation" in the lowest mode will occur. The S_c can be found by calculating the upper limit in the number of quanta in excited state N^* with ($K \neq 0$) and $\mu = (\omega_0 - \hbar\theta)$ in the high-temperature approximation

$$N^* = \sum_{K(k \neq 0)} n_k(\mu) \cong \left(\frac{1}{\beta\hbar} + \frac{\phi(\omega_0 - \theta_0)}{\chi Z} \int_{\omega_0^+}^{\omega_0} \frac{D(\omega_k) \, d_{\omega k}}{\omega_k - \theta_k - \omega_0 + \theta_0} \right)$$

where ω_0^+ is the next to lowest frequency and is given by

$$[(\omega_0^{+2} + \Delta^2\pi^2)/Z^2)]^{1/2} \cong \omega_0 + \Delta^2\pi^2/2\omega_0 Z^2$$

S_c is now associated with the critical energy in the polarization mode, which must be exceeded and is given by

$$\epsilon_t = \left[\left(\frac{\hbar\omega_0}{\gamma} \right) \frac{4\omega_0^2 Z}{\Delta^2\pi^2\beta} \right] \Big/ \left(1 + \frac{c^2\epsilon_l}{\sigma^2\omega_0^4 L} \right)$$

The rate for N^* differs from the usual pumping rate by a factor due to elastic restoring force, equal to $(1 + C^2\epsilon_t/\sigma^2\omega_0^4 L)^{-1}$. It is interesting to note that the elastic restoring force lowers the S_c, and that is proportional to ϵ_t and as θ_0 increases, the mode become softer and the frequency becomes imaginary at certain levels, and the material may be torn apart. This system is stabilised, however, by a nonlinear term which produces a restoring force which is quartic in the polarization field. The effective potential energy that includes this stabilization is given by

$$V_{\text{eff}}(P_0) = \tfrac{1}{2} \frac{C^2\epsilon_t^2}{\sigma^2 L\omega_0^4} + \tfrac{1}{2} \left(\omega_0^2 - \frac{C^2\epsilon_t}{\sigma^2 L\omega_0^2} \right) P_0^2 + d^2 P_0^4$$

P_0 is the contribution of the lowest mode to the polarization field, and d is the strength of stabilizing interaction. At $\epsilon_t \geqslant \dfrac{\sigma^2 L\omega_0^4}{C^2}$ the mode goes soft. Below this, P_0 is zero and therefore P_0 at equilibrium is

$$(P_0)_{\text{equil}} = \pm \frac{1}{2d} \left(\frac{c^2}{\sigma^2} \frac{\epsilon_t}{L^2\omega_0^2} - \omega_0^2 \right)^{1/2}$$

Above discussion indicates the possibility of metastable state displacement appearing like a ferroelectric phase transformation. Thus either Bose condensation or a "ferroelectric" state occurs, the former at a lower pumping rate. The critical flux without restoring force from adenosine is calculated to be 0.4 mW/cm². This is already in the order of magnitude observed in experiments with Escherichia coli. In addition to the experiments with E. coli there has also been the enhancement of specific activity of chymotrypsinogen when in low concentration, by laser irradiation [23]. Phytochrome induced root growth can be caused by laser irradiation as far away as ¼ mile [24]. Bhaumik et al. [25–28] have tried to explain this "switching on" of an enzyme into a giant dipolar state, and have formulated two separate enzyme kinetic schemes [28] including both excitation and collisional de-excitation in the two schemes:

one parallel case, in which the enzyme is activated by a conventional channel as well as by laser irradiation and the other, in series, in which only the excited-state enzyme participates in the reaction. The two channels suffer de-excitation by various processes. In the series formulation it is excited by collisions and de-excited. Excitation by irradiation leads to excited enzyme substrate complex formation and the usual resultant products. Relevance of this analysis to photoexcitation of chlorophyll and to such occurrences as thymine dimerization, photochemical reaction [30–32], photoviscosity effects [33], and photopyroelectricity [24] is evident.

(C) *Bose-Einstein-like coherence—A way of living state*: The above viewpoints, however, beg some questions. It may be possible that the "condensation" or "alignments" or 'coherences' occur but is it the way of Nature? It is quite evident that the interior of bigger organisms is not available to a varying of radiations relevant to the concept. Other points to consider are: high body temperature (300°K) in homeo-iotherms, the fluxes and flow of ions, presence of a material of variable dielectric, heterogeneity of molecular species, as well as the identification of bosons. Some of these are unnecessary and others demand extension of the concept as provided below.

As stated above the restoring force reduces the critical pumping rate of 0.4 mW/cm^2. Effects of electrical fields are sometimes observable in chemical systems like field effect in liquid crystals [35]. Even so the molecular force field with a mosaic of effective residual electronic charges, in a bed of gegenions located on atoms and bonds on ATP and spectral data of adenosine studied by us, shows that the molecules can provide such a force field, using the narrow-band approximation. Mutual distortion mechanisms [32, 33], whether photoinduced or otherwise, are cases in point. Using the procedure by Bykov [36], Jain [37] calculauted atomic and bond charges in molecules. We may suggest the possibility of damped oscillation of charge on the bond during the period of bond formation under the influence of electronegativity which could be detected by appropriate resonance spectroscopy. Even luminescence can result as is already well known in the cases of the luciferin-luciferase system. The net result in the case of enzyme substrate, drug receptor, or antigen-antibody union will be reduce E of the whole system.

Judging from the orientational and rotational freezing of a pair of molecules like pentane [17] or others, by the full molecular force field of pentane [17, 17a, 12a, 12b, 38–41] the entire subject of the conformational hyperspace, intramolecular coupling by mutual interactions [42–45] variation in charge densities on molecularization [44–45], lengthening of CN bond in haem-haem interaction [46], cascading of conformations in solutions [47–48], coenzyme conformation [49], lowering of barriers of internal rotation [50] and intramolecular acyl migration in lipids point to the effectiveness of molecular and atomic force fields at distances relevent to living state. *Many such interactions are strengthened in a hydrophobic environment. Lowering of the internal energy of the system by Bose pumping strengthens the notion of its feasibility.* If it is true as a general mechanism, then the consequences are recognition of a specific substrate or an antigen (in case of an antibody). By holding the molecule for a sufficiently long period for necessary transformation, the molecule is transformed, pumping ceases, and the activity is reduced. Thus the gradients are maximised and the substrates seek such "sinks" to which they are directed.

(D) *Boson ordering—A unified and widespread phenomenon*: In the above analysis London-van der Waals interaction, photons and elastic recoil have been considered. The London-van der Waals interation can be considered excitonic in nature due to uncertainty-generated electron excitation. To consider the structural basis of such a view, let us consider, as in Lamb shift, trapped and detrapped electron. This would mean electron in a hole, or a zero radius exciton. We now consider the interaction between the Lamb shift type "excitons", the photons emitted or absorbed in such transitions and the phonons [62]. The question then arises is what would be the energy profile under this treatment. Ultimately it will be a step in the direction of defining the interaction responsible for the stability of living matter.

We discuss the question of energy change in such excitations and other excitations. The interaction between two electrons trapped under potential fields may be viewed, apart from the coulombic interaction, as interaction between two excitonic states through the simultaneous exchange of photons. The excitonic states arise out of the elementary fluctuations in the electron field, which could otherwise be regarded as electron, behaving in the limit as Frenkel excition.

The interaction matrix element for this interaction is

$$M_{eh} = \langle i_e | \mu \cdot \epsilon | j_e \rangle \langle j_e g_h | V_{eh} | g_e f_h \rangle$$

A further correction to this interaction in energy space would be provided by the coupled motion of the potential fields, and thus in a dense media where the two potential fields are situated, in a harmonic approximation, this could be visualised as an exciton (trapped electron) phonon interaction which is given by

$$G_{eh} = \langle g_h \chi(\epsilon_q) | V_{ev} | i_e \chi(\epsilon_q - \Delta) \rangle \langle j_e \chi(\epsilon_q - \Delta) | V_{ev} | f_h \chi(\epsilon_q - \Delta) \rangle$$

(for phonon energy $\epsilon_q \rangle \Delta$)

where $|i_e\rangle, |j_e\rangle, |g_e\rangle$ are electron states, $|g_h\rangle, |f_h\rangle$ (coupled to trapped electron states through photon/phonon) are exciton states and $|\chi\rangle$ are phonon states. μ is dipole moment and ϵ is electromagnetic field operator, V_{eh} is exciton-trapped electron interaction, V_{ev} trapped electron vibronic coupling and $\Delta = \epsilon_i - \epsilon_j$, the trap energy with ϵ_i energy of electron.

In second quantized notation these interactions can be written as

$$M_{eh} = \sum_{j_e, g_e, g_h, f_h} \sum_k \widetilde{M}_{eh} a_{ie}^\dagger c_{gh}^\dagger c_{fh} a_{ge} (b_k + b_k^\dagger)$$

$$G_{eh} = \sum_{g_h f_h} \sum_q \widetilde{G}_{eh} c_{g_h}^\dagger c_{f_e} (b_q + b_q^\dagger)$$

$$\widetilde{M}_{eh} = \langle \bar{i}_e | \bar{\mu} \cdot \bar{\epsilon}_0 | \bar{u}_k \bar{j}_e \rangle \langle \bar{j}_e \bar{g}_h | \left(\frac{V_{eh}}{\epsilon_j - \epsilon_j} \right) | \bar{g}_e f_h \rangle$$

$$\widetilde{G}_{eh} = \langle \bar{g}_h \chi(\epsilon_q) / \bar{V}_{ev} | \bar{i}_e \chi(\epsilon_q - \Delta) \rangle \langle \bar{j}_e \chi(\epsilon_q - \Delta) | \bar{V}_{ev} | f_h \chi(\epsilon_q - \Delta) \rangle$$

$a_i^\dagger$, $b_k^\dagger$, $b_q^\dagger$, $c_g^\dagger$ are creation operators for electron, photon, phonon and exciton. Similarly, a_l, b_k, b_q, c_f are annihilation operators for electron, photon, phonon and exciton, $\bar{u}_k$ describes photon wavefunction. The term $\bar{V}_{ev}$ is given by

$$\bar{V}_{ev} = \sqrt{\frac{h}{2MN\omega_q}} \frac{\partial V}{\partial X} \bigg|_{X=R_0}$$

We take quinone as an example. The vibronically coupled exciton (triplet) state matrix element $\widetilde{G}_{eh}$ comes out for $C=0$ stretching mode in quinone group at 1740 cm^{-1} as

$$\widetilde{G}_{eh} = 0.0316 \text{ (eV)}$$

at vibronic energy $\epsilon_q = 0.216$ (eV).

The exciton (trapped electron) is coupled to photon through an interaction of the order of

$$\widetilde{M}_{eh} = 0.80 \text{ (eV)}$$

and self-energy contribution $= 0.27$ (eV) $= 6.3$ kcal/mole at 300 mμ photon wavelength for triplet-triplet transition. The relation for $\widetilde{M}_{eh}$ in terms of exciton radius r_2 is

$$\widetilde{M}_{eh} = 0.0834 e^2 \alpha_h (\alpha_h/\alpha_e) S_e S_e S_h [15 - e^{-2\alpha_e r_2} \{ 4\alpha_e^5 r_2^5 + 10 \alpha_e^4 r_2^4 + 20 \alpha_e^3 r_2^3$$
$$+ 30(\alpha_e^2 r_2^2 + \alpha_e r_2)(1 - \alpha_e^3/\alpha_h^3) + 15 \alpha_e^3/\alpha_h^3 \}]$$

$$S_e = \langle \bar{i}_e | \bar{u}_k \bar{j}_e \rangle$$

$$S_e = \langle \bar{j}_e | \bar{g}_e \rangle$$

$$S_h = \langle \bar{g}_h | \bar{f}_h \rangle$$

and $\alpha_h = 1.2 a_0^{-1}$ (in terms of Bohr radius a_0)

for triplet exciton states and $\alpha_e = \sqrt{\frac{2m\epsilon_l}{\hbar^2}}$, where ϵ_l is the energy of electron state.

It is thus interesting to note that the energy of excitation in this formalism is 6.3 kcal/mole, close to energy involved in biological interaction, i.e.

a few to 7 kcal/mole. We also see that London attraction due to charge fluctuation can be viewed in excitonic framework as done earlier.

Rhythmic variation in crystallisation or oscillation in enzyme activity are explainable as being due to elastic field coupling.

III. *Internal Pump and "Self-organization"*

Comfirmation has often been sought by energy changes under the influence of laser [23]. This is rarely the condition in life. It is not necessary for self-organisation. The pump, in order to subserve the needs of self-organisation must be within the system. It is known that transitions between rotational quantum levels are always occurring at body temperature and there is continuous energy input maintaining this radiation. The formation or breakage of bonds also leads to changes in energy [44] and thus a radiative field always exists at the conditions of body temperature. The *rate* of pumping differs from molecule to molecule and determine the specificity.

All the above discussion is aimed to pinpoint the *connectivity in a system which is far from equilibrium, with input of mass and energy and thermalising bath, both in microsystems as well as in macro ones and establish the basic framework in which self-organisation, self-reference, can operate, as per Prigogine.*

The extended generalisation has also been pointed out by Haken, who has coined the word synergetics [54–55], Walter B. Cannon [56] had drawn up an analogy between various feedback systems in cities, economic flows, etc. and suggested a widespread utility of these concepts. The concept of synergistic action, permissive action, or conditioning action in biology emanates from the work of Selye [57–59]. The matter has been reviewed very remarkably by Jantsch [59]. In the next papers we give some more basis for instabilities [60, 61].

Acknowledgements

The author wishes to acknowledge very helpful discussions with Dr. G. C. Shukla, Dr. K. Bhaumik, Dr. G. Subha Rao, S. K. Dube and R. S. Tyagi whose work has been cited in this paper and with whom it was a pleasure to have extremely useful discussions on this paper. The author is also grateful to Prof. H. Frohlich and Prof. K. P. Sinha on some aspects of the paper.

References and Comments

1. Pirschle, K. (1934), *Planta*, **23**, 177–224.
2. Lendle, A. (1936), *Chem. Ztg.*, **60**, 933–935.

3. Thatcher, R. W. (1934), *Science*, **79**, 463–466.
4. Frey-Wyssling, A. (1935), Naturwiss, **23**, 767–769.
5. Steinberg, R. A. (1939), *J. Agriculture Res.*, **57**, 251–258.
6. Wald, G. (1962), "Life in the Second and Third Periods", in; *Horizons in Biochemistry*, Kasha, M. and B. Pullman (eds.), Academic Press, N. Y., p. 127.
7. Rich, R. (1965), *Periodic Correlations*, W. A. Benjamin, N. Y.
8. None of the following factors that influence chemical trends of periodic table can account for the chemical behaviour of the elements: electron configuration of atoms, atomic number and orbital energy levels, "penetration" effects and shielding, "spin pairing" and "exchange" effects, spin dominated or van der Waals dominated aggregation, "coincidence" effect, electron correlation, formation of hybrid orbitals, atomic radii, covalent and metallic radii, "exposure" of inner orbitals in ions, orbital assymetries, ionization potentials, electron affinity, orbital electronegativity, electrostatic or Sanderson or Mulliken electron-negativities, valence state electron affinities, polarizability, polarising strength, thermodynamics of atomisation, acidity, basicity in solvents, hydration of cations, redox behaviour, self-association, precipitation, solubility, coordination and complex formation.

    Na^+, Cu^+ have similar radii, but different chemistry. d electrons differentiate Na^+ and Cu^+, S^{6+} and Se^6 have dissimilar radii, but several similarities. Chemically Ag^+ and K^+ are of nearly same size but have different solubilities.

    Cu^+ is smaller in X and Y axis than in the Z axis. It has vacancy in dx^2-y^2 orbital, but has a spherically symmetrical d_{10}. Chemical reactivity of compounds is not easily predictable. Silane and phosphine take fire in air easily but other reducing agents, methane and ammonia are relatively inert. H_2O_2 can be preserved longer than H_2S_2. Atmosphere has large amounts of O—O inspite of the fact that there are many good reducing agents on Earth.

9. Free energy of hydration would put Na^+ superior to K^+, if that is the necessary requirement for biological activity. Li^+ should be even superior but it is toxic and so are Rb^+ and Cs^+ which have larger G. Colour, redox reactivity, strength for complexing or precipitation have little to contribute. Perhaps a whole group of properties determines what should be suitable for any specific function which is to be subserved. Perhaps what one notices is the atavism of the evolutionary past.

    The first function and the approximation of the composition of sea water, and the preponderance of Na^+ over K^+, although K^+ has more to do intracellularly, is clearly a reminder of the marine past. But this function is a very vital one looking at the elaborate and sensitive handling of water and ionic balance. Indeed, this attracted the attention of Claude Bernard, Cannon [56] and Selye [57, 58]. This, along-with separation and transport of gases, appears as one of the central "goals" of the living state. Both, the ionogenic potentials, propagation of altered electrical state and the production and dissemination of required molecular species, enzymatic or hormonal devices pressed into play and are responsible for fine titration and modulation of affinity and activity of reactive groups.

The role of carbon is truly extraordinary. It shows the sharpest difference between terrestrial abundance and organisms abundance, being 350 times more than in the earth's crust. But once selected it has unique role. It is the key element in "Frey-Wyssling line" which is the more functionally drawn up periodic table marking the biologically essential from biologically non-essential elements, at least in plant nutrition.

One of the important properties of second period is the ability to form multiple bonds, as pointed out by Wald [6]. This has the important consequence that a spectrum of bond strengths and lengths now becomes available, allowing for the generation of asymmetry out of random association as we shall show below. The second most important property of carbon is its hybridisability which gives rise to a tetrahedral arrangement which by substitutions leads to helices, sheets, pleated sheets, balls. In spite of being otherwise hydrophobic, its tetrahedrally disposed bonds are best able to correlate with water and very many molecules which show this structure. As Pople has shown if all orbitals are equally occupied it does not matter mathematically whether the completely antisymmetric wavefunction is based on atomic orbitals or on localised, hybrid, "equivalent" orbitals. The third most important property of carbon because of its tetrahedral bond disposition is the ability to form emphiphiles, perhaps the most important single property required for the Living State. Solvophobic force, partitions, surfaces can now be generated, with suitable selective gates.

The arguments for S and P for group and energy transfer reactions are already presented [6]: (1) they form open and usually weaker bonds than their congeners in the 2nd period i.e. N and O, (2) they possess $3d$ orbitals, so that their valence can increase beyond four, (3) the capacity to form multiple bonds, (4) their tendency to add lone pair of electrons in unoccupied 3rd orbitals. These lead to metastabilities and vulnerability to exchange reactions. In short, they provide mutability of shapes and can provide energy pumps.

The capability of interaction with the lone pair of electrons in oxygen, leading to oscillation in bond lengths and in affinity and participation of nitrogen in electron-handling cyclic structeres provides for "bussing" the electrons and the participation in enzyme reactions.

The reference of Mn, Fe, Co, Cu and Zn also deserves comment.

For a good hybrid orbital the component orbitals must overlap well and not have disparate energies. In the early nd elements where n may be 3, 4 or 5, the energy of $(n+1)P$ orbitals is rather too high. In the later period they are somewhat lower. Accordingly d^6 case of Fe provides the most numerous structure, while d_{10} Zn has the least numerous.

Perhaps a remark on Si is well in order. A life based on Si is not possible on Earth, since S—Si has half the strength of C—C bond and $1\frac{1}{2}$ times its length. It is thus vulnerable to attack by water. This accounts for natural selection of carbon over silicon, although it is its congener. It also suggested that carbon-based life on the Earth must have been precedeed by the existence of water. If silicon has to play a role of structures similar to those by carbon in terrestrial living forms, it would perhaps be possible, if at all, in planets like Mars.

10. Selye, H. (1956), Personal Communication.
11. Mishra, R. K. and G. S. Rao, *Conformational Atlas, Molecular Fragments*. Lists essential molecular fragments from which physiologically or pharmacologically molecules can be constructed. (Unpublished).

List of Small Molecules
In addition to the small list of functional groups, the small molecules that are found in living systems are: glycol, erythritol, sugars, fatty acids, cellulose, digitoxin, substituted amines, adrenaline, ephedrine, urea, guanidine, creatine, barbituric acid, sulphanilamide, catechol, quinol, cyclopentane, cyclohexane, benzene. naphthalene and naphthol, carbazole, furane, thiophene, pyrrole, thiazole, imidazole, pyridine, quinoline, coumarone, indole, acridine, acriflavine, -propyl piperidine, tropic acid, nicotinamide, pyridoxin.

List of Molecular Fragments
Hydrocarbons, straight or branched, unsubstituted or substituted, bearing hydroxyl, NH_2, S, SH and natural and artificial amino acids, nucleic acid bases sugars (ribose etc.).

12. Selye, H. and R. K. Mishra (1951), *Acta Pharmacol. et Taxicol.*, **14**, 359.
12a. Mishra, R. K. (1972), Fluctuating liquid crystallinity.
12b. Mishra, R. K. (1975), *Mol. Cryst. Liquid Cryst.*, **29**, 201–224. The condition for the Living State; IV International Cong. Biophys.,

Moscow, Abstract (EXXX a2/5).
12c. Brown, G. H. and R. K. Mishra (1971), *Agriculture and Food Chem.*, **19**, 645.
13. Dzyaloshinskii, I. F., E. M. Lifshitz and L. P. Pitaevskii (1960), *Soviet Physics, TETP*, **37**, 161.
14. Mishra, R. K. (1970), *Mol. Cryst. Liquid Cryst.*, **10**, 85–114.
14a. Mishra, R. K., A. Shrivastava and K. Bhaumik (1980), "Validity of the two-body model of van der Waals interaction between 'loose' structure in the body system" (under publication).
15. Axilrod, B. M. and E. Teller (1943), *J. Chem. Phys.*, **11**, 299.
16. Sinanoglu, O., S. Abdulnur and N.R. Kestner, in: *Electronic Aspects of Biochemistry*, B. Pullman (Ed.), Acad. Press, N.Y., p. 301.
17. Rao, G. S., R. S. Tyagi and R. K. Mishra (1970), International Conference on Liquid Crystals, Kent, Ohio, Abstract.
17a. Mishra, R. K. and G. Subha Rao, *A Study of Intermolecular Association in the Living State*, V (under publication).
18. Rao, G. S., R. S. Tyagi and R. K. Mishra (1981), *J. Theor. Biol.*, **90**, 377–389.
19. Mishra, R. K. and G. C. Shukla, "Intermolecular forces and solvent effect for few bodies" (unpublished work).
20. Longuet Higgins, H. C. (1965), "Intermolecular Forces", Spiers Memorial Lecture, Faraday Soc. Symp., p. 7.
21. Mishra, R. K., K. Bhaumik, A. Srivastava and S. S. Chaudhury (1981), *Intern. J. Q. Chem.*, **20**, 377–383.
21a. Maclachlan, A. D. (1963), *Proc. Royal Soc.*, **4**, 171, 387 and 274.
22. Mishra, R. K., K. Bhaumik, S. C. Mathur and S. Mitra (1979), *Int. J. Q. Chem.* **16**, 691–706.
23. N. Kolias and R. Melander (1976), *Phys. Lett.*, **57A**, 102.
24. Paleg, L. G. and D. Aspinall (1970), *Nature*, **228**, 970.
25. Bhaumik, P., K. Bhaumik and B. Dutta-Roy (1976), *Phys. Lett.*, **56A**, 145.
26. Bhaumik, D., K. Bhaumik and B. Dutta-Roy (1976), *Phys. Lett.*, **59A**, 77.
27. Bhaumik, D., K. Bhaumik, B. Dutta-Roy and M. M. Engineer (1977), *Phys. Lett.*, **A 62**, 197.
28. Bhaumik, D., K. Bhaumik, A. K. Roy and B. Dutta-Roy (1978), *Bull. Math. Biol.* (in press).
29. Brown, G. H. (ed.) (1971), *Photochromism*, Wiley-Interscience, New York, p. 687.
30. Galanian, M. D. (1951), *Zh. Eksp. Theor. Fiz.*, **28**, 485.

31. Lovrien, B. and T. Linn (1967), *Biochemistry*, **6**, 2281.
32. Lovrien R. and J. C. B. Waddington (1964), *J. Am. Chem. Soc.* **86**, 2315.
33. Lovrien, R. (1967), Proc. Natl. Acad. Sci. USA, **57**, 236.
34. Moeckel, P. (1968), *Z. Chem.*, **8**, 382.
35. Schadt, M. and W. Helfrich (1971), *Appl. Phys. Lett.*, **18**, 187.
36. Bykov, G. V. (1974), *Electronic Charges on Bonds in Organic Compounds*, Pergamon, London.
37. Jain, V. K. (1970), *J. Chem.* **8**, 298.
38. Walter, P. and B. Pachofen (1977), Preprint.
39. Yos, M. T., W. L. Rade and H. Jehle (1957), Proc. Natl. Acad. U.S.A., **43**, 314.
40. Mishra, R. K. and N. K. Roper (1974), *Liq. Cryst. Ordered Fluids*, **2**, 743.
41. Mishra, R. K. and R. S. Tyagi (1974), *Liq. Cryst. Ordered Fluids*, **2**, 759.
42. Kuhn, K. (1972), *Angew. Chem. Int. Ed.*, **11**, 798.
43. Smillie, L. F., A. C. Enenkel and C. M. Kay (1966), *J. Biol. Chem.*, **241**, 2097.
44. Boyd, D. B. (1970), *Theor. Chim. Acta*, **18**, 184.
45. Boyd, D. B. (1973), *J. Am. Chem. Soc.*, **94**, 64.
46. Perutz, M. F. (1972), *Nature*, **237**, 495.
47. Grell, E., Th. Funck and F. Eggars (1972), Proccedings of the Grandata Symposium on Molecular Mechanisms of Antibiotic Action on Protein Biosynthesis and Membranes, D. Vazzney (ed.), American Elsevier, New York.
48. Funck, Th., F. Eggers and E. Grell, (1971), First European Biophysics Congress (Wiener Acad. Med., Vienna).
49. Balasubramanian, D. and D. L Wetlaufer (1966), Proc. Natl. Acad. Sci. USA, **55**, 762.
50. Boyd, D. R. (1973), *Theor. Chim. Acta.*, **30**, 137.
51. Brumberger, M. (1970), *Nature*, **227**, 490.
52. Shnoll, S.E. (1965), In: *Molecular Biophysics*, **Nauka,** Moscow, Vol. 56.
53. Eigen, M. (1960), *Fast Fundamental Transfer Process in Aqueous Blmolecular System*, MIT, Cambridge, Mass., pp. 18–19.
54. Haken, H. (1964), *Z. Phys.* **181**, 96.
55. Haken, H. (1976–77 and 1978), *Synergetics*, Springer-Verlag, Berlin-Heidelberg-New York.
56. Cannon, W. B. (1932 and 1939), *The Wisdom of the Body*, Routledge and Kegan faul, London, pp. 305–324.

57. Selye, H. (1936), "The alarm reaction", *Canad. M. A. J.*, **34**, 706.
58. Selye, H. (1950), *Stress,* Acta Inc., Montreal.
59. Jantsch, E. (1980), *The Self Organising Universe,* Pergamon Press, based on Gaither Lectures in Systems Science, May, 1979, University of California at Berkeley.
60. Mishra, R. K. and G. C. Shukla (Present volume).
61. Shukla, G. C., K. Bhaumik, R. K. Mishra (present volume).
62. Mishra, R. K., S. K. Dubey and R. S. Tyagi (unpublished work).

Metastable States and the Soft Mode in Liquid Crystalline Matter

R. K. Mishra and G. C. Shukla

Department of Biophysics
All India Institute of Medical Sciences
New Delhi-110029, India

1. Introduction

Despite the fact that the molecular aggregate ("lattice") model of liquid-crystal does not clearly depict the physical situation of liquid crystaline state, it is still useful to study specially its dynamical behaviour in search of a state the stability which could be changed by altering the temperature. In an earlier work Novakovic and Shukla (1) considered a lattice of liquid-crystal with random phase approximation, wherein the orientational ordering of the system was specified with the components of pseudo-spin ($S=1$). The collective behaviour of elementary excitations in the liquid-crystal was studied leading to soft-mode in the system at which the liquid-crystal underwent transition to isotropic liquid phase. It has been pointed out by De Gennes (2, 3) that any second rank tensor should suffice to describe the orientational ordering in liquid-crystal. We test this and develop this in this communication by taking the quadrupole moment tensor. For uniaxial liquid-crystal there are only two nonvanishing quadrupole operators, namely Q_0 and Q_2 defined below:

$$Q_0 = 3S_z^2 - S(S+1) = 3S_z^2 - 2, (S=1)$$
$$Q_2 = (S_x^2 - S_y^2)$$

The next section we write down the Hamiltonian of the system i.e. liquid-crystal of nematic type in term of the operators Q_0 and Q_2. The Hamiltonian is simplified in mean-field-approximation. We transform the resulting Hamiltonian, with a set of transformations, *to Boson representation*. In section 3, the elementary excitations of the Hamiltonian are worked out leading to the soft-mode in the system at T_L, the temperature at which nematic-liquid transition occurs. Section 4 briefly concludes the behaviour of this soft-mode.

2. Hamiltonian

The Hamiltonian of the uniaxial liquid-crystal of nematic type is described by (4, 5):

$$H = - \sum_{i=0,2} J_{ij} Q_i Q_j,$$

$$\sum_i J_{ij} = \sum_j J_{ji} = J > 0 \qquad (1)$$

In mean field approximation (MFA) the Hamiltonian reduces to:

$$H_{MFA} = H_0 - \tfrac{1}{2} \langle H_0 \rangle, \qquad (2)$$

$$H_0 = -J \sum_i [\beta_0 \langle Q_0 \rangle Q_0^i + \beta_1 \langle Q_2 \rangle Q_2^i] \qquad (3)$$

where β_0 and β_1 are normalisation constants, namely

$$\beta_0 = \tfrac{1}{3}; \quad \beta_1 = 1$$

As we are interested in the dynamical aspect leading to elementary excitations in the systems so $\tfrac{1}{2} \langle H_0 \rangle$, being a constant, is dropped out of $\mathcal{H}_{MFA}$. Thus,

$$H_{MFA} = -J \sum_i [\langle Q_2 \rangle Q_2^i + \tfrac{1}{3} \langle Q_0 \rangle Q_0^i] \qquad (4)$$

defines the Hamiltonian of the system.

The Q_0 and Q_2 are determined from the following self-consistent equations:

$$\tfrac{1}{3} \langle Q_0 \rangle = \frac{K_B T}{J} \frac{\delta \log Z_0}{\delta \langle Q_0 \rangle} \qquad (5a)$$

$$\langle Q_2 \rangle = \frac{K_B T}{J} \frac{\delta \log Z_0}{\delta \langle Q_2 \rangle} \qquad (5b)$$

where K_B is Boltzmann's constant and

$$Z_0 = e^{-\beta H_{MFA}}, \quad \beta = \frac{1}{K_B T} \qquad (6)$$

Q_0 and Q_2 are subjected to the following set of transformations:

$$S_i^\pm = \tfrac{1}{2}(S_i^x \pm i S_i^y),$$

$$S_i^+ = \sqrt{2}\,(1 + b_i^\dagger b_i)\, b_i + \ldots \qquad (7a)$$

$$S_i^- = \sqrt{2}\, b_i^\dagger (1 - b_i^\dagger b_i) + \ldots$$

$$S_i^z = i - b_i^\dagger b_i \qquad (7b)$$

3. Elementary Excitations

With the help of the transformations (7a) and (7b) the Hamiltonian (4) takes the following form:

$$\mathcal{H} = \mathcal{H}_2 + \mathcal{H}_3 + \ldots \qquad (8)$$

However, we only write down $\mathcal{H}_2$ as the higher order terms in Boson operators (H_2 etc.) are not needed for the present study. H_2 takes the following form:

$$H_2 = \sum_i 2\omega b_i^\dagger b_i - \sum_{i,i} \omega(1)[b_i b_i' + b_i^\dagger b_i'^\dagger] \tag{9}$$

where

$$2\omega = 2J\langle Q_0\rangle, \quad \omega(1) = 4J\langle Q_2\rangle \tag{10}$$

We take the Fourier transform for H_2 by letting

$$b_\mathbf{q} = \frac{1}{\sqrt{\omega}} \sum_j e^{i\mathbf{q}\cdot\mathbf{r}j} b_j$$

$$b_\mathbf{q}^\dagger = \frac{1}{\sqrt{\omega}} \sum_j e^{-i\mathbf{q}\cdot\mathbf{r}j} b_j^\dagger$$

so $\mathcal{H}_2$ becomes.

$$H_2 = \sum_\mathbf{q} [\omega n_\mathbf{q} - \omega(\mathbf{q})(b_\mathbf{q} b_\mathbf{q} + b_\mathbf{q}^\dagger b_\mathbf{q}^\dagger)] \tag{12}$$

where

$$\omega(\mathbf{q}) = 4J(\mathbf{q})\langle Q_2\rangle, \quad n_\mathbf{q} = b_\mathbf{q}^\dagger b_\mathbf{q} \tag{13}$$

$$J(\mathbf{q}) = \frac{1}{N} \sum_{ij} e^{i\mathbf{q}\cdot(\mathbf{r}i-\mathbf{r}j)} J_{ij} \tag{14}$$

We apply the following Bogoliubov transformation:

$$b_\mathbf{q} = \mu(\mathbf{q})B_\mathbf{q} + V(\mathbf{q})B_{-\mathbf{q}}^\dagger$$

$$b_\mathbf{q}^\dagger = \mu(\mathbf{q})B_\mathbf{q}^\dagger + V(\mathbf{q})B_{-\mathbf{q}}$$

$$\mu^2(\mathbf{q}) - V^2(\mathbf{q}) = 1 \tag{15}$$

$$\mu^*(\mathbf{q}) = \mu(\mathbf{q}) = \mu(-\mathbf{q})$$

$$V^*(\mathbf{q}) = V(\mathbf{q}) = V(-\mathbf{q})$$

with which $\mathcal{H}_2$ is transformed into

$$H_2 = \sum_\mathbf{q} \omega(\mathbf{q}) B_\mathbf{q}^\dagger B_\mathbf{q} \tag{16}$$

$$\omega(\mathbf{q}) = 2\sqrt{\omega^2 - w^2(\mathbf{q})} \tag{17}$$

Thus (17) yields the collective modes in liquid-crystal.

Sustituting the values of ω and $\omega(\mathbf{q})$ it reduces to

$$\omega(\mathbf{q}) = 2J\langle q_0\rangle \left[1 - \left(\frac{4J(\mathbf{q})\langle Q_2\rangle^2}{J\langle Q_0\rangle}\right)^2\right]^{\frac{1}{2}} \tag{18}$$

Now, we have to determine $\langle Q_0\rangle$ and $\langle Q_2\rangle$. It can be seen that the solutions (5a) and (5b) are, which yield lowest free energy of the system.

(i)
$$\langle Q_0\rangle = \langle Q_2\rangle = Q\left(\frac{K_B T}{J}\right) \tag{19}$$

(ii) $$\langle Q_2 \rangle = 0, \langle Q_0 \rangle = -2Q\left(\frac{K_BT}{J}\right) \tag{20}$$

where $Q\left(\frac{K_BT}{J}\right)$ is a positive function (see equations (15) and (16) in reference (4)).

However, a rotation which brings the y-axis into the z-axis (tumbling) transforms the solution (19) into the solution (20). Thus these two solutions are equivalent. Taking the second solution (20) and substituting the values of $\langle Q_2 \rangle$ and $\langle Q_0 \rangle$ in (18) $\omega(q)$ becomes:

$$\omega(q) = -4Q\left(\frac{K_BT}{J}\right) = \text{constant}, \tag{21}$$

as it is indepedent of q. So we disregard this solution. Now we substitute the first solution (19) $\omega(q)$ becomes

$$\omega(q) = 2JQ\left(\frac{K_BT}{J}\right)\left[1 - \frac{16J^2(q)}{J^2}\right]^{\frac{1}{2}} \tag{22}$$

Note that the Hamiltonian (1) is a general Hamiltonian describing the situation wherein the system has quadrupolar interaction alone (9) whether it be magnetic system (4, 6), solid Hydrogen (5) or liquid-crystal. The only difference is in the assumption of coupling term. In case of liquidcrystal J_{ij} is continuosly varying function for a liquid-crystal.

$$J_{ij} = \mathcal{F}(\gamma), \gamma = \gamma_j - \gamma_i$$

One could relate J_{ij} with Lennard-Jones type interaction (1, 7, 8) or with the more general expression as proposed by Mc-millan (9).

Let T_Q be the transition temperature at which a magnetic system with the Hamiltonian (1) undergoes a phase transition from the ferroquadrupolar to paramagnetic phase. Then, near T_Q, one has

$$Q\left(\frac{K_BT}{J}\right) = \tfrac{1}{2} + D(T_q - T), T \to T_q \tag{23}$$

with

$$D = \frac{K_B}{J}\frac{(\log 4)^2}{(3 - 2\log 4)} \tag{24}$$

We assume that T_L coincides with T_Q where T_L be the temperature at which nematic liquid-crystal goes over to liquid phase. The only difference will be that in expansion (23) and (24) J should be for the liquid-crystal. Further we assume that near $T = T_L$, $J(q)$ can be expressed as

$$J(q) = J[1 - \alpha(q - q_L)^2] \tag{25}$$

This gives

$$\left[1 - \frac{16J^2(q)}{J^2}\right]^{1/2} = [-15 + 16\alpha(q - q_L)^2]^{1/2}$$

$$\sim \sqrt{15}\left[1-\frac{8\alpha}{15}(q-q_L)^2\right] \times \sqrt{-1}$$
$$= i\sqrt{15}[1-C(q-q_L)^2] \qquad (26)$$

Substituting the values from (26) and (23) into (18) we get,

$$\omega(q) = i\sqrt{15}J[-1 + C(q-q_L)^2 + 2D(T-T_L) + \ldots]$$

or $\quad i\omega(q) = \omega = J\sqrt{15}[-1 + C(q-q_L)^2 + 2D(T-T_L)] \qquad (27)$

Neglecting -1 in the bracket, it becomes

$$\omega = C'(q-q_L)^2 + D'(T-T_L), \ T \to T_L \qquad (28)$$
$$= 0, \ T > T_L \qquad (29)$$

The relation (29) follows from the property of the function

$$Q\frac{(K_B T)}{J}$$

Thus (28) and (29) describe the desired soft-mode in liquid crystal.

4. Results

We have been able to derive the soft-mode in liquid crystal at which the system undergoes a phase transition from nematic to liquid on the basis of quadrupole–quadrupole interaction. Now, we discuss the features of this soft-mode as described by (28) and (29).

(i) as $\omega = \mathrm{Im}\,\omega q \neq 0$, whereas $\mathrm{Re}\,\omega(q) = 0$, so the soft-mode is dissipative made.

(ii) $\omega = 0$ for $T > T_L$. implies that the soft-mode is non-propogating.

(iii) The relation (29) implies that the soft-mode occurs at $T = T_L$ and $q = q_L$. Similar mode has been described for classical liquids (10, 11, 12) and for a liquid-crystal by Kobayashi et al (13). It would seem that the system is suitable for application of concepts of Prigogine (14).

In conclusion the lattice model of liquid-crystal is equally capable of reproducing the soft-mode in liquid-crystal.

References

1. L. Novakovic and G. C. Shukla, Fizika, 4, 29 (1972).
2. P. G. de Gennes (Personal communication).
3. P. G. de Gennes, Phys. Lett. 30A, 454 (1969).
4. H. H. Chen and P. M. Levy. Phys. Rev. 7B, 4267 (1973).
5. T. Nakamura, Prog. Theor. Phys. 14, 135 (1955).
6. S. Strassler and C. Kittel, Phys. Rev. 139, 758 (1965).
7. N. Hijikuro, K. Miyakawa and H. Mori, Phys. Lett. 45A, 257, (1973).

8. S. Chandrasekhar, D. Krishnamurti and N. V. Madhusudana, Mol. Crystal Liq. Crystal **8**, 45 (1969).
9. W. L. McMillan, Phys. Rev. **4A**, 1238 (1971).
10. K. K. Kobayashi, J. Phys. Soc. Japan **27**, 1116 (1969).
11. T. Schneider, R. Brout, H. Thomas and J. Feder, Phys. Rev. Lett. **25**, 1423 (1970).
12. S. K. Mitra and G. C. Shukla, Phys. State Solidi **62b**, K 39 (1974).
13. K. K. Kobayashi, W. H. Franklin and D. S. Morol, Phys. Rev. **7A**, 1781 (1973).
14. I. Prigogine, G. Nicolis. J. Chem. Phys. **46**, 3542 (1967).

An Aspect of Self-Organization and Symmetry Breaking in Biological Systems

G.C. Shukla, K. Bhaumik and R.K. Mishra

Department of Biophysics
New Delhi, 110029, India

An obvious problem regarding biological systems is to explain the macroscopic observable behaviour. To answer this problem the effects of input of mass and momentum have to be understood as leading to collective behaviour of biosystems even though comprised of molecules[1]. This has been Fröhlich's central idea of dynamics of biosystems[2]. For proper understanding he has, indeed, investigated the dielectric behaviour of biological systems in some detail[3]. Nicolis and Prigogine[4] have approached this from the point of far-from-equilibrium thermodynamics while Haken[5] has generalised a theory of cooperative behaviour. These approaches cover behaviour of molecules and chemical waves and upto large systems like communities, metereology etc. The concept of a class of biosystems which do not passes the *spatial order* but a dynamic *order due to motion* is elaborated here.

At the temperature of the body the constituents of biosystems oscillate or vibrate. The interatomic and intermolecular distances change. Since they are coupled they show instability or stability as a function of time. In an earlier work[6] biosystems were regarded as consisting of dipolar molecules. This may be a fair approximation as a system of dipoles may have a net dipole moment. Their resultant charge displacement variables, x and y, influence a class of Van der Waal's interaction. Because of oscillation or vibration they show mode softening in quadratic displacements which may be stabilized by including quartic changes. The Lagrangian for such a systems is:

$$L = \frac{\dot{x}^2}{2} + \frac{\dot{y}^2}{2} - \frac{\omega^2}{2}(x^2+y^2) + g'xy - \frac{\epsilon}{4}(x^4+y^4) \tag{1}$$

To render the potential a positive definite for ensuing stability one requires

$$g' > \omega^2 \tag{2}$$

The dynamical analysis of this composite system shows the existence of isotropic stable point $x(t)=y(t)$. Spatial fluctuation around this bifurcation may lead to a fold type of deformation during the interaction due to potential function. This in turn conforms to Thom's theorem[7] that for conservative and discrete systems the instability is of fold type. However, such biosystems are generally dissipative and open in an environment including all electromagnetic radiations. We assume that such wide environmental effect is unimportant as compared to local intermolecular or intersystem interactions. The biosystems are also affected by medium, mostly water, containing ionic charges, amphiphiles etc. In a later work, effect of medium had been taken into account[8].

The system of earlier work is generalized and now the attractive van der Waal's force is considered to arise from interaction of multipoles of the two systems, a more realistic approach. The coupling term with $x^n y^m$ where n, m depict order of multipoles, is included in the Lagrangian:

$$L = \frac{\dot{x}^2}{2} + \frac{\dot{y}^2}{2} - \frac{\omega^2}{2}(x^2+y^2) + g'_{nm} x^n y^m - \frac{\epsilon}{p}(x^p + y^p) \tag{3}$$

The stabilizing term has been chosen with pth order of charge displacement with the restriction

$$p \geqslant m+n \tag{4}$$

if $\quad p = m+n, \quad$ then $\quad g' < \epsilon/p \tag{5}$

The equations of motion for displacements are:

$$\ddot{x} = -\omega^2 x + n g'_{nm} x^{n-1} y^m - \epsilon x^{p-1}, \tag{6}$$

$$\ddot{y} = -\omega^2 y + m g'_{nm} x^n y^{m-1} - \epsilon y^{p-1} \tag{7}$$

For internal stability of this composite system, the force acting on its constituents should vanish provided corresponding acceleration term vanishes (force = mass × acceleration; mass = 1, so force corresponds numerically to acceleration), that is,

$$\omega^2 x + \epsilon x^{p-1} = n g'_{nm} x^{n-1} y^m \tag{8}$$

$$\omega^2 y + \epsilon x^{p-1} = m g'_{nm} x^n y^{m-1} \tag{9}$$

Now, we state a theorem without proof:

Theorem: The system of coupled Eqns. (8) and (9) yield isotropic real solution $[x(t)=y(t)]$ whenever these equations remain unaltered by the transformation $x(t) = y(t)$ otherwise it yields anisotropic real solution that is, if $[x(t) \neq y(t)]$. Guided by the above theorem and searching for real isotropic stable points we get the generalized Lagrangian:

$$L = \frac{\dot{x}^2}{2} + \frac{\dot{y}^2}{2} - \frac{\omega^2}{2}(x^2+y^2) + g'_{nn}(xy)^n - \frac{\epsilon}{p}(x^p+y^p) \tag{10}$$

Consider two cases only, namely $n=1$ or 2 and $p=4$.

The equation governing isotropic stable point is:

$$ng'_{nn} x^{2n-2} - \epsilon x^2 - \omega^2 = 0 \tag{11}$$

with solutions as

$$n=1, \quad x_0 = y_0 = \sqrt{g'_{11} - \omega^2} \tag{12}$$

$$n=2, \quad x_0 = y_0 = \frac{\omega}{\sqrt{2g'_{22} - \epsilon}} \tag{13}$$

The strain energy at the isotropic stable point is:

$$E_{\text{strain}} = \omega^2 x_0^2 - g'_{nn} x_0^{2n} + \frac{\epsilon}{2} x^4 \tag{14}$$

$$n=1, \quad E_1 = \frac{-(g'_{11} - \omega^2)^2}{2\epsilon} \tag{15}$$

$$n=2, \quad E_2 = \frac{-\omega^4}{2(\epsilon - 2g'_{22})} \tag{16}$$

Thus other kind of isotropic stable point and strain energy are obtained than discussed in earlier work[6]. It should be noted that the coupling term $(x^2 y^2)$ for $n=2$, accounts for *quadrupole-quadrupole interactions*. From (13) and (14) one has

$$E_1 > E_2 \text{ if } \epsilon > 2g'_{22} \text{ and } g'_{11} < \omega^2 \tag{17}$$

Combining these one has

$$g'_{22} < g'_{11} \text{ and } E_1 > E_2 \text{ if } g'_{22} < \epsilon/2. \tag{18}$$

Thus given the choice, the system will choose the conformation of E_2 rather than E_1, i.e. prefering quadrupole-quadrupole interaction whenever $g'_{22} < \epsilon/2$. This result also follows by consideration of positive definiteness of potential of the system with quadrupole interactions. Now, a Lagrangian is discussed which can yield anisotropic stable points. The simple situation is of biomolecules in which one has dipole (displacement x) and other quadrupole (displacement y^2). The Lagrangian is

$$L = \frac{\dot{x}^2}{2} + \frac{\dot{y}^2}{2} - \frac{\omega^2}{2}(x^2 + y^2) + g'_{12} xy^2 - \frac{\epsilon}{4}(x^4 + y^4) \tag{19}$$

$$g'_{12} = \frac{\beta}{R^4} \tag{20}$$

The acceleration equations are:

$$\ddot{x} = -\omega^2 x + g'_{12} y^2 - \epsilon x^3 \tag{21}$$

$$\ddot{y} = -\omega^2 y + 2g'_{12} xy - \epsilon y^3 \tag{22}$$

Stable points are determined by:

$$\omega^2 x + \epsilon x^3 = g'_{12} y^2 \tag{23}$$

$$\omega^2 y + \epsilon y^3 = 2g'_{12} xy \qquad (24)$$

One of the solution of coupled equations (23) and (24) is:

$$x^\circ = y^\circ = 0 \qquad (25)$$

Other bifurcation solutions are given by the equations (23) and (25):

$$\omega^2 + \epsilon y^2 = 2g'_{12} x \qquad (26)$$

Substituting for

$$y^2 = \frac{2g'_{12} x - \omega^2}{\epsilon}$$

in (23) yields a cubic equation for x:

$$x^3 + \frac{x}{\epsilon^2}(\omega^2 \epsilon - 2g'^2_{12}) + \frac{\omega^2 g'_{12}}{\epsilon^2} = 0 \qquad (27)$$

Rewriting it as

$$x^3 + 3px + 2q = 0 \qquad (28)$$

$$3p = \omega^2/\epsilon - 2g'^2_{12}/\epsilon^2, \qquad (28a)$$

$$q = \frac{\omega^2 g'_{12}}{2\epsilon^2} \qquad (29)$$

In order that solutions of cubic equation be all real, one must have

$$q^2 + p^3 = 0 \qquad (30)$$

with

$$g'^6_{12} - \frac{3}{2} g'^4_{12} \omega^2 \epsilon - \frac{3}{32} \omega^4 \epsilon^2 g'^2_{12} - \frac{\omega^6}{8} \epsilon^3 = 0 \qquad (31)$$

Letting $g'^2_{12} = x$, it is transformed into

$$x^3 + ax^2 + bx + c = 0,$$

$$a = -\frac{3}{2}\omega^2 \epsilon, \; b = -\frac{3}{32}\omega^4 \epsilon^2 \cdot c = \frac{\omega^6}{8}\epsilon^3 \qquad (32)$$

This is transformed into

$$y^3 + 3p_1 y + q_1 = 0, \qquad (33)$$

$$3p_1 = -\frac{a^2}{3} + b, \; q_1 = \frac{a^3}{27} - \frac{ab}{2} + \frac{c}{2} \qquad (34)$$

We have only one real solution of g'^2_{12},

for

$$q_1^2 + p_1^3 = 0, \; g'^2_{12} = 1 \cdot 606 > \omega^2 \epsilon \qquad (35)$$

This is the critical value of g'_{12}, $(g'_c) = 1.2676 \, \omega \sqrt{\epsilon}$

Now allow some fluctuations to this value of g_c and discuss the real solutions, i.e. anisotropic stable points governed by Eqn. (26) and (27).

Let us put $g'^2_{12} = (1.6067 + \delta)\omega^2 \epsilon$, so as allow the fluctuations in g'_{12}. The equation (28), then becomes

$$x^3 + 3px + 2q = 0 \qquad (36)$$

$$p = -(2.2135 + 2\delta)\frac{\eta^2}{3} \qquad (37)$$

$$q = \sqrt{1.6067 + \delta\eta^3/2} \tag{38}$$

where
$$\eta = \frac{\omega}{\sqrt{\epsilon}} \tag{39}$$

Consider three cases: *case I*
$$\delta = 0 \quad \text{or} \quad g_{12}^{'2} = g_c^{'2}$$

The results, without giving lengthy algebra, are as follow:
There appear two identical anisotropic real solutions:

Case I $\qquad x_0 = 0.8589\,\eta, \quad y_0 = 1.0852\,\eta \tag{40}$

Case II $\qquad \delta < 0 \quad \text{or} \quad g'^2 < g_c^{'2}.\ \text{Let}\ \delta = -1.$

One finds real anisotropic stable point of the system is non-existent.

Case III $\qquad \delta > 0 \quad \text{or} \quad g_{12}^{'2} > g_c^{'2},\ \text{Let}\ \delta = 1.0.$

The two real anisotropic solutions are

$$x_1 = 1.8244\,\eta;\quad y_1 = 2.2116\,\eta \tag{41}$$
$$x_2 = 0.3982\,\eta;\quad y_2 = 0.2857\,\eta \tag{42}$$

Now linearize the equations (23) and (24) around the resting stable points whether isotropic ($x_0 = y_0 = 0$) or anistropic one governed by equations (40), (41) and (42).

Letting $x = x_0 + \xi$, $y = y_0 + \eta$, resulting equations are

$$\ddot{\xi} = -(3x_0^2 + \epsilon)\,\xi + 2y_0 g_{12}'\,\eta \tag{43}$$
$$\ddot{\eta} = -(3y_0^2\epsilon + \omega^2 - 2x_0 g_{12}')\,\eta + 2g_{12}' x_0\,\xi \tag{44}$$

Corresponding frequences are

$$\lambda_+^2 = \frac{\omega^2 + (A+B) - \sqrt{(A+B)^2 - 4(AB - g_{12}'^2 y_0^2)}}{2} \tag{45}$$

$$\lambda_-^2 = \frac{\omega^2 + (A+B) + \sqrt{(A+B)^2 - 4(AB - g_{12}'^2 y_0^2)}}{2} \tag{46}$$

$$A = 3x_0^2 + \epsilon,\ B = 3y_0^2\epsilon - 2g_{12}' x_0 \tag{47}$$

For linear branch with resting point $x_0 = y_0 = 0$ frequencies are:

$$\lambda_+ = \lambda_- = \omega,\quad \text{independent of } g_{12}' \tag{48}$$

Thus interaction potential for

$$r = \frac{R_0}{R},\ R_0^4 = \frac{g_{12}' B}{\omega\sqrt{\epsilon}}$$

has the following classical limit and extreme quantual limit respectively

$$I_c = K_B T \ln\left(\frac{\lambda_+ + \lambda_-}{\omega^2}\right) \tag{49}$$

$$I_Q = \frac{\hbar}{2}(\lambda_+ + \lambda_- - 2\omega) \tag{50}$$

both identically vanish.

Thus both limits have identical result for free energy of the system whether it be classical or quantal. The numerical values for λ_+, λ_-, for other resting points, have been worked out. At $R = R_c$ defining

$$\mu_c = \frac{R}{R_c}, \quad R_c = \left(\frac{g'_c}{\omega\sqrt{\epsilon}}\right)^{1/4} \tag{51}$$

one has

Case I: $R = R_c : g'(R_c) = 1.2675 \, \omega\sqrt{\epsilon}$

$$\lambda_+ = -66.8924 \, \omega, \quad \lambda_- = -2.2630 \, \omega \tag{52}$$

$$V_{(R_c)} = \frac{-\hbar\omega}{2}(4.9319) + \frac{\hbar\omega}{2}(0.1583)\rho \tag{53}$$

$$\rho = \frac{\omega^2}{\epsilon\hbar} \tag{54}$$

Case II: $R < R_c$ (slightly less), say $g'_{12}(R) = 1.6145 \, \omega\sqrt{\epsilon}$

$$\lambda_+ = 4.1893 \, \omega, \quad \lambda_- = 1.4890 \, \omega \tag{55}$$

$$V_{(R)} = -\frac{\hbar\omega}{2}(3.3784) - 1.3470 \, \hbar\omega\rho \tag{56}$$

Case III: $R \ll R_c$; say $g'_{12}(R) = 10.08 \, \omega\sqrt{\epsilon}$

$$\lambda_+ = -39.3670 \, \omega, \quad \lambda_- = -24.0138 \, \omega \tag{57}$$

$$V_{(R)} = \frac{-\hbar\omega}{2}(39.3670) - 12.0167 \, \omega\hbar\rho \tag{58}$$

As done in earlier work (6), an expansion of $V(r)$ was obtained for $R > R_c$ which yielded the effective Van der Waal's interaction for rigid structure. Here such expansion does not appear and interaction potential is defined only for $V(R)$ ($R < R_c$) Also the plot of $V(R)$ for R is as follows.

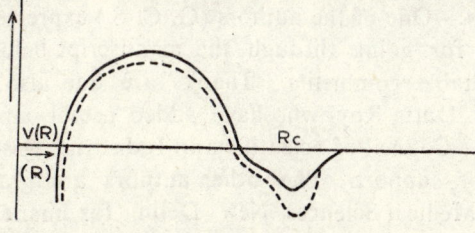

Note that one of the anisotropic stable points occurring in case III is unstable so one has only one anisotropic stable point. However, the interaction energy at linear branch corresponding to isotropic stable point

$(x_0 = 0 = y_9 = 0)$ is higher than for corresponding anisotropic stable point at R_c. This shows that the anisotropic state is favoured as compared with the isotropic one. This interesting result is an example *of symmetry-breaking where in* the system transits to a lower symmetry for stability reason. The interaction competing against the present dipole-quadrupole $(g'_{12} xy^2)$ is the quadrupole-quadrupole form $(g'_{22} x^2y^2)$. As in later case the strength of interaction (g'_{22}) varies as $\dfrac{1}{R^5}$. The analysis of $V(R)$ for for later model is straightforward. The result is:

$$V(r) = 0, r > r_c \qquad (59)$$

$$\gamma = \frac{\gamma_0}{R},\ R_0 = \left(\frac{\epsilon}{\beta}\right)^{1/5}$$

$$V(\gamma) = -2\left[2 + \frac{1}{r5}\left(2 - \frac{1}{r5}\right)^{-1}\right]_{\gamma \leqslant \gamma_c}, \qquad (60)$$

The plot of $V(R)$ verses R now does not have a cusp as in a dipole quadrupole coupling case. And again, $V(r) = 0$, $r > \gamma_c$ implies the short-range of interaction potential. The present $V(R)$ is more short-range than that of dipole-quadrupole interaction. This is due to fact that in defining R_c, the ratio of g to $\omega/\sqrt{\epsilon}$ is required, as compared with dipole-dipole case[6] wherein g/w^2 was required. This is itself indicative of smallness of V_c. To conclude, multipole-multipole interactions due to nonlinear charge displacements in biosystems becomes weaker beyond dipole-dipole or dipole-quadrupole and therefore uninteresting; *however, the peculiar feature is that symmetry breaking is possible only in dissimilar multipoles like dipole-quadrupole and not in similar ones like dipole-dipole.* The present study has been confined to two molecules. Its extension to several interacting oscillators as well as van der Waal's attracting forces with more complicated structure and type will be discussed in a separate paper.

Acknowledgements:—One of the authors (G. C. S.) expresses his gratitude to Prof. Frohlich for going through the manuscript before presentation and offering valuable comments. Thanks are due also to Prof. K. P. Sinha and Prof. B. Dutta Roy who have added useful comments. He is also grateful to the Council of Scientific and Industrial Research, Government of India, for support. The other authors are grateful to the All India Institute of Medical Sciences, New Delhi, for financial support for the work.

References

1. Mishra, R. K.: The Living State (this Volume).
2. Fröhlich, H.: Riv. Nuovo Cimento, **7**, 399 (1977).
3. Fröhlich, H.: Proc. Nat. Acad. Sci. (USA), **72**, 4211 (1975).
4. Nicolis, G. and Prigogine, I.: Self-organization in non-equilibrium systems. Wiley-Interscience, New York (1977).
5. Haken, H.: Synergetics (Springer-Verlag, Berlin, Heidelberg, New York) (1978).
6. Bhaumik, D., Dutta-Roy, B. and Lahiri, A.: Physics Letters, **68A**, 131 (1978).
7. Thom, R.: in "Synergetics" (H. Haken, ed.); Springer Verlag, Berlin (1977), p. 26.
8. Bhaumik, D., Dutta-Roy, B. and Lahiri, A.: Physics Letters A, **69A**, 68 (1978).
9. Mishra, R. K., Bhaumick, K. and Shukla, G. C.: Paper presented in Proceedings of the International Congress on "Application of Physics to Medicine and Biology", I. C. T. P., Trieste, 30 March-3 April, 1982. Alberi, G., Bajzer, Z. and Baxa, P. (Eds), World Scientific Publishing Co. Priv. Ltd., Singapore, 1983.

Cryptobiosis: A Phase of the Living State

James S. Clegg

Laboratory for Quantitative Biology
University of Miami
Coral Gables
Florida, USA

Introduction
There exist severely dried but viable cells which show no overt metabolic activity, but which can be revived by the addition of proper amounts of water. In these the dynamics of metaboblic transformation which characterise living state are not observable, only structure provides proof of its biological nature. Because of uncertainty of the metabolic status of the dried but viable cells many previous students of the phenomenon have assigned various terms, the latest being "Cryptobiosis"[1]. The dried cryptobiote represents a unique state of biologic organisation, but only if it is reversibly ametabolic and with quantitative and qualitative differences[2]. These matters have been discussed for over 150 years (see review by Keilin[1], 1959 and by Crowe and Clegg[2]).

I

For the study of this state, cysts of Artemia salina, an anostracan crustacean were examined as a model of a cryptobiotic state as a dry biological system extensively studied by us in the past [4,5,6].

Because the dry cryptobiote can be "revived" by the addition of necessary amount of water and because "life" and metabolism are related as stated above, it is clear that the dry biological state undergoes initiation to active living state by hydration. Such hydration-dependent metabolic transitions are, in a manner of speaking, a mark of the initiation of the dynamics of cellular life, which represents the Living State.

It is also obvious that such systems of suspended animation will provide information on structure, properties and role of intracellular water which it is very important to know [7-12]. However, this subject has not received the attention that is due to it.

In our studies on Artemia cysts referred to above two kinetic probes, NMR and dielectric studies, and two thermodynamic ones, isotherms and differential scanning calorimetry relationships were obtained between metabolic state, hydration, and variation in physical parameters as revealed by the probes employed. These latter likely reflect physical state of water (see Fig. 1).

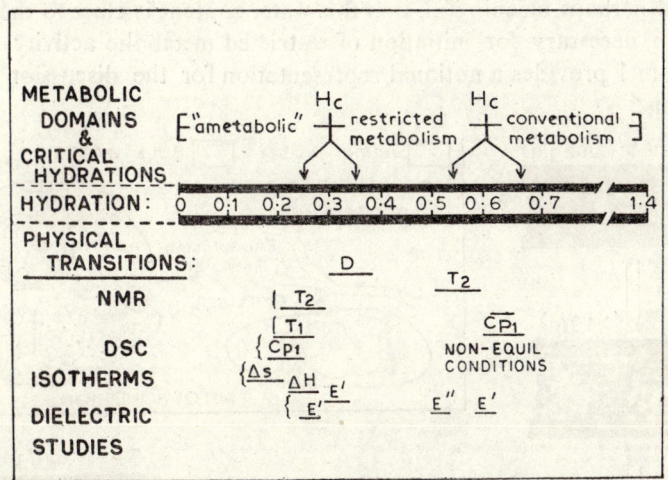

Fig. 1 Summary and integration of hydration dependence of metabolic and physical parameters in Artemia cysts.

A number of other studies on Artemia cysts indicate that major cellular structures in the cysts remain intact inspite of severe dehydration and no great processes for reassembly are necessary for reactivation by rehydration. This is similar to findings on a structure of other animal cryptobiotes (13). Does it mean if metabolism is to be reactivated sufficient water must be supplied so that preformed dehydrated ultrastructural elements incorporate sufficient water and start functioning metabolically? Again for this purpose transfer of metabolites between various organelles and intracellular compartments must be provided for in addition to essential dehydration of enzymes and substrates.

Let us start with the $CaSO_4$—dried system which has only "vicinal water". Now whatever water is present, is available only to hydrogen binding by primary hydration. It cannot be used for any other process. We suggest that a large percentage of this vicinal water could exist hydrogen bonded directly to glycerol. Since glycerol dehydrates macromolecules such as proteins, nucleic acids and polysaccharides at low prevailing a_w's[14] that is environmental water activities, this point of view may be supportable. Presumably hydroxyl groups of glycerol can compete with and prove better than nonionic polar groups of the macromolecule and may even be for the second hydration layers of at least certain ionic groups. If each glycerol molecule will hold only one single water molecule in all the vicinal water of cysts which have been dried by $CaSO_4$, then it can be shown that only 28% of the total glycerol would be actively thus engaged. The glycerol content is about 0.04 g/g dried cysts. However, a glycerol molecule may bind as many as six water molecules (2 per hydroxyl) and thus immobilise an amount of water equivalent to about 0.24 gm H_2O/g cysts. It is remarkable, and perhaps meaningful, that this water content is close to the critical hydration necessary for initiation of restricted metabolic activity.

Diagram 1 provides a notional representation for the discussion of our hypothesis.

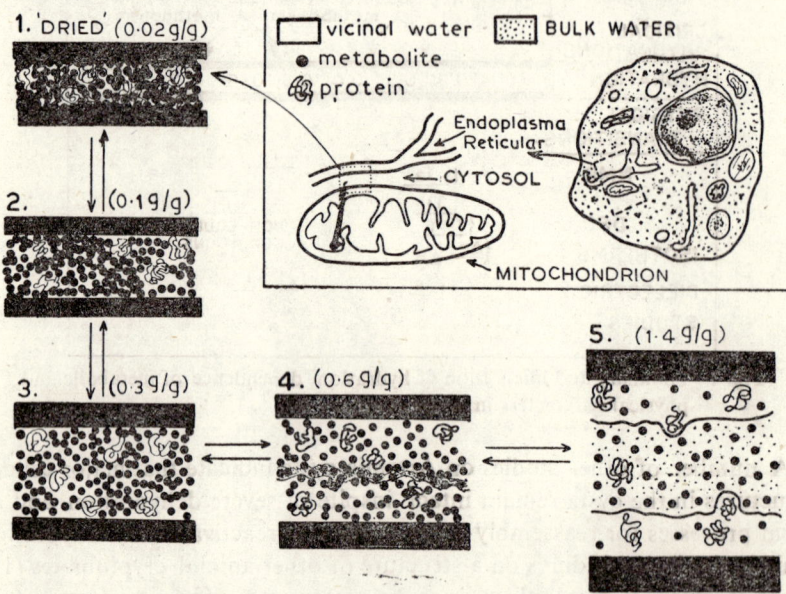

Diagram I Diagrammatic description of the proposed relationships between cellular water and metabolism in Artemia cysts (see text for explanation).

We now present a working hypothesis on the relationships between metabolism and water in the cysts. We suggest that all water upto the hydration levels close to 0.3 g/g in areas 1–3 in Diagram I is engaged in the primary hydration of cyst components in $CaSO_4$ dried cells. Bulk water does not exist below this level of hydration and therefore there is no possibility of transfer of metabolites and other substances between various intracellular compartments and in essence these are functionally cut off. As per our data several metabolic pathways start functioning when the hydration is very close to 0.3 g/g. Enzymes of the pathways of metabolism must therefore have achieved enough hydration to assume conformation of active state which can bring about enzyme activity. It is natural to suppose that sufficient water must exist to allow the carriage and movement of metabolites which would provide energy sources, coenzymes, substrates and products. These conditions are satisfied when a critical amount of water that is 0.3 g/g is provided as stated above. The communication pathways must be restricted to interfacial locations at various sites since an appreciable amount of water apparently does not yet exist and all water is close to various surfaces as described by Drost-Hansen[12, 15, 16]. This water may be called vicinal water. This supports restricted metabolic activity since long range transfer of metabolites between various organelles and other compartments in the cells at a sufficiently rapid rate is not possible. The deduction we can make from this is that the enzymes constituting the operating pathways must exist in highly localised areas in the cryptobiotic cells.

However the vicinal water cannot support the full traffic of metabolites to lead to the "conventional metabolism" when the cells are in the restricted hydration zone 0.3—0.6 g/g. However, as water is imbibed over this range interactions between water and hydration sites on intracellular components and organelles become increasingly weak and at the upper limit of this range, all such sites are maximally hydrated (Diagram 1, area 4). At this point and above greater than 0.6 g/g, water-water interactions become much more frequent and appreciable bulk aqueous phase is generated. We would like to suggest that sufficient water is now available to provide channels of continuity in the transfer of metabolites and other ingredients necessary for metabolism. Again these transfers should occur at sufficiently rapid rate so that "conventional metabolism" within cysts may be initiated. Of course it is still not certain if the water in excess of 0.6 g/g is ordinary bulk water or not.

It must be stated that when critical hydration is achieved further addition of water does not qualitatively alter metabolism and there is reduced effect on the properties and behaviour of water in the cyst. We therefore suggest a possibility that even in fully hydrated cells (Diagram 1, area 5) what is happening at the interface dominates metabolic activity even if fair amounts of water is present in solution phase in the cells. On the basis of information obtained from the Artemia cells and a review of literature of intracellular water we have proposed[5] that "soluble metabolic pathways" may be actually occurring in or on the water immediately adjacent to intracellular surfaces, or vicinal water and this situation could prevail in all caryotic cells. Or in other words the "soluble pathways" whether they are in the cytosol or in the nuclear nucleoplasm or in interior of intracellular organelles may not be in free solution in the living cells. It can be shown that restriction of soluble metabolic pathways to the interfacial water altogether with channels of continuity in transfer and communication of metabolites, enjoins more efficient function than would be if the pathways are operated by mass action or diffusion based processes in free solution. Since partitions in the cells play a large role in regulating the metabolism by providing metabolites whose rate and direction and transport can be controlled in the face of large differences in water content that occurs during the process of dehydration and hydration. These concepts have been invoked in detail[5] in our formulation of "vicinal water network model" in the intracellular environment.

II

A theory to account for the phenomenon of cryptobiosis was advanced by Crowe[17], following the work of Levitt[18], Webb[19], Warner[20]. This may be called the water-replacement hypothesis. It recalls that glycerol and other polyhydroxy alcohols and sugars are often present in very high concentration in organisms which exhibit cryptobiosis by desiccation. This is perhaps because glycerol and these other compounds replace structural water of cellular components as intracellular water is reduced and they prevent damage to cell by preventing several lethal events[21,22]. Indirect support for this is provided by NMR data, considered along with data on differential dielectric constant ($\Delta\epsilon'/\Delta\omega$).

NMR results in Figure 2 show a decrease in relaxation times T_1 and T_2, perhaps due to decrease in the total water and increase in proportion of water that interacts strongly with water binding sites. By this process translational and rotational mobility is reduced until at very low contents the relaxation times becomes immeasurable as in case of glycerinated skeletal muscle which were made glycerol-free before measurement. In case of Artemia cysts this does not happen. As cysts are dehydrated below 0.2 g/g and to about 0.3 g/g for T_1 and T_2 respectively the relaxation times lengthen, indeed T_2 values go upto 50% longer than is fully hydrated cases.

A plausible explanation, amongst possibly others, could be the following: sitebound water is replaced by glycerol and removed by hydrogen-bonded solvation of glycerol molecules and/or dissacharide trehalose. Trehalose is about 10-15% of the dry cyst mass[23-25]. The diffusion coefficient of water increases somewhat simultaneously as T_1 and T_2 increase at water content of 0.25 g/g. The data actually suggest that the increase in T_1 and T_2 arises from increase in rotational mobility of water in the hypothesized glycerol-water complex as compared with that sitebound water on membranes and other structures. Since diffusion coefficient increases slightly but T_1 and T_2 markedly as water content decreases it might mean that the translational mobility of water bound to glycerol or trehalose is not much different from that water bound to macromolecules and intracellular organelles.

It is to be emphasized that this is just one possible explanation for anamalous NMR data and others may be equally tenable, pending further experimentation.

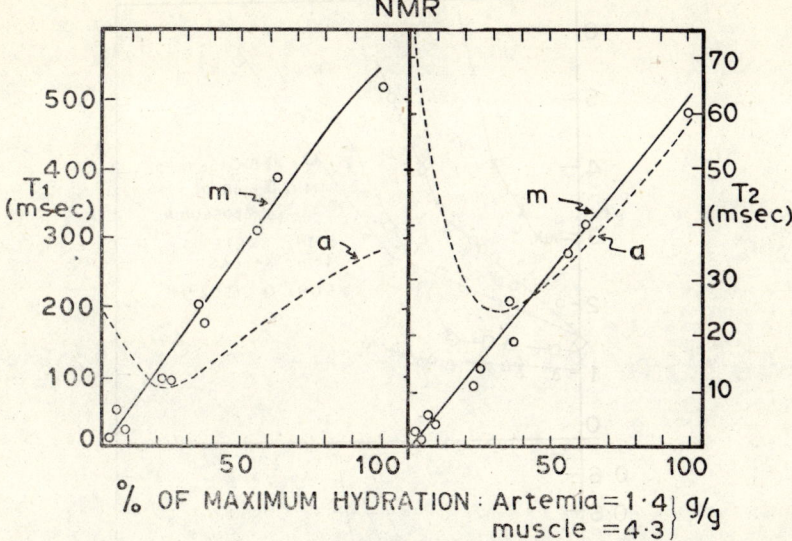

Fig. 2 Comparison of relaxation times, T_1 and T_2, for Artemia cysts (dashed line) and skeletal muscle (open circles) as a function of water content. Data for muscle were recalculated from Cooke and Wien (32) in terms of % maximum water content to facilitate comparison with Artemia.

As regards the unexpected differential dielectric increment and with reference to Figure 3 it is known that ϵ' increases at critical water concentration corresponding to completion of the bound water layer which may be less restricted. The unexpected values of $\Delta\epsilon'/\Delta\omega$ induced by virtue of hydration level may be explainable by water replacement hypothesis.

Figure 3 represents studies on a variety of dried materials [26,27,28] using vapour phase hydration as was done for Artemia cysts and these led to the belief that increase in ϵ' and ϵ'' is due to increase in rotational mobility in water which is sufficiently away from protein surface or other surfaces as compared to site-bound water. This differs in Artemis cysts where as discussed in regard to NMR data, water bound to glycerol or may be to trehalose, below hydration level of 0.25 g/g, exhibits greater polarisability. When hydration exceeds this level a longer fraction of water is site bound on macromolecules and organelles and that leads to corresponding decrease in polarisability and $\Delta\epsilon'/\Delta\omega$ is reduced. Again the dielectric measurements possibly refer to rotations rather than translations and therefore parallel NMR data. Of course, Maxwell-Wagner effect and higher frequency response is to be studied to be definite about this. In any case, the NMR and dielectric increment data so far seem to be consistent with water replacement hypothesis, even if these may not be able to "explain" them.

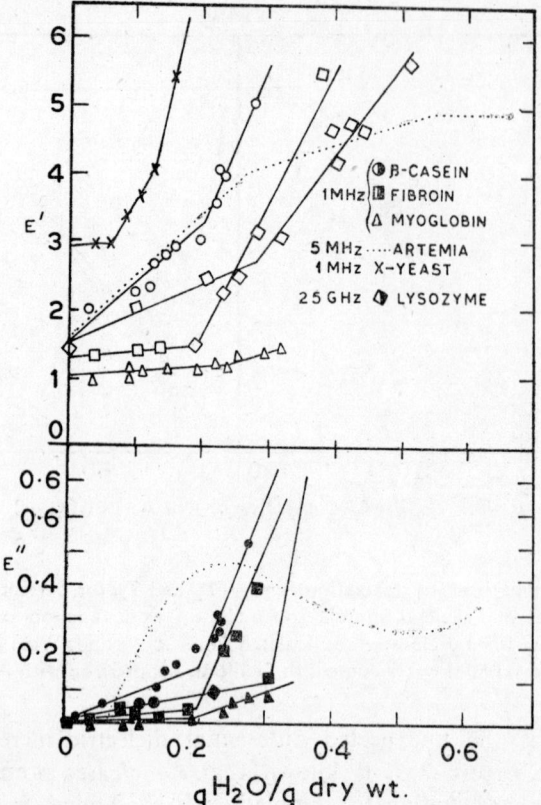

Fig. 3 Comparison of the real (ε′) and imaginary (ε″) parts of the dielectric constant for various dried materials hydrated from the vapor phase. The dotted line represents my results on Artemia. Data for casein, fibroin, and myoglobin from Rosen (26), for yeast from Koga et al., (27), and for lysozyme from Harvey and Hoekstra (28).

A simple calculation of relative proportion of glycerol and water in Artemia cysts as per Table I yields further support to water hydration hypothesis. Correcting for shell water, the amount of water and glycerol at hydration of 0.25 g/g is equivalent to a weight percent glycerol solution of about 17. In this it is assumed that water and glycerol are distributed uniformly in the cyst and both are available to interact with each other and not bound to other structures. This assumption may not be true. But these data along with those from water vapour pressure of glycerol solutions as in Newman's work[14] are able to express the degree to which glycerol and water may co-exist simultaneously. If we chose water vapour pressure activity (a_w) in environment that will be in equilibrium with glycerol concentration (weight percent) to produce hydration of 0.25 g/g of cyst then this a_w will be 0.840 and a glycerol concentration of about

Table I Glycerol-water Relationships as a Function of Hydration in Artemia Cysts

Cyst Hydration (gH_2O/g dried wt.)	% Cyst Water in cells	gH_2O in cells	wt. % glycerol*
0.05	77	0.0385	98
0.10	79	0.0790	47
0.20	83	0.166	22
0.30	85	0.255	15
0.40	88	0.352	11
0.50	90	0.450	8.3
0.60	91	0.546	6.8
0.70	92	0.644	5.8

*The glycerol content of these cysts is 3.72% of the dried weight and the location is exclusively in the cells. The calculation assumes all glycerol and water are equally distributed and in the same compartments.

45 weight percent. This is close to 6 water molecules per glycerol molecule which are otherwise also the maximum bondable to a glycerol molecule. More hydration means some additional molecules of water which will now be available for reaction.

III

In the definition of cryptobiosis by Keilin (1959) [1] two elements are important : (1) metabolic activity is hardly measurable, or (2) organism comes reversibly to a standstill. The first one indicates a quantitative slowing down, the second a qualitative difference; it indicates the organism is ametabolic. The latter is unique [13, 29] and has interesting consequences. We have said earlier [3] that for any organism in which water is dried upto 5% or less of its mass, it would not be possible to carry on enzyme reactions. These implications do not change even if this estimate is revised upwards to 0.1 g/g or 9% water by weight which we have done [6]. This level is chosen because over the frequency rate of 0.01—100 MH [5] the cysts do not respond suggesting that molecular rotation, reorientation of dipole, relaxation, ion transferance are forzen due to insufficient water. This is for small molecular species let alone enzymes. Hence ametabolic state would exist. A rough estimate even justifies the paucity of water if 0.1 g/g is adopted as the cutoff level. In previous work [4] we stated that on an average $CaSO_4$–dried cyst weighs $3.64/\mu g$, and there are 3864 cells/cyst (SD=212, n=12). Thus 0.1 g/g dried cysts have 3.2×10^{12} water molecules per average cell. At this level of hydration 9% of cyst water is within the cell[4, 6]. It will be seen that resultant number 2.5×10^{12} water molecules/cell is rather small considering the components requiring hydration. For example, McClean and Warner[30] found close to 65% of the total RNA of cyst was rRNA and therefore out of total RNA (7.83 mg RNA/g cysts SE=0.25, n=15) is 0.051 g/g dried cyst on 4.8×10^{-11}g rRNA/cell. Each ribosome contains rRNA equivalent to a molecule of 1.94×10^6 Daltons. Artemia cell contains on an average $4.8 \times 10^{-11}/1.94 \times 10^6 \times$ Avogadro number i.e. 15×10^6 ribosomes per cell. Average Artemia cyst ribosome has a hydrated mass of 7.3×10^{-18}g containing 82% water (2×10^5 molecules). So only the ribosomes in Artemia requires 3×10^{12} water molecules which exceed 2.5×10^{12} actually estimated to be present in cyst at hydration level of 0.1 g/g and question of concerted enzyme activity does not arise.

Table II Approximations for the Relative Contributions of Cyst Components to the Total Amount of Water Sorbed at Water Activity of 0.9 (23°C)*

Component	Amount present (g/g dried cysts)	gH$_2$O sorbed per g of component	gH$_2$O sorbed in 1 g dried cysts	% of total water sorbed
Proteins	0.48	0.40	0.192	51.3
Lipids	0.19	0.27	0.051	13.6
Trehalose	0.13	0	0	0
Polysaccharides	0.051	0.37	0.019	5.1
Glycerol	0.04	1.86	0.074	19.8
Nucleotides	0.036	0.65	0.023	6.2
Amino acids	0.007	0.75	0.005	1.3
TOTAL	0.95	—	0.374	100

*Sorption Isotherm Measurements: 0.361 gH$_2$O/g dried cysts (± 0.022, n = 6), for cyst homogenates: 0.397 g/g (± 0.036).

In Table II for sorbent capacities the a_w accounted for is 0.9 (23°C). This is the level at which most workers have found water in bound state using spectroscopic, dielectric and calorimetric evidence[9, 31]. Thus if at cyst hydration of 0.37 g/g free water is almost absent then it can be regarded to be so absent above at 0.1 g/g.

On the above grounds we conclude that at 0.1 g/g. hydration in a cryptobiote like Artemia cyst global metabolism cannot occur. This metabolism alongwith various pathways and control mechanism should be distinguished from isolated enzyme catalysed reaction "Adventitious Chemistry and Physics".

Acknowledgement

I thank Drs. Patsy Seitz and Carlton Hozlewood for permission to use their unpublished NMR data.

References

1. Keilin, D., Proc. Roy. Soc. Lond. B. 150, 149 (1959).
2. Hinton, H.E., Proc. Roy. Soc. Lond. B., 171, 43 (1968).
3. Clegg, J.S., in "Anhydrobiosis" (J. Crowe, and J. Clegg, eds.), p. 141. Dowden, Hutchinson, and Ross, Publ., Stroudsburg, Pa. (1973).
4. Clegg, J., J. Cell Physiol., in Press (1978).
5. Clegg, J.S., in "Cell-Associated Water" (W. Drost-Hansen, ed.), Academic Press, New York, in Press (1978).
6. Clegg, J.S., in "Dry Biological Systems" (J. Crowe, and J. Clegg, eds.), p. 117. Academic Press, N.Y. (1978).
7. Tait, J.J., and Franks, F., Nature (Lond.) 230, 91 (1971).
8. Hazlewood, C.F., (Ed.), Ann. N.N. Acad. Sci. 204, 1–631, 1973.
9. Cooke, R., and Kuntz, I.D., Ann. Rev. Biophys. Bioeng. 3, 95 (1974).
10. Alfsen, A., and Berteaud, A.J., (eds.) "L'eau et les Systemes Biologiques", 1/7 pp. Editions de Centre Nationale Recherche Scientifique, Paris, (1976).
11. Ling, G.L., Mol. Cell. Biochem. 15, 159 (1977).
12. Drost-Hansen, W., in "L'eau et les Systemes Biologiques" (A. Alfsen and A.J. Berteaud, eds.), p. 177, Editions due Centre Nat. Recherche Scientifique, Paris (1976).
13. Crowe, J.H., and Clegg, J.S. (eds.) "Anhydrobiosis" 477 pp. Dowden, Hutchinson and Ross, Inc., Stroudsburg, Pa., (1973).
14. Newman, A.A., "Glycerol" 298 pp. C.R.C. Press, Cleveland (1968).
15. Drost-Hansen, W., in "Chemistry of the Cell Interface, Part B." (H.D. Brown, ed.), Chapter 6, Academic Press, Now York (1971).
16. Drost-Hansen, W., (ed.) "Cell-Associated Water", Academic Press, New York, in Press (1978).
17. Crowe, J.H., Amer. Nat. 105, 563 (1973).
18. Levitt, J., J. Theoret. Biol. 3, 355 (1962).
19. Webb, S.J., "Bound Water and Biological Integrity"., 187 pp., C.C. Thomas, Publ., Springfield, Ill. (1965).
20. Warner, D.T., Ann. Rep. Med. Chem. 4, 256 (1969).
21. Clegg, J.S., Trans. Amer. Microscop. Soc. 93, 481 (1974).

22. Madin, K.A.C., and Crowe, J.H., J. Exp. Zool. 193, 335 (1975).
23. Dutrieu, J., Arch. Zool. exp. gen. 99, 1 (1960).
24. Clegg, J.S., J. Exp. Biol. 41, 879 (1964).
25. Boulton, A.P., and Huggins, A.K., Comp. Biochem. Physiol. 57A, 17 (1977).
26. Rosen, D., Trans. Faraday Soc. 59 2), 2178 (1963).
27. Koga, S., Echigo, A., and Nunomura, K., Biophys. J. 6, 665 (1966).
28. Harvey, S.C., and Hockstra, P., J. Phys. Chem. 76, 2987 (1972).
29. Hinton, H.E., Proc. Roy. Soc. Lond. B., 171, 43 (1968).
30. McClean, D.K., and Warner, A.H., Develop. Biol. 24, 88 (1971).
31. Kuntz, I.D., and Kauzmann, W., Adv. Protein Chem. 28, 239 (1974).
32. Cooke, R., and Wien, W., Ann. N.Y. Acad. Sci. 204, 197, (1973).

Time, Probability and Dynamics*

B. Misra and I. Prigogine**

Faculte des Sciences
Universite Libre des Bruxelles
1050 Bruxelles, Belgium

Introduction

In classical (as well as quantum) mechanics time appears in the most elementary form: as a numerical parameter serving to label the states occurring in the course of temporal dynamical evolution. Morover, the laws of dynamical evolution (with the single possible exception of the law relating to the decay of K-meson, which is not directly relevant in the context of the following discussion) are themselves invariant with respect to the transformation $t \to -t$ of the time parameter. The dynamical laws are also strictly deterministic. Classical (as well as quantum) mechanics thus envisage a timeless reality in which there is no true change and no intrinsic distinction between past and future.

By contrast, both our conscious experience of time and (thermodynamically) irreversible physical phenomena seem to point to a far richer and subtler concept of time that is endowed with an intrinsic direction, "time's arrow," in Eddington's phrase. Further, it is, in general, meaningless in both classical and quantum mechanics to say that one state is more "aged" than the other. As opposed to this, it is the very essence of the second law governing irreversible thermodynamic evolution to enable one to attribute such "age" to physical states on the basis of their entropy. Time in thermodynamics thus seems to refer to a concept of "internal time" associated with a certain class of physical systems to which the second law applies. Does this concept of oriented and internal time represent a fundamental feature of the physical world or is it merely a phenomenological concept which is ultimately subjective in origin?

---

*Combined version of the two talks also presented by the authors at the Workshop on Long-Time Prediction in Nonlinear Conservative Dynamical Systems.
**Also, Center for Studies in Statistical Mechanics, University of Texas, Austin, Texas, 78712.

The conventional viewpoint in this respect is that, the fundamental lws of physics being symmetric with respect to the time-reversal transformation, there can be no fundamental physical basis for oriented time. The appearance of irreversibility in physical processes and the related "arrow of time" are only the result of statistical averaging (or "coarse graining") which is necessitated not by any objective aspect of physical phenomena but simply to take into account our ignorance (or lack of interest) of the *exact* dynamical state of the system. Thus Born, for example, asserts that "irreversibility is a consequence of the explicit introduction of ignorance into the fundamental laws" [1].

Recent developments in physics and chemistry, however, make it increasingly difficult to maintain such a view of irreversibility. The progress towards a unified theory of particle interaction, which seems to indicate that all matter is unstable, the discovery of the background black-body radiation and the associated "Big Bang" theory of the origin of the universe, theories indicating that the so-called black holes are subject to thermodynamic laws, and finally the study of dissipative structures that can emerge in far-from-equilibrium situations, all testify to a growing recognition of the essential and constructive role of irreversibility in fundamental physical processes. In view of these developments, it becomes necessary to reconsider the relation between the dynamical description with its reversible and deterministic laws of evolution on the one hand and the thermodynamic description with its law of monotonic increase of entropy on the other.

The problem, obviously, is that of the existence of a suitable "mechanism" for breaking the time-reversal symmetry of dynamics. Not all forms of violation of time-reversal invariance can, however, lead to the irreversibility expressed in the second law. For this one needs a special form of symmetry breaking that entails a transition from the unitary group describing the reversible dynamical evolution to a semigroup describing an irreversible evolution to which one can associate an *H*-function or a Lyapounov functional. As is well known, probabilistic Markov processes (with an invariant measure) provide suitable models of such irreversible evolution admitting an H-function. Indeed, if the evolution $\tilde{\rho}_0 \to \tilde{\rho}_t$ (of distribution functions $\tilde{\rho}$) is induced from such a Markov process, then the usual expression for (negative) entropy

$$\int_\Gamma \tilde{\rho}_t \log \tilde{\rho}_t d\mu$$

(as well as any other convex functional of $\tilde{\rho}_t$) is an *H*-function [2]. The important question, thus, is how, and for what class of dynamical systems, can a symmetry-breaking transition from dynamics to probabilistic Markov processes occur. This is the question we consider in this paper in the light of modern developments in the theory of (classical) dynamical systems.

In this connection, let us recall that it is in Boltzmann's work that one first sees the deep link between irreversibility and probability. The infusion of probabilistic concepts into dynamics plays so central a role in Boltzmann's work that Gibbs has remarked that in reading Boltzmann "we seem rather to be reading in the theory of probabilities" [3]. But in Boltzmann's time dynamics and probability theory were not sufficiently advanced to support Boltzmann's physical insight and, as is well known, the validity of Boltzmann's approach was soon the subject of violent controversy. We need not follow the course of this debate here. Instead, let us turn our attention to the recent developments in the theory of classical dynamical systems and probabilistic processes that have made the situation somewhat clearer than it could be in Boltzmann's time.

One of the most important outcomes of recent advances in the theory of dynamical systems is a growing recognition of the limited scope of the concept of phase space trajectories. Classical mechanics starts by *idealizing* temporal evolution of dynamical systems in terms of the deterministic motion of phase points along phase space trajectories. Implicit in this idealization is the supposition that phase points in (initial conditions) can be determined with *infinite* precision. This idealization is unobjectionable as long as one deals only with simple dynamical motion (such as the periodic planetary motion, etc.) for which the phase space trajectories depend on the initial conditions in a continuous manner. In this situation initial imprecision about initial phase points does not get amplified in an unbounded manner in the course of time. However, as the present symposium testifies, interest has shifted now to dynamical systems exhibiting strong trajectory instability. For such systems *arbitrarily close* initial conditions can give rise to exponentially diverging or qualitatively distinct types of trajectories. Obviously, in this situation the very concept of deterministic motion along phase space trajectories ceases to be a physically meaningful idealization.

In our opinion, this limitation of the concept of phase space trajectories implied by instability of motion is not just a practical limitation but is of conceptual character. This makes it *conceptually* necessary to give up the unphysical idealization of phase points and trajectories and to go to a probabilistic description of physical states in terms of open regions of phase space, or, more generally, Gibbs distribution functions.

Obviously, such a step beyond phase points and trajectories to a probabilistic description of physical states in terms of distribution functions is *necessary* for irreversibility implied in the second law to occur. This by itself is, however, not sufficient. The evolution $\rho_0 \mapsto \rho_t$ of distribution functions is governed by a (unitary) group U_t (or in the infinitesimal form by the Liouville equation) which is induced by the dynamical motion of the phase points. And, of course, no symmetry breaking can occur in this transition from the Hamiltonian description of dynamics in terms of phase space trajectories to the Liouvillian description in terms of the evolution of distribution functions. The usual expressions for (negative) entropy such as

$$\int_\Gamma \rho_t \log \rho_t \, d\mu$$

remains constant in the course of the time evolution of the system.

The passage from the Hamiltonian description to the Liouvillian description is, however, not without certain important gains. First, it permits the formulation of concepts in ergodic theory that do express at least some aspects of irreversible thermodynamical evolution. For instance, the concept of mixing dynamical systems expresses that initial conditions are forgotten with time going on. Moreover, Gibbs, who was the first to have arrived at the concept of mixing, has also observed that the very use of the probability concept in the description of physical states implies a sort of time asymmetry. Despite the reversibility, in a mathematical sense, of the evolution of distribution functions, as Gibbs says, "it should not be forgotten when ensembles are chosen to illustrate the probabilities of events in the real world, that while the probabilities of subsequent events may often be determined from the probabilities of prior events, it is rarely the cace that probabilities of prior events can be determined from those of subsequent events, for we are rarely justified in excluding the considerations of the antecedent probability of the prior events" [3].

Nevertheless, concepts of ergodic theory and the above mentioned time asymmetry implicit in the use of the probability concept do not provide the dynamical basis for the characteristic feature of irreversible thermodynamic evolution, i.e., the *monotonic* character of the evolution expressed in the second law. As said before, this aspect can be described most naturally in terms of a symmetry breaking that causes the unitary group U_t of dynamics to be "realized" as a dissipative semigroup associated with a probabilistic Markov process.

Such a symmetry breaking would occur, for example, if for some reason not all distribution functions, but only a certain proper subset of them, could be physically observed or realized. Then the physically realized evolution would be described not by the unitary group U_t, but by the "restriction" or projection of U_t to this subset. Further, if the subset of physically realizable distributions were asymmetric with respect to time reversal transformation in an appropriate manner, one might expect the projected evolution to reduce, in one time-direction (say ($t \geqslant 0$)), to the semigroup evolution associated with a probabilistic Markov process. Mathematically, the question of the possibility of such a symmetry breaking is the question of the existence of a suitable operator Λ such that the evolution $\Lambda_{\rho 0} \equiv \hat{\rho}_0 \longmapsto \Lambda u_{t\rho 0} \equiv \hat{\rho}_t$ of the transformed states is described for $t \geqslant 0$ by a strongly irreversible Markov semigroup W_t^* (for details see section 2).

The idea of a possible non-unitary equivalence between dynamic and stochastic evolutions has been introduced earlier by one of us and his coworkers in the study of kinetic theory. This study implies an N-body problem and the taking of the thermodynamic limit $N \to \infty$, $V \to \infty$, which leads to complicated mathematical problems. Therefore it is important that a new approach to the problem summarized in this paper establishes this possibility rigorously for a well-defined class of dynamical systems.

The symmetry breaking can take two different forms according to the assumption made on the operator Λ which relates ρ to $\hat{\rho}$. First, Λ may be invertible, in which case the dynamical group u_t is *non-unitarily equivalent* or *similar* to a strongly irreversible Markov semigroup. The other is the case case where Λ is a projection, say, P_0. In this case the projected evolution $P_0 U_t P_0$ is a strongly irreversible Markov semi-group for $t \geqslant 0$.

At this point let us emphasize that the symmetry breaking projection operator that we are considering is distinct from the projection operator employed in the usual scheme in nonequilibrium statistical mechanics for deriving the so-called generalized master equation from dynamics. This latter projection is taken to correspond to an operation of "coarse-graining" or contraction of description which is *independent of dynamics and does not break the time reversal symmetry*. The resulting projected evolution then obeys the so-called generalized master equation which is, in general, non-Markovian. To restore the Markovian character of the evolution, one needs to resort to further asymptotic approximation schemes (e.g., the weak coupling limit, etc.).

In contrast, the symmetry-breaking mechanism we are discussing asks for a projection operation which itself breaks the time reversal symmetry and leads directly to the master equation of a Markov process without the necessity of resorting to special approximation schemes. In this sense, we are considering the possibility of deriving an *exact* Markovian master equation which does not depend on special approximation schemes.

Naturally, one must ask if such a procedure of symmetry breaking (either through an invertible Λ or a projection) leading from deterministic dynamics to probabilistic Markov processes is at all possible for conservative dynamical systems. As already mentioned, the possibility of a nonunitary equivalence or similarity between dynamical groups and dissipative semigroup associated with irreversible processes was already considered in the "subdynamics" approach to the foundation of kinetic theory [4]. A rigorous proof of this possibility was given, however, only recently for a class of dynamical systems called the K-flows in modern ergodic theory [5, 6]. Subsequently it was found that the class of dynamical systems for which a symmetry breaking *projection* leading to an *exact* master equation (for all $\rho \epsilon L_\mu^2 \wedge L_\mu^1$) exists is identical with the class of K-flows [7, 8]. The K-flows are known to include several systems of physical interest, such as the system of hard spheres within a box, the Lorentz gas model and geodesic flow on a manifold of negative curvature, etc. [9-11].

The existence of a link between instability of dynamical motion and irreversibility has been intuitively grasped in several early works, notably that of the Russian physicist Krylov [12]. The results mentioned above make this link precise and are of obvious interest in nonequilibrium statistical mechanics. They show how (thermodynamic) irreversibility can be the manifestation of a special form of symmetry breaking entailed by limitations on realizable physical states of the dynamical system.

The possibility of symmetry breaking as described here is based on the existence of an internal time operator* introduced in ref [13]. Briefly, this is an operator acting on the distribution functions of the phase space which is canonically conjugate to the generator of motion (Liouvillian). Its existence permits one to attribute (average) internal time (or age) to individual distribution functions in such a manner that advance in internal age corresponds to decrease in *H*-function of the system (or increase of entropy). The internal time operator may thus serve as a microscopic model of the phenomenological concept of thermodynamic time (for a discussion of thermodynamic time, the reader may see [14].

Independently of the "thermodynamic" context, the significance of these results seems to extend beyond their application in statistical mechanics. As is evident from contributions to this symposium, there is at present a growing interest in what may be called randomness of dynamical origin. Our result on non-unitary equivalence between dynamical motion and stochastic Markov processes contributes towards an understanding of the precise nature of this "dynamic randomness." It shows that in the presence of suitably strong forms of instability (expressed, for example, by the K-flow condition) the (deterministic) dynamical motion can be transformed into the stochastic evolution of a Markov process simply through a "change of representation" that involves no loss of information. Such systems may hence be said to be *intrinsically random*. The demonstration of the existence of intrinsically random dynamical systems in this sense shows that the appearance of the probability concept in the description of physical phenomena need not involve the "introduction of ignorance into the fundamental laws."

---

*In contrast, the symmetry breaking as described in the subdynamics approach is based on the study of the analytical continuation of the resolvent of the Liouville operator.

Finally, let us briefly turn to the question of the fundamental physical basis for the so-called "arrow of time." The second law of thermodynamics has two aspects. First, it implies a symmetry breaking transition from dynamics to irreversible probabilistic processes with which one can associate an H-function. The theory of symmetry breaking described in this paper fully accounts for this aspect of the second law. But the second law also endows time with a "preferred" direction. This aspect of the second law or the existence of a "preferred" direction of time is, however, not described by the existence of a symmetry breaking transition from dynamics to Markov processes, for the very formulation of the idea of symmetry breaking as described here assumes the "correct" or "preferred" direction of time being given. Note that we require the symmetry breaking P or Λ to be such that a strongly irreversible Markov process results for $t \geqslant 0$. If one wished, one could also consider a *necessarily different* symmetry-breaking operation that would yield an *exact* master equation for the reverse direction of time. The reversibility of the dynamical group implies that if symmetry breaking is possible for one direction of time then it is also possible for the reverse direction.

Let us note that in this respect the situation is exactly similar in the usual approach to deriving the Boltzmann equation or a master equation. In the usual approach to deriving the master equation, the symmetry breaking occurs at the stage of a special asymptotic limit (e.g., the so-called $\lambda^2 t$ limit); and one could consider a suitable limit to yield a master equation in either the "forward" direction ($t \geqslant 0$) or "backward direction ($t \leqslant 0$) of time. Thus neither the usual approach nor the theory of symmetry breaking presented here answers the question of the "preferred" direction of time or "time's arrow." This fundamental question is related to the problem of *intrinsic* physical distintion (if any) between the two possibility of symmetry breaking corresponding to the opposite directions of time. We shall not discuss this question in detail here but shall be content only with offering a few brief remarks in section 4.

We now proceed to a more precise and detailed formulation of the ideas described above.

Formulation of the Problem

Consider an abstract dynamical system (Γ, B, μ, S_t). Here Γ denotes the phase space of the system equipped with a σ-algebra B of measurable subsets, S_t a group of measurable transformations mapping onto itself and preserving the measure μ. For example, Γ could be the energy surface of a classical dynamical system, S_t the group of dynamical evolution and μ the invariant measure whose existence is assured by Liouville's theorem. For convenience we shall assume the measure μ to be normalized $\mu(\Gamma) = 1$. As is well known, the evolution $\rho \longmapsto \rho_t$ of density functions under the given deterministic dynamics is described by the unitary group U_t induced

by S_t:
$$\rho_t(\omega) \equiv (U_t\rho)(\omega) = \rho(S_{-t}\omega).$$

The generator L of the unitary group U_t is called Liouvillian operator of the system: $U_t = e^{-iLt}$. It is given by
$$L\rho = i[H, \rho]_{\text{P.B.}}$$

if the evolution is generated by the Hamiltonian function H. Here $[,]_{\text{P.B.}}$ denotes the usual Poisson bracket.

Every measure preserving deterministic evolution S_t thus defines a unitary group. Conversely, (under certain mild assumptions about the measure space (Γ, B, μ)) every unitary group which preserved positivity (i.e., maps nonnegative functions to nonnegative functions) and leaves the constant functions unchanged is induced by a group S_t of measure preserving transformations on Γ [15].

On the other hand, stochastic Markov processes on the state space Γ, preserving μ, are associated with contraction semigroups of $L^2_\mu(\Gamma)$ [16]. In fact, let $P(t, \omega, \Delta)$ denote the probability of transition from the point $\omega \in \Gamma$ to the region Δ in time t. Then the operators W_t defined by

$$(W_t f)(\omega) = \int_\Gamma f(\omega') P(t, \omega, d\omega')$$

form a contraction semigroup for $t \geqslant 0$.

Moreover, W_t has the following properties:
(i) W_t preserving positivity (i.e., $f \geqslant 0$ implies $W_t f \geqslant 0$ for $t \geqslant 0$).
(ii) $W_t \cdot 1 = 1$

The evolution of the distributions $\tilde{\rho}$ under the Markov process is described now by the adjoint semigroup W_t^* which also preserves positivity since W_t does: $\tilde{\rho}_0 \longmapsto \tilde{\rho}_t \equiv W_t^* \tilde{\rho}_0$. Since the measure μ is an invariant measure for the process (or equivalently the *micro canonical distribution function* 1 is the equilibrium state of the process), we also have

(iii) $W_t^* \cdot 1 = 1$.

Every Markov process on Γ with stationary measure μ is thus associated with a contraction semigroup satisfying the conditions (i) through (iii). Conversely every contraction semigroup W_t on L^2_μ satisfying the above conditions comes from a stochastic Markov process, the transition probabilities $P(t, \omega, \Delta)$, being given by

$$P(t, \omega, \Delta) = (W_t \varphi_\Delta)(\omega).$$

Here φ_Δ denotes the characteristic (or indicator) function of the set Δ.

In the following we are interested in a special class of Markov processes whose semigroups W_t satisfy (in addition to the condition (i)-(iii)) the condition:

(iv) $\|W_t^* \tilde{\rho} - 1\|^2$

decreases strictly monotonically to 0 as $t \to +\infty$; for all states $\tilde{\rho}$ (i.e., all nonnegative distribution functions $\tilde{\rho}$ with $\int_\Gamma \tilde{\rho} d\mu = 1$). This condition expresses the requirement that any initial state $\tilde{\rho}$ tends *strictly monotonically* in time to the equilibrium distribution 1. For such processes the functional

$$\int_\Gamma \tilde{\rho}_t \log \tilde{\rho}_t \, d\mu; \quad \tilde{\rho}_t \equiv W_t^* \tilde{\rho}_0$$

and indeed any other convex functional $\tilde{\rho}_t$ is an *H*-function.

Semigroups satisfying the above conditions (i) through (iv) will be called strongly irreversible Markov semigroups.

The problem before us is to determine the class of dynamical systems for which one can construct a bounded operator Λ having the following properties:

(i) Λ preserves positivity
(ii) $\Lambda \cdot 1 = 1$
(iii) $\int_\Gamma \Lambda \rho \, d\mu = \int_\Gamma \rho \, d\mu$
(iv) The dynamical group $U_t = e^{-iLt}$ satisfies the intertwining relation

$$\Lambda U_t = W_t^* \Lambda \quad (t \geqslant 0)$$

with a *strongly irreversible Markov semigroup* W_t.

The meaning of conditions (i)–(iii) is obvious. They express the requirements that the transformation $\rho \longmapsto \Lambda \rho$ maps states to states and leaves the equilibrium state unchanged. The intertwining condition (iv), on the other hand, expresses that the transformation $U_t \rho \equiv \rho_t \longmapsto \Lambda \rho_t \equiv \tilde{\rho}_t$ brings about the desired form of symmetry breaking: The transformed evolution $\tilde{\rho}_0 \longmapsto \tilde{\rho}_t$ obeys for $t > 0$ the master equation corresponding to a strongly irreversible Markov semigroup W_t.

We shall consider two cases: In the first case, Λ has a densely defined inverse Λ^{-1}. Existence of such a Λ means that the dynamical group U_t is *similar* (non-unitarily equivalent) to a strongly irreversible Markov semigroup $\Lambda U_t \Lambda^{-1} \equiv W_t^*$ for $t \geqslant 0$. Dynamical systems admitting such a Λ may be said to be *intrinsically random*. For such systems a "change of representation" of dynamics $\rho_t \longmapsto \Lambda \rho_t = \tilde{\rho}_t$ involving "no loss of information" (expressed by the invertibility of Λ) can convert the dynamical motion into that of a stochastic Markov process.

The other case we consider is: Λ is a projection operator P. The transformation $\rho_t \longmapsto P \rho_t \equiv \tilde{\rho}_t$ may now be considered as a generalized form of "coarse graining" that eliminates from ρ_t physically unobservable or "uncontrollable correlations." The intertwining condition now is equivalent to the requirement that the restriction PU_tP of the dynamical group

U_t to the subspace of physically observable states (of the form $P\rho$) is a strongly irreversible Markov semigroup for $t \geqslant 0$. Let us reemphasize that we are not considering here the usual form of "coarse graining" employed in the derivativation of, say, non-Markovian generalized master equations but are considering the existence of a symmetry breaking projection that leads directly to an *exact* master equation.

The time reversibility of the dynamical motion U_t implies, of course, that if there exists a symmetry breaking projection P_+ (or a similarity Λ_+) such that $W_t^{(+)} \equiv P_+ U_t P_+$ (or $\Lambda_+ U_t \Lambda_+^{-1}$) is a strongly irreversible Markov semigroup for the "forward" direction of time ($t \geqslant 0$), then there also exists another projection P_- (or a similarity Λ_-) for which $W_t^{(-)} \equiv P_- U_t P_-$ (or $\Lambda_- U_t \Lambda_-^{-1}$) will be a strongly irreversible semigroup for the "backward" direction $t \geqslant 0$. The important point, however, is that the symmetry breaking projections P_+ and P_- (or Λ_+ and Λ_-) corresponding to the two directions of time must necessarily be distinct. In fact, it can be easily verified that if $P_+ U_t P_+$ is a strongly irreversible Markov semigroup for $t \geqslant 0$ then $P_+ U_t P_+$ is unitary (in the subspace of P_+) for $t < 0$, and hence $P_+ U_t P_+$ cannot represent, in the backward direction of time, an irreversible stochastic evolution. Similarly, if $\Lambda_+ U_t \Lambda_+^{-1}$ represents the irreversible stochastic evolution of a Markov process for $t \geqslant 0$, then $\Lambda_+ U_t \Lambda_+^{-1}$ cannot even be positivity preserving, and hence cannot represent any kind of physical evolution for $t < 0$ [15].

The symmetry between the two directions of time is thus broken by the choice of a symmetry breaking projection or similarity. But as we have seen, this choice between P_+ and P_- (or between Λ_+ or Λ_-) and the corresponding semigroups $W_t^{(+)}$ and $W_t^{(-)}$ presupposes the "correct" direction of time being known. Thus, while the existence of a symmetry breaking projection P (or similarity Λ) expresses, at the dynamical level, the characteristic features of thermodynamic irreversibility, it cannot serve to define an intrinsic direction of time or the so-called "arrow of time." As we have said, this is the further question of *intrinsic* physical distinction (if any) between P_+ and P_-.

Internal Time and Symmetry Breaking

If the projected evolution $P U_t P$ to the subspace of states of the form $P\rho$ is to behave asymmetrically with respect to the two directions of time, then one expects that the states $P\rho$ themselves must, in some suitable sense, contain the time asymmetry. In order to give meaning to such a notion of time asymmetric states, one needs the notion of an "internal time" operator that permits to attribute (average) "age" or "internal time" to individual states. In this section we define this concept and discuss the close connection between the existence of a symmetry breaking transition from dynamics to probabilistic processes and the existence of an internal time operator for the system. For further detailed see references [5–8, 13].

Let $H_{-\infty}$ denote the one-dimensional subspace of constant functions on the phase space Γ, $P_{-\infty}$ the projection onto $H_{-\infty}$ and $H_{-\infty}^{\perp}$ the subspace of $L_{\mu}^2(\Gamma)$ that is orthogonal to $H_{-\infty}$. By an internal time operator T of the (abstract) dynamical system with dynamical group U_t, we mean a self-adjoint operator T on $H_{-\infty}^{\perp}$ satisfying the two following conditions:

(a) $$U_t^* T U_t = T + tI$$
(The infinitesimal form of this relation is the familiar canonical commutation relation $[T, L] = iI$ (on $H_{-\infty}^{\perp}$) between the Liouvillian L and T.)

(b) The projections $P_\lambda = E_\lambda + P_{-\infty}$ of L_μ^2 preserve the positivity of functions: i.e., $P_\lambda \rho \geqslant 0$ a.e. if $\rho \geqslant 0$ a.e. Here E_λ (λ real) denotes the spectral projections of T (i.e., E_λ is an increasing family of projections such that: (i) $E_\lambda \to 0$ ($\lambda \to -\infty$) (ii) $E_\lambda \to I_{H_{-\infty}^{\perp}}$ ($\lambda \to -\infty$), and (iii) given functions f and g (in the domain of T)

$$\langle f, Tg \rangle = \int_{-\infty}^{\infty} \lambda d \langle f, E_\lambda g \rangle.$$

We have used here, as elsewhere in this paper, the usual inner product notation which denotes, for example, the integral $\int_\Gamma \bar{f} g d\mu$ by $\langle f, g \rangle$.

If such an internal time operator T exists, one can consistently interpret the quantity

$$\langle T \rangle_\rho \equiv \frac{\langle \bar{\rho}, T \bar{\rho} \rangle}{\langle \bar{\rho}, \bar{\rho} \rangle}$$

(with $\bar{\rho} = \rho - 1$ being the departure from microcanonical equilibrium ensemble 1), as the average internal time or "age" of the (Gibbsian) ensemble 1. The condition (a) on T, then expresses the desirable consistency requirement that in the course of dynamical evolution the system's *average age* advances in step with the increase in the external time parameter t: i.e., if $\rho_t = U_t \rho_0$ then $\langle T \rangle_{\rho_t} = \langle T \rangle_{\rho_0} + t$.

As in the case of quantum mechanical observables, the time operator T does not in general permit to attribute a *definite* "age" to a distribution function. Definite "ages" can be attributed only to the eigenfunction of T. Corresponding to the fact that a distribution function ρ is, in general, the "superposition" of several eigenfunctions of T, there will be associated with ρ not a definite "age" but a statistical distribution of possible "ages", the average being given by $\langle T \rangle_\rho$. The projection operator $P_\lambda = E_\lambda + P_{-\infty}$ associated with T may be interpreted as the projection to subspace of states, all whose possible "ages" lie in the interval $(-\infty, \lambda)$. The projection P_λ may, thus, be called the "projection to the past of λ." In particular, the projection P_0 will simply be called the projection to the past.

The positivity condition (b) on P_λ means that the projection operation to this past maps states (i.e., nonnegative and normalized distribution) to states. Besides being a natural requirement, this condition is indispensible if P_λ is to be used for constructing symmetry breaking transformations Λ or P with properties stated in the previous section.

The condition (b) is independent of the condition (a). In fact, the existence of a T satisfying this condition (a) alone is known to be equivalent to a spectral property of the Liouvillian L: viz. that L (restricted to $H^\perp_{-\infty}$) has absolutely continuous spectrum of uniform degeneracy extending over the entire real line. The condition (b), on the other hand, cannot follow from merely *spectral properties* of L. The further restriction on dynamics that the existence of T satisfying both (a) and (b) imposes will be discussed below. Here, let us study the connection between the possibility of constructing an internal time operator and the existence of symmetry breaking transitions from dynamics to stochastic processes. This connection is described by the following:

Theorem 1
(a) Let the dynamical group admit a symmetry breaking projection P (i.e., a projection P mapping states to states and such that PU_tP is a strongly irreversible Markov semigroup for $t \geqslant 0$). Assume further that P is sufficiently large in the sense that, given a state ρ and $\epsilon > 0$ there exists a state ρ' and $t > 0$ such that $\|\rho - U_t P \rho'\| < \epsilon$. In other words, the set of states in the subspace of P evolve in time to give rise to all possible states.) Then the dynamical system admits an internal time operator T whose "projection operator to the past" coincides with P.

(b) Conversely, if the dynamical system U_t admits an internal time operator T then the associated "projection operator to the past" P_0 is a symmetry breaking projection for U_t.

(c) Moreover, if an internal time operator T exists, then the system is *intrinsically* random, i.e., there exists a state preserving operator Λ with densely defined inverse Λ^{-1} such that $\Lambda U_t \Lambda^{-1}$ a strongly irreversible Markov semrigroup for $t \geqslant 0$.

A proof of the last part (c) of this theorem is essentially contained in [4]. Let us only mention that the symmetry breaking invertible transformation can be constructed as an operator function of T. More specifically, Λ is of the form:

$$\Lambda = \int_{-\infty}^{\infty} h(\lambda)\, dE_\lambda + P_{-\infty},$$

where E_λ denotes the spectral projections of T and $h(\lambda)$ is a function satisfying the following conditions:

(i) $h(\lambda)$ is strictly monotonically decreasing with $\lim_{\lambda \to +\infty} h(\lambda) = 0$ and $\lim_{\lambda \to -\infty} h(\lambda) = 1$.

(ii) $\dfrac{h(\lambda + s)}{h(\lambda)} \equiv \tilde{h}_s(\lambda)$ is monotonically decreasing function of λ for every $s \geqslant 0$. The function $h(\lambda)$ can, for example, be the function $e^{-(e^\lambda)}$.

The proofs of parts (a) and (b) are fairly straightforward. As an illustration, let us consider the proof of (b). Let P_0 be the "projection to the past" associated with the internal time operator T. We have to show that $P_0 U_t P_0$ has the semigroup property for $t \geqslant 0$ and satisfies the strong irreversibility condition $\| P_0 U_t P_0 \rho - 1 \|^2 \to 0$ as $t \to \infty$. (The positivity preserving property of $P_0 U_t P_0$ follows automatically from the fact that P_0 and U_t do so.) $P_0 U_t P_0$ will obviously form a semigroup for $t \geqslant 0$ if $P_0 U_t Q_0 = 0$ for $t \geqslant 0$; where $Q_0 = I - P_0$. Consider now $P_0 U_t Q_0 U_t^*$:

$$P_0 U_t Q_0 U_t^* = P_0 - P_0 U_t P_0 U_t^*.$$

At this point one may use the fact that the relation $U_t^* T U_t = T + tI$ (on $H_{-\infty}^\perp$) between T and U_t is equivalent to the following so-called *imprimitivity relation*

$$U_t P_\lambda U_t^* = P_{\lambda+t}$$

between the "projections to the past λ" P_λ associated with T. Thus $P_0 U_t P_0 U_t^* = P_0 P_t$. Since $P_t \geqslant P_0$ if $t \geqslant 0$ it follows that $P_0 P_t = P_0$ for $t \geqslant 0$. Thus $P_0 U_t Q_0 U_t^*$ and hence also $P_0 U_t Q_0$ vanish for $t \geqslant 0$. Similarly, the strong irreversibility condition follows because

$$\| P_0 U_t P_0 \rho - 1 \|^2 = \| U_t^* (P_0 U_t P_0 - 1) \|^2$$
$$= \| P_{-t} P_0 \rho - 1 \|^2$$
$$= \| P_{-t} \rho - 1 \|^2$$

where the last equality follows from the fact that $P_{-t} P_0 = P_{-t}$ for $t \geqslant 0$. The last term tends monotonically to 0 with $t \to \infty$ because P_{-t} decreases monotonically to the projection $P_{-\infty}$ on to the equilibrium state 1 as $t \to \infty$.

The proof of part (a) is essentially a reversal of the steps of the previous argument.

Before discussing the properties of dynamical systems admitting an internal time operator, let us briefly comment on the possible physical meaning of internal time. We have remarked in the Introduction that the internal time operator seems to have the right properties to serve as a microscopic model for the phenomenological concept of thermodynamic time. Moreover, one can show that, in a certain suitable sense, internal times associated with states ρ reflect the "degree of indistinguishability" of ρ from the equilibrium state. The more advanced the "age" the more indistinguishable is the state from equilibrium state. But one may still raise the question of the *microscopic* meaning of statistically distributed possible internal times or "ages" that are associated with individual states.

Now the description of physical states in terms of a (Gibbs) distribution function ρ is a mathematical shorthand for two types of information: the information about the velocity distribution of the particles (in our kinetic theory this is called the "vacuum of correlations") and the hierarchy of correlations between the particles which depends *both* on the positions and the velocities.

We have therefore to expect some relation between the internal time and the correlations which may be the result of past events ("post-collisional" correlations) or the cause for future events ("pre-collisional" correlations). For instance, it is to be expected that the *projection* operation P_0 to the *past* eliminates all *future directed* ("pre-collisional") correlations that would cause future "collisions." We shall come back to this question in section 4 where we shall find, at least in a qualitative manner, such a link.

K-Flows and Symmetry-Breaking Transitions to Markov Processes

It can be shown that necessary and sufficient condition for the existence of an internal time operator is that the dynamical system be a K-flow. In view of the theorem above, this means that the K-flow condition is also both necessary and sufficient for the existence of a symmetry breaking projection leading to an exact master equation. The existence of an internal time operator for K-flows also implies that K-flows are *intrinsically random*, i.e., the dynamical group U_t induced from a K-flow is (for $t \geqslant 0$) non-unitarily equivalent, through a state preserving and invertible similarity transformation, to the semigroup associated with a strong irreversible Markov process.

We shall not stop here to supply the proofs of the above statements (the interested reader may see refs. [5–8 and 13]). Let us only describe briefly the concept of K-flow [17] and the construction of the symmetry breaking projection in terms of associated K-partitions.

A K-flow is by definition a dynamical system (Γ, B, μ, S_t) for which there exists a distinguished (measurable) partition ξ_0 of the phase space into disjoint cells such that:

(i) $\qquad\qquad\qquad S_t \xi_0 \equiv \xi_t \geqslant \xi_s \text{ if } t \geqslant s$

Here ξ_t is the partition into which the original partition ξ_0 is transformed under dynamical evolution in time t. The notation $\xi_t \geqslant \xi_s$ signifies that the partition ξ_t is finer than ξ_s (i.e., every cell of ξ_t is entirely contained in some single cell of ξ_s).

(ii) The (least fine) partition $\bigvee_{t=-\infty}^{\infty} \xi_t$

which is finer than each ξ_t ($-\infty < t < \infty$) is identical with the partition of the phase space into individual phase points.

(iii) The (finest) partition $\bigcap_{t=-\infty}^{\infty} \xi_t$ that is *less fine* than each ξ_t is the trivial partition consisting of a cell of full measure 1. (A partition ξ_0 with the stated properties is called a K-partition).

Another mathematically equivalent characterization of K-flows is that they have completely positive Kolmogorov entropy. Without going into a definition of Kolmogorov entropy, let us simply mention that complete positivity of Kolmogorov entropy means that the knowledge about the past history of the system obtained from an infinite repetition of any realistic measurement that corresponds to a partition of the phase into a *finite* number of disjoint cells is insufficient to predict the future outcome of the same experiment. In this sense, the *observed* behavior of K-flows is non-deterministic in character.

The simplest example of K-flow, or rather a K-system (discretized time), is the so-called baker's transformation. It may be described as the transformation B of the unit square onto itself that is the result of two successive operations (see Fig. 1): (a) First, the unit square is squeezed in the vertical direction to half its width, and it is at the same time stretched along the horizontal direction to double the length; (b) Next, the resulting rectangle is cut in the middle and the right half is stacked upon the left half.

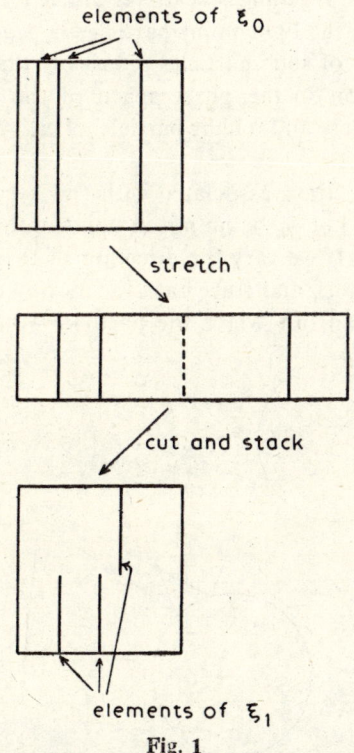

Fig. 1

The iterates B^n of the baker's transformation can be considered as describing the evolution of an *abstract* dynamical system at unit intervals of time. It is clear that the partition of the unit squares into vertical lines is a K-partition for the baker's system. In fact, under the operation of B^n ($n \geqslant 0$), the partition in question becomes successively finer and the reader can easily verify that the defining conditions (i) through (iii) of K-partitions hold.

Besides this rather artificial example, many systems of physical interest, such as the system of Lorentz gas and hard spheres in a box [9], the geodesic flow on a compact Riemann manifold of negative curvature [10, 11], etc. are known to be K-flows. Although the existence of K-partitions for these more realistic systems have been established, their construction (and description) cannot yet be given analytically with a sufficient degree of explicitness.

As an illustration, and also for future comment, let us geometrically describe the elements (individual cells) of a K-partition for the (two-dimensional) Lorentz model of gas. The model is the following: we have a fixed configuration of discs (scatterers) and light point particles which do not interact among themselves but move between the scatterers freely with constant velocity and on reaching a scatterer are reflected elastically. Since the interaction between the light point particles is neglected, the study of the behavior of a beam of such particles reduces to the study of the motion of a distribution function on the phase space of the one-particle system: the fixed convex scatterers and a light particle. This system is known to be a K-flow [9].

The cells of the K-partition associated with this system may be obtained as follows (see Fig. 2): Let ω_0 be an initial point in the phase space and ω_t the point after a time t. If we vary the direction of the particle (but not the spatial position) around ω_t and trace back the motion to $t = 0$ to determine the initial phase point from which the particle would have started, we

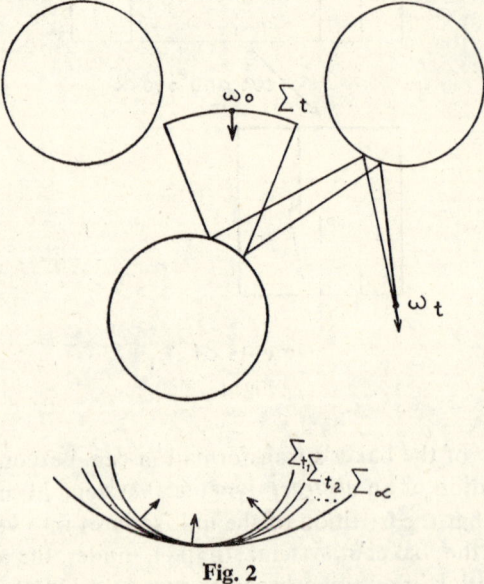

Fig. 2

obtain a curve Σ_t containing the given point ω_0. The curve Σ_t will be smooth at least if one considers a reasonably small neighborhood of ω_0. Now one can construct Σ_t in the manner described above, for various values of t (but for the same initial points).

A typical element (cell) of the K-partition associated with the system will be the limiting curve Σ_∞ to which the family of curves converge as $t \to \infty$. In other terms, the elements of K-partition for the system under consideration are curves in phase space such that all particles having initial conditions on one such curve are so intricately correlated that after being scattered repeatedly by the scatterer, they all converge as $t \to \infty$ towards a single position. Thus, their correlation is similar to the correlations obtained in an incoming wave front which will converge in the infinitely distant future to a single point. Later, we shall make use of this feature of K-partitions to suggest a possible distinction between the two semigroups (corresponding to the two directions of time) that result from the symmetry breaking mechanism discussed in this paper.

The interest in explicit construction of K-partitions comes from the fact that for K-flows the symmetry breaking projections leading to exact master equations (and equivalently the projection to the past associated with the internal time operator of the K-flow) can be constructed as the *projection operation of "coarse-graining" with respect to K-partitions*. More explicitly, if ξ_0 is a K-partition associated with the K-flow (Γ, B, μ, S_t), then the orthogonal projection P_0 of $L^2\mu(\Gamma)$ onto the subspace $L^2(a(\xi_0), \mu)$ is a symmetry breaking projection, i.e., $P_0 U_t P_0$ is a strongly irreversible Markov semigroup for $t \geq 0$. Here we have denoted by $a(\xi_0)$ the σ-subalgebra of B formed by only the (measurable) subsets Δ *that are unions of elements (cells) of the partition* ξ_0. And $L^2(a(\xi_0), \mu)$ denotes the subspace of $L^2\mu(\Gamma)$ that consists of functions that are measurable with respect to $a(\xi_0)$. For instance, in the example of the baker's transformation, $a(\xi_0)$ is the σ-algebra of subsets formed by all *vertical* rectangular strips only.

The projection P_0 onto $L^2(a(\xi_0), \mu)$ has the following properties :

(i) $\quad f \geq 0 \Rightarrow P_0 f \geq 0$

(ii) $\quad \int_\Delta f d\mu = \int_\Delta (P_0 f) d\mu \quad$ for all $\Delta \epsilon a(\xi_0)\quad$ and $\quad f \epsilon L^2\mu(\Gamma)$.

(iii) $P_0 f$ (being measurable with respect to $a(\xi_0)$) assumes constant values on each element (individual cell) of the partition ξ_0.

In the language of probability theory, the function $P_0 f$ is the *conditional expectation* of the function (random variable) f, given (the outcomes of the measurement represented by) the partition ξ_0. The symmetry breaking projections for K-flows may thus be viewed as representing a generalized process of "coarse graining" which, in a sense, averages over the cells of the K-partition. To put it differently, the symmetry breaking projection

P_0 eliminates from dynamical considerations the degree of freedom and the "correlations" associated with the cells of K-partitions.

When the k-partition ξ_0 is of simple structure, one can give an explicit analytic formula for the symmetry breaking projection P_0 onto $L^2(a(\xi_0), \mu)$ and display in explicit form the *exact* master equation that results from symmetry breaking. Such explicit calculations have recently been made for the infinite ideal gas [18] using the explicit form of K-partitions described in [19].

To conclude this section, let us briefly return to the question of the choice between the two semigroups (corresponding to the two different directions of time) that can result from symmetry breaking. As discussed above, one semigroup (for $t \geqslant 0$) comes from "coarse-graining" with respect to the K-partition. The other semigroup for $t \leqslant 0$ would correspondingly come from the projection associated with the "time reversed" k-partition, and such a partition, of course, exists as the originally given dynamics is assumed to be reversible. (The *"time reversed"* K-partition is a partition ξ_0' such that the partition $\xi_t' \equiv S_t \xi_0'$ becomes *less fine* as t increases: $\xi_t' \leqslant \xi_s'$ if $t \geqslant s$). The question then is whether there exists any physical distinction between the two semigroups to recommend one over the other as the physically realized temporal evolution. It seems to us that such a distinction can be made on the basis of the type of "correlations" that are eliminated by the two projections respectively. As an illustration, "coarse-graining" projections associated with the K-partition of the Lorentz gas model would eliminate from the description of dynamical states precisely the future directed correlations of the type present in "incoming waves" that would converge to a point in the *infinite future*. The projection associated with the "time inverted" K-partition would, on the other hand, eliminate the "outgoing wave" type correlations but retain the "incoming wave" type correlations. It is the physical unrealizability of the "incoming wave" type correlations that recommends the former semigroup arising from the elimination of such unrealizable correlations as the physically preferred semigroup over the other.

These remarks, although qualitative, indicate how an intrinsic asymmetry between the two time directions can be sought in terms of the question of physical realizability and unrealizability respectively of "outgoing wave" type and "incoming wave" type correlations.

Similar considerations can also be advanced in the "subdynamics" approach to the problem of irreversibility. In brief, in the subdynamics approach the distinction between the semigroups can be formulated as follows: Between particles situated at large distances there may exist only post-collisional correlations but no pre-collisional correlations. Collisions can give rise to correlations and correlations to collisions, but the relation between them is not a symmetrical one. However, these considerations will be developed in a separate paper.

Internal Time and Geodesic Instability in Cosmology

An interesting application of the foregoing ideas occurs in certain simple cosmological models described below. In this interesting lecture at this workshop, Professor Szebehely has exhibited in a striking manner the stochasticity of motion in the restricted three body problem. It is interesting that in the cosmological model we consider stochasticity appears already at the level of a one-body problem.

The model we consider is the usual model of an expanding universe (with Robertson-Walker metric) where three-dimensional spatial hypersurfaces of simultaneity are assumed to be of negative curvature. Furthermore, we shall suppose that the spatial hypersurfaces are compactified by suitable identification of points. It is known that this is possible to do without changing the metric structure of the space.

A free test particle in the universe moves, of course, along the four-dimensional space-time geodesics. Because of the symmetry of the Roberson-Walker metric, however, one can show that the *three* spatial coordinates (in the co-moving frame) of the freely moving test particles follow geodesic curves in (any given) three-dimensional spatial hypersurface of simultaneity. In other terms, the four-dimensional geodesic motion of test particles when projected to a three-dimensional spatial hypersurface defines a geodesic flow in the three space. The important point to note is that this affine parameter λ of the projected geodesic flow on the three space turns out not to the proper time of the test particle nor the *cosmic time parameter t*, but a nonlinear function $\lambda(t)$ of the from:

$$\lambda(t) = \lambda(t_0) + A \int_{t_0}^{t} \frac{ds}{R(s)}$$

for massless particles, and

$$\lambda(t) = \lambda(t_0) + A \int_{t_0}^{t} \frac{ds}{R(s)\,[\alpha^2 + R^2(s)]^{1/2}}$$

for massive test particles. Here $R(t)$ denotes the scale factor or expansion parameter of the universe. It is to be determined, of course, from the Einstein equation in conjunction with an equation of state for the cosmological fluid and the equation of conservation of energy. The quantity α is a constant related to the speed of the particle. The above-mentioned reduction of the four-dimensional geodesic motion to geodesic flow in three space is possible in any model with Robertson-Walker metric irrespective of the sign of spatial curvature. But it is only in the case of models with negative spatial curvature and compactified three space that this possibility has interesting implications.

This is because in this case the reduced geodesic flow in three space associated with free motion of test particles is known to be K-flow (and, in fact, Bernoulli flow) and hence a fortiori mixing and ergodic. Thus in such a universe arbitrary initial beams of test particles distributed in space and direction (or more precisely, beams corresponding to square integrable distribution functions on the space of line elements of three-space) will tend toward the uniform (microcanonical) distribution as time progresses. There is thus a natural mechanism leading to homogeneity and (local) isotropy in such a universe. In particular, photons from different regions of the early universe would form the isotropic backgroud radiation after a sufficient lapse of time.

It is interesting that in the cosmological model of the universe under consideration, distributions or beams of massive test particles behave somewhat differently than beams of massless test particles. Though the "mixing mechanism" is present for massive particles also, an initial distribution of massive particles may not be able to reach the uniform distribution. This difference in behavior between photons and massive test particles comes from the fact that the parameter $\lambda(t)$ for massive test particles, in contrast to that of photons, stays bounded even as $t \to \infty$: the physically admissable values of λ (for massive test particles) cannot be made arbitrarily large, whereas the mixing property of three-dimensional geodesic flow would lead to a uniform distribution in general only in the asymptotic limit $\lambda \to \infty$.

Let us now turn to the internal time operator T associated with the geodesic motion of test particles in the cosmological model under consideration. Its existence is assured because the "projected" geodesic flow on the fixed three-dimensional hypersurface of simultaneity is a K-flow. The internal time operator T under discussion will thus be an operator acting on the distribution functions on the space of line elements of the three-dimensional hypersurface and will satisfy the relation

$$U_\lambda^* T U_\lambda = T + \lambda(t) I.$$

Here U_λ denotes, of course, the unitary group induced by the projected three-dimensional flow.

The important point to note is that the "time parameter" of the projected flow is not the cosmic time parameter t but is a nonlinear function $\lambda(t)$ given in the preceeding section, and it is $\lambda(t)$ rather than t that must occur in the defining relation of T. As a result, the (average) age of distribution functions on the space of line elements changes as the test particles move freely, keeping step not with t, but with $\lambda(t)$. The time scale defined by internal time is thus distinct from the cosmic time scale and corresponds to that of $\lambda(t)$. It is interesting that (see the expression for $\frac{d\lambda}{dt}$ given in the previous section) in very early epochs of the universe the internal time flows more rapidly compared with the cosmic time $\frac{d\lambda}{dt} + \frac{1}{R(t)} \to +\infty$ as $t \to 0$. Similarly, as the universe ages the internal time scale gets dilated relative to the cosmic time: $\frac{d\lambda}{dt} \to 0$ as $t \to +\infty$. The physical meaning of these relative rates of flow of internal time and cosmic time is that the mixing rate (i.e. rate of approach to equilibrium) *with respect to change in t* approaches ∞ as one nears the singularity: whereas there is practically no mixing in a sufficiently aged universe.

The existence of an internal time operator T in the cosmological model under discussion has important implications. As discussed in this paper, it allows one to associate a Lyapounov variable or H-function with the geodesic motion of test particles. Moreover, the existence of T implies that the free geodesic motion test particles are intrinsically random. This illustrates how irreversibility and randomness could emerge as essential features of dynamical systems embedded in a suitable cosmological model. A further interesting feature of the cosmological model under discussion follows from the uncertainty relation

$$(\Delta T)_\rho (\Delta L)_\rho \geq \tfrac{1}{2}$$

which is a consequence of the canonical commutation relation

$$[T, L] = iI$$

between the Liouvillian L and T. Now $(\Delta L)_\rho$ represents the dispersion of frequencies of periodic components into which the (Gibbs) distribution ρ can be decomposed. The above uncertainty relation thus shows that for a ρ in the early epochs of the universe (when $(\Delta T)_\rho$ is necessarily small), the distribution function ρ must have a wide dispersion of frequencies and hence its motion must have been very chaotic.

For more details on these questions, the reader may see the forthcoming publication [20].

It is amusing to recall that Einstein cherished the belief that "God does not play dice." A serious challenge to this point of view comes from quantum mechanics. In our opinion, an equally important challenge comes from the recent studies of classical systems exhibiting strong forms of trajectory instability. As was said before, such systems are intrinsically random, and we find here that Einstein's own theory allows cosmological models in which the simplest and most fundamental of all motions, the geodesic motion of test particles, has this feature of nondeterminism.

Concluding Remarks

The preceding considerations thus lead us to the viewpoint that irreversibility expressed in the second law results from a special form of symmetry breaking at the dynamical level that causes the dynamical group to be "realized" as a dissipative semigroup associated with a probabilistic process admitting an H-function. The physical origin of the symmetry breaking in question is a limitation on physically observable states. Such a limitation comes, in the first place, from (strong) instability of dynamical motion as a consequence of which the concept of phase space trajectories ceases to be physically meaningful and the physically realizable states of the system need to be described in terms of (Gibbs) distribution functions. But the existence of symmetry breaking under consideration is the expression of a further limitation: not all distributions but only a suitable proper subset of them can correspond to physically realizable states. We have presented arguments indicating that this second limitation is a consequence of the fact that certain types of "future directed" correlations can not exist in physical systems so that only those distributions which do not contain such "future directed" correlations can represent physically realizable states.

Thus the second law, which implies at the macroscopic level a limitation on the possibilities of "manipulation" of matter (e.g., the impossibility of perpetual machines of the second kind), implies a limit to our manipulation also at the *microscopic level*. To put it differently, the second law makes explicit on the macroscopic level a basic structure referring to the microscopic level. It expresses an essential new element foreign to the laws of dynamics but, of course, compatible with them. The analogy with quantum statistics may perhaps clarify what we want to say. The limitation to symmetrical (or anti-symmetrical) wave functions is of course not a consequence of the Schrödinger equation. However, once a restriction on the symmetry of wave functions is formulated it is propagated by the laws of quantum mechanics.

In the present cultural context, entropy plays a considerable role from basic physics to economics and political thought. It is therefore gratifying that we begin to understand somewhat better the microscopic meaning of the second law. Boltzmann's famous conclusion that increase of entropy means the evolution to the "most probable" state still remains basically correct. However, it hides some rather complex features. Indeed, we have first to express the dynamical condition which lead to a Markov process, as it is only for such processes that Boltzmann's statement is correct. Moreover, we have to find physical reasons to introduce a selection principle which would permit us to choose the right semigroup. Anyway, the wide gap which existed between the far-ranging macroscopic applications of dissipativity and the microscopic theory of irreversible processes is beginning to narrow, and further progress can be expected in the near future.

Acknowledgements

We thank Dr. C. M. Lockhart for his help in preparing this paper. This work was partly supported by the Robert A. Welch Foundation of Houston, Texas and the Institut Internationaux de Physique et Chimie (Solvay), Brussels.

References

1. M. Born, *Natural Philosophy of Cause and Chance* (Oxford, Clarendon Press, 1949).
2. K. Yosida, *Functional Analysis* (Springer, New York, 1974).
3. J. W. Gibbs, *Elementary Principles in Statistical Mechanics* (Dover, New York, 1960).
4. I. Prigogine, C. George, F. Henin and L. Rosenfeld, Chem. Scripta **4** (1973) 5-32.
5. B. Misra, I. Prigogine and M. Courbage, Physica **98A** (1979) 1-26.
6. S. Goldstein, B. Misra and M. Courbage, J. Stat. Phys. **25** (1981) 111-126.
7. B. Misra, in *Proceedings of the Special Session in Mathematical Physics of the American Mathematical Society Meeting* (March 1980, Boulder, Colorado) Plenum Press, 1981.
8. B. Misra and I. Prigogine, Suppl. Prog. Theor. Phys. **69** (1980) 101-110.
9. Ya. G. Sinai, Usp. Mat. Nauk. **27** (1972) 137.
10. Ya. G. Sinai, Sov. Math. Dokl. **1** (1960) 335-339.
11. D. Anosov, Proc. Steklov Inst. No. 90 (1967).
12. N. S. Krylov, *Works on the Foundations of Statistical Physics* (Princeton Univ. Press, New Jersey, 1979).
13. B. Misra, Proc. Natl. Acad. Sci. USA **75** (1978) 1627-1631.
14. I. Prigogine, *Etudes Thermodynamiques des Phenomenes Irreversibles* (Dunod, Editeurs Paris, 1947).
15. K. Goodrich, K. Gustafson and B. Misra, Physica **102A** (1980) 379-388.
16. E. B. Dynkin, *Markov Processes* (Springer, New York, 1965).
17. V. I. Arnold and A. Avez, *Ergodic Problems of Classical Mechanics* (Benjamin, New York, 1968).
18. M. Theodosopulu, C. Coutsomitros (to appear).
19. Ya. G. Sinai, Funct. Anal. Appl. **6** (1972) 35.
20. C. Lockhart, B. Misra and I. Prigogine, "Geodesic Instability and Internal Time in Relativistic Cosmology" (to appear).

Dissipative Structures in Multiple Unit Systems

A. Babloyantz

Chimie-Physique II
Campus Plaine U. L. B.
C. P. 231
1050 Bruxelles, Belgium

1. Introduction

The theory of dissipative structures based on bifurcation analysis has been applied with great success to several fields of research such as hydrodynamics, chemistry, ecology, and biology[1]. Domain of application of the theory is still expanding as more and more chemical reactions exhibiting selforganization are found[2]. On the other hand recent development of the theory of bifurcation has shown a great variety of selforganizing solutions thus extending the domain of application of the theory.

Recent advances in the problem of slime mold aggregation and chemical waves may be found in the original papers[3-4].

All these systems are usually modelled in terms of either two coupled first order nonlinear differential equations or two second order partial differential equations if the diffusion of the dynamical variables are considered. In this case the reaction diffusion equations are

$$\dot{x} = f(x, y) + D_x \nabla^1 x$$
$$\dot{y} = g(x, y) + D_y \nabla^2 y \tag{1}$$

These equations are quite appropriate for description of continuous chemical systems. Reaction diffusion equations have also been applied successfully to the study of various aspects of morphogenetic fields formed from a large number of individual cells[5-6]. In these models one neglects the cellular membranes that separate various cytoplasms; and the system is assimilated to a continuous chemical vessel.

However the reaction diffusion mechanisms cannot account for the regulative properties of morphogenetic fields. Therefore they fail to explain the phenomenon of size invariance[7] of morphogenetic patterns. On the other hand reaction diffusion mechanisms are unable to predict the onset of polarity in cleaving eggs in the absence of embryonic growth, unless one assumes, the unlikely event, that an appropriate and well defined change in cell metabolism occurs.

Moreover the investigation of multiunit network properties in terms of various types of cellular connectivity cannot be performed even in the framework of discontinuous diffusion processes.

The intent of this paper is to introduce a discontinuous approach to selforganizing ensembles formed of multicellular and multiunit assemblies. It will be shown that in this framework complex networks such as neuronal systems may be investigated. We show also that if in a morphogenetic field, cell individuality is conserved, a solution to above mentioned unsolved problems may be found.

2. Pattern Formation in Multicellular Systems

Let us consider a developmental system composed of N identical cells. Complex metabolic reactions proceed inside each cell, two morphogens x^1 and x^2 are formed by auto or cross catalytic processes. These substances influence the concentration of morphogens in other cells via activation of surface receptors, gates, channels or gap junctions. Some of these processes must be described in terms of non linear functions. In morphogenetics fields the variables of a single unit are influenced by their immediate neighbours only.

We assume that the centers of N cells form a cubic lattice and are contained in a cubic volume (see figure 1).

The kinetic equations describing the rate of change with time of the concentrations x^1 and x^2 in the i, j, k cell are given by

$$\dot{x}^1_{ijk} = f(x_1, x_2) + DC^1(x^1_{i-1jk} + x^1_{i+1jk} + x^1_{ij-1k} + x^1_{ij+1k}$$
$$+ x^1_{ijk-1} + x^1_{ijk+1})$$
$$\dot{x}^2_{ijk} = g(x_1, x_2) + DC^2(x^2_{i-1jk} + x^2_{i+1jk} + x^2_{ij-1k} + x^2_{ij+1k}$$
$$+ x^2_{ijk-1} + x^2_{ijk+1}) \qquad (2)$$
$$i = 1,...N/3, j = 1,...,N/3, k = 1,...,N/3$$

Functions f and g refer to the non linear chemical reactions proceeding inside a given cell whereas C^1 and C^2 describe the exchange processes between the i, j, k cell and its first neighbours. These functions will be designated as contact operators. Coefficient $D = \frac{sd}{n}$ is the product of the fraction of the surface s of the cell i, j, k covered by the neighbouring cells, times the density d of transport channels. As we shall be concerned mainly with large systems, therefore periodic boundary conditions will be considered.

With the present discrete description of multicellular systems the nature of the mathematical problem changes completely. Now we must solve a set of first order coupled non linear differential equations with many variables. Analytical approaches to the solution of these systems is extremely tedious. Computer evaluation of solutions remains the best way to deal with this kind of systems.

Numerical integration of the set of equations (2) for a one dimensional system, and specific contact operators shows[8] that the set of equations (2) have the same type of solutions as the reaction diffusion systems. Temporal, statio-temporal and nonhomogeneous steady state patterns analogous to reaction diffusion systems[1] are found.

However numerical evaluation of solutions of the set of equations (2) may become extremely complex and costly if we do not have a clue as to the range of parameter values for which the homogeneous steady state of the system becomes unstable.

Fortunately by performing a linear stability analysis[9] of the non linear system (2) we may find easily the range of parameter values for which the homogeneous steady state becomes unstable. Moreover the nature of the unstable points may be guessed from this analysis. In some instances this information is sufficient for discussion of some aspects of biological systems such as the nature of various emerging patterns.

In order to perform a linear stability analysis we linearize the system around a given steady state and find

$$\dot{x}^1_{ijk} = f_{x^1} x^1_{ijk} + f_{x} x^2_{ijk} + DC^1_{x1} \{x^1_{i-1,jk} + x^1_{i+1,j,k} + x^1_{ij-1,k}$$
$$+ x^1_{ij+1k} + x^1_{ijk-1} + x^1_{ijk+1} - 6x^1_{ijk}\} + 6DC^1_{x^1} x^1_{ijk}$$

$$\dot{x}^2_{ijk} = g_{x^1} x^1 + g_{x^2} x^2 + DC^1_{x_2} \{x^2_{i-1jk} + x^2_{i+1jk} + x^2_{ij-1k} + x^2_{ij+1k}$$
$$+ x^2_{ijk-1} + x^2_{ijk+1} - 6x^2_{ijk}\} + 6DC^2_{x^2} x^2_{ijk}$$

$$C^1_{x^1} = \frac{\partial C^1}{\partial x^1_{i-1jk}} = \frac{\partial C^1}{\partial x^1_{i+1jk}} = \ldots \tag{3}$$

The characteristic equation of such a large system is a polynomial of order 2N and thus again must be solved by numerical methods. However the matrix of coefficients of linearized system may be diagonalized readily for periodic boundary conditions if the contact operator shows some symmetry in interactions. This is the case for the present problem as is seen from figure 1.

The eigenvalue $(-\lambda_k)$ of the contact operator may be evaluated easily and reads

$$\lambda_k = -12 \sin^2 \frac{\pi k}{N} \tag{4}$$

With this eigenvalue the matrix of order 2N is diagonalized and the stability problem thus reduces to the resolution of a second order polynomial

$$\omega^2 - T\omega + \Delta = 0 \tag{5}$$

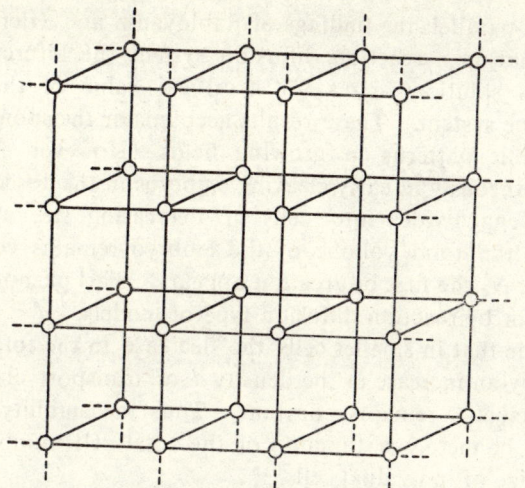

Fig. 1. A cubic lattice with six immediate neighbours.

where Δ and T are defined as: $\text{Det } \Lambda_k = \Delta$ and $T_r \Lambda_k = T$ and $\rho = T^2 - 4\Delta$. Λ_k is the Jacobian Matrix

$$\Lambda_k = \begin{bmatrix} f_x - C_{x'}^1 \lambda_k & f_y \\ g_x & g_y - C_{x^2}^2 \lambda_k \end{bmatrix} \quad (6)$$

Therefore

$$T = f_{x^1} + g_{x^2} - C_{x^1}^1 \lambda_k^{x^1} - C_{x^2}^2 \lambda_k^{x^2} = T_s + T_k$$

$$\Delta = (f_{x^1} g_{x^2} - f_{x^2} g_{x^1}) + C_{x^1}^1 C_{x^2}^2 \lambda_k^{x^1} \lambda_k^{x^2} - f_{x^1} C_{x^2}^2 \lambda_k^{x^2} - g_{x^2} C_{x^1}^1 \lambda_k^{x^1}$$

$$= \Delta_s + \Delta_k$$

$$\rho = (f_{x^1} - g_{x^2})^2 + 4 f_{x^2} g_{x^1} + (C_{x^1}^1 \lambda_k^{x^1} - C_{x^2}^2 \lambda_k^{x^2})^2$$
$$+ 2 C_{x^1}^1 \lambda_k^{x^1} (g_{x^2} - f_{x^1}) + 2 C_{x^2}^2 \lambda_k^{x^2} (f_{x^1} - g_{x^2})$$

$$= \rho_s + \rho_k \quad (7)$$

where the subscripts s and k refer to those parts of each expression independent of and dependent on λ_k respectively.

The remaining part of the discussion follows that of references 9 and 10 and will not be given here.

We shall note that if the uniform steady state for the unconnected ensemble is a stable focus and N is small the steady state remains homogeneous. As N increases and reaches a critical value N_c, the homogeneous state may become unstable and the solution will bifurcate into a non homogeneous pattern. The nature of the pattern is dependent on the value of k of expression (4).

If the volume of the individual cells does not change then the total volume of the system must reach a critical size before a bifurcation point is reached.

These results parallel the findings of Babloyantz and Hiernaux[5]. They have shown that in a reaction diffusion system the bifurcation into an inhomogeneous solution arizes for a critical value of the geometrical dimension of the system. These results account for the spontaneous onset of morphogenetic patterns in growing fields. However gradients may generate spontaneously in early cleaving embryos in the absence of growth. The fertilized egg divides into cells of decreasing size and increasing number, while the total volume of the embryo remains constant. At a critical number N_c the first bifurcation appears. This phenomenon cannot be accounted for by reaction diffusion type of models.

Let us assume that in smaller cells the decrease in the total cell surface S is followed by an increase in the density d of transport channels, therefore the coefficient D remains constant. Thus the stability properties of the system will be merely a function of the total cell number N independently of the size of individual cells.[11]

Recently we have shown the possibility of size invariance in cell-cell contact models. The details of the model will be found elsewhere[12].

3. Connectivity Number as Bifurcation Parameter

The discrete contact models are suitable for the study of systems with various topologies. To see this we assume that the N identical units form a closed ring, therefore again the ensemble is submitted to periodic boundary conditions.

Let us again consider the set of equations (2) but with various connective topologies, whereby a given unit is in direct contact with two other units m step to its left and to its right as seen in figure (2).

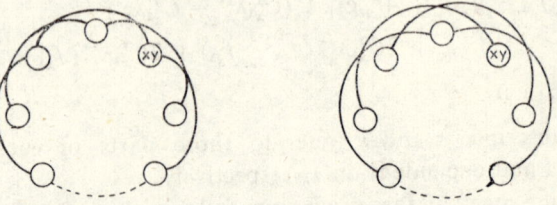

Fig. 2. Two connective topologies for $m=2$ and $m=3$ steps.

Presently we ask if the bifurcation properties of the unique steady state of all these configurations will change as m takes different values.

Linearized contact operator is of the form $C[x_{l+m} + x_{l-m} - 2x_l]$. Therefore one verifies easily that eigenvalue of this operator is

$$\lambda_k = 2\left[1 - \cos m \frac{2\pi k}{N}\right] \qquad k = 1, \ldots, N$$

For large N we find $\lambda_k = \left(\frac{m2\pi k}{N}\right)^2$. Therefore T_k, Δ_k and ρ_k are all functions of m and are different from one configuration to the next. Thus the bifurcating solutions are also determined by m and may present different characteristics. This fact shows how by keeping all other characteristics of the network constant, the mere change of connections influences self-organising properties of the system.

In more complex network systems a given unit is usually in direct contact with many other units that are not necessarily in its immediate neighbourhood. This type of situation may arize in ecosystems, economic ensembles or neural networks.

Most interesting cases arize when a given unit with fixed output and input per unit is connected to $2n$ neighbours. (see figure 3). From preceeding paragraphes we expect that for each value of n different bifurcation properties will arize.

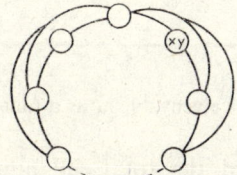

Fig. 3. A multiply connected network with $n=3$ neighbours. Only the connectivity of a single unit is shown.

The linearized set of differential equations describing the dynamics of the network is:

$$\dot{x}_i = f_x x_i + f_y y_i + C_x \left[\frac{1}{n}(x_{i-n} + \ldots + x_{i-1} + x_{i+1} + \ldots + x_{i+n}) - 2x_i \right] + \frac{2C_x}{n} x_i$$

$$\dot{y}_i = g_x x_i + g_y y_i + C_y \left[\frac{1}{n}(y_{i-n} + \ldots + y_{i-1} + y_{i+1} + \ldots + y_{i+n}) - 2y_i \right] + \frac{2C_y}{n} y_i$$

(8)

Again for periodic boundary conditions it can be shown that the eigenvalues $(-\lambda_k)$ of the contact operator are given by

$$\lambda_k = 2 \left[1 - \frac{1}{n} \frac{\cos((n+1)\omega_k/2) \sin(n\omega_k/2)}{\sin(\omega_k/2)} \right]$$

$$\omega_k = \frac{2\pi k}{N} \quad k = 1, \ldots, N \qquad (9)$$

Figure 4 shows the first four values of λ_k as a function of n for a 40 unit ensemble. It is seen that for every k, λ_k increases with n. This property may also be seen analytically by expanding λ_k for large values of N. We must note that the nth eigenvalue is always zero and therefore independent of n.

From expressions (7) cf T, Δ and ρ we see that the connectivity number n enters these quantities and therefore affects the bifurcation points of the set of equations (8). If all chemical parameters, coefficients related to

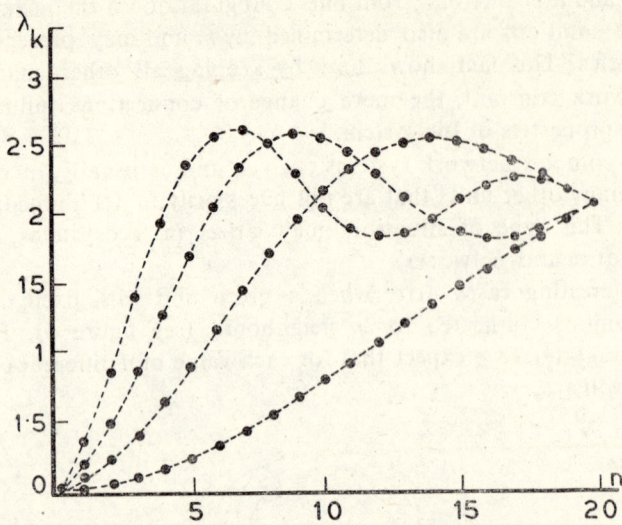

Fig. 4. The eigenvalue λ_k as a function of n.

exchange between units and the total cell number N are held fixed, the connectivity number n may become a new bifurcation parameter. The latter will determine the nature of the newly emerging solutions.

An extensive discussion of the influence of n on the stability properties of the set of equations (8) may be found elsewhere[10]. Let us only note that the nature of self-organizing phenomena depends strongly on the functional form of f_i and g_i. It appears that increasing the connectivity of an ensemble of units with multiple contact may transform the spatio-temporal patterns generated by the ensemble. In general increasing connectivity results in greater stability. Temporal periodicity nearly always disappears if the number of connections is sufficiently increased. However, under certain conditions, purely spatial patterns may arise though an increased number of interconnections. Thus, by merely changing the way a unit is connected to its surrounding, a system may present successively all different forms of self organization and even become completely stable.

In the next section, we shall use these ideas for the study of the self-organizing properties of a neutral network. In particular we shall be concerned with the epileptic seizure in a cerebral cortex.

4. Spatio Temporal Patterns in Epileptic Seizures

Figure (5a) shows a microelectrode recording of cortical neurones made under normal conditions in the absence of external stimuli. Usually no correlation is seen either between individual units in the same cortical domain or between any one unit and the local gross E. E. G. activity. However under the influence of specific drugs or lesions this situation may

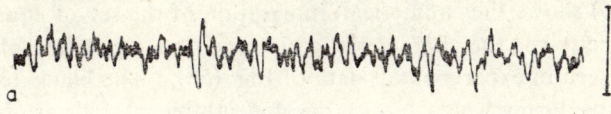

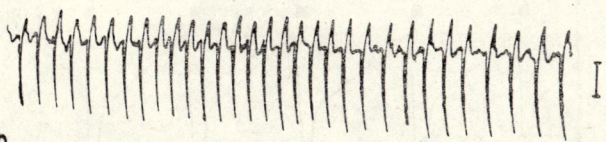

Fig. 5. (a) Normal cortical electro-encephalographic activity (E.E.G.)
(b) Activity during a seizure.

change dramatically. Most of the neuronal cells picked up by a microelectrode start to oscillate and fire in phase. One sees a large increase in the amplitude of the E. E. G. which shows extremely regular sharp waves with a characteristic biphasic profile as seen in Figure (5b). This is the phenomenon of epileptic seizure.

A model for epilepsy may be constructed in the framework of bifurcation analysis if we use the cell-cell contact models of the preceeding section.

The network is composed of $2N$ neurons, divided equally between excitatory and inhibitory neurons. The variables are the set of mean membrane potentials $\{x_i\}$ and $\{y_i\}$ across the electrically inexcitable dendritic membranes of respectively excitatory and inhibitory neurons.

The time evolution of these variables is given by the following equations

$$\frac{dx_i(t)}{dt} = k(V - x_i(t)) + (\varepsilon - x_i(t)) \sum_{j=1}^{N} a_{ij} f(x_j(t-\tau))$$
$$+ (E_{Cl} - x_i(t)) \sum_{j=1}^{N} b_{ij} f(y_j(t-\tau)),$$
$$\frac{dy_i(t)}{dt} = k(V - y_i(t)) + (\varepsilon - y_i(t)) \sum_{j=1}^{N} c_{ij} f(x_j(t-\tau))$$
$$i = 1...N. \tag{10}$$

The first term on the right hand side represents the passive properties of the membrane in the absence of any synaptic input. V is the resting value of membrane potentials. $1/k$ is the relaxation time constant. Function f describes the contact and exchange process between two neurons. Physiological requirements necessitate the introduction of a time delay τ in the contact functions. The functional form of f may be deduced from experimental data.

This network has been studied extensively analytically as well as by numerical methods by Kaczmarek and Babloyantz[13].

Figure (6) shows the numerical integration of the set of equations (10). We have recovered the characteristic biphasic time oscillation of excitatory potentials seen in experimental data of Fig. (5b). The black bars indicate the time laps during which excitatory and inhibitory cells are firing. The

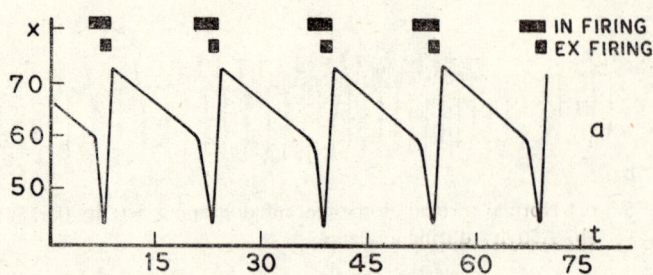

Fig. 6. A plot of sharp biphasic oscillating solution of equations (10).

simplified model therefore accounts satisfactorily for the known experimental data of epileptic seizures. Different connective topologies were also considered. Each connectivity gives rize to new selforganizing behaviour. These could be divided into spatially inhomogeneous steady states, and time dependent solutions such as standing waves, travelling waves or chaotic behaviour. Figure (7) represents such a chaotic solution. Details of other connective topologies may be found in the original paper[13].

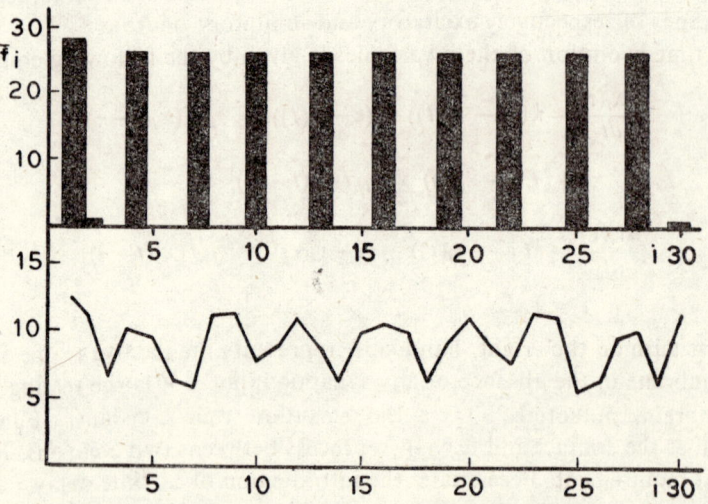

Fig. 7. A chaotic solution of equation (10). The upper diagram gives the mean firing rates for a spatio temporal pattern in membrane potentials that may be described as a regular travelling wave with nodes.

Conclusions

We have shown that cell contact models may improve the results obtained from reaction diffusion mechanism for understanding of selforganizational properties of multiunit assemblies.

Cell-cell contact models based on bifurcation analysis extends the range of the theory of dissipative structures into the domain of multiunit systems with various connective topologies where reaction diffusion systems cannot be used.

This discrete approach is most suitable for the study of ecosystems, social systems and also economic problems. Or in other domains where the properties of the system as a whole is dependent on the cooperation of an ensemble of separate units.

References

1. G. Nicolis and I. Prigogine, (1977) Self-Organization in Non equilibrium Systems. Wiley-Interscience.
2. P. De Kepper (1981) Nonlinear Phenomena in Chemical Dynamics. C. Vidal and A. Pacault Editors Springer Verlag, Berlin.
3. A. Goldbeter and L.A. Segal (1980) Differentiation **17**, 127.
4. D. Walgraef, G. Dewel and P. Borckmans (1982) Advances in Chemical Physics, Vol. **49**, I. Prigogine and S. Rice Editors, John Wiley and Sons.
5. A. Babloyantz and J. Hiernaux (1975), Math. Biol. **37**, 437.
6. A. Gierer and H. Meinhardt (1972), Kybernet. **12**, 30.
7. L. Wolpert (1969) J. Theor. Biol. **25**, 1.
8. A. Babloyantz (1977) J. Theor. Biol. **68**, 551.
9. N. Minorski (1962) Nonlinear Oscillations. Van Nostrand, Princeton.
10. A. Babloyantz and L. K. Kaczmarek (1979) Bul. Math. Biol. **41**, 193.
11. A. Babloyantz (to appear)
12. A. Babloyantz (to appear)
13. L. K. Kaczmarek and A. Babloyantz, (1979) Biol. Cybernetics. **26**, 199.

Irreversible Thermodynamics of Living State

R.P. Rastogi

Chemistry Department, Gorakhpur University
Gorakhpur, India

1. Introduction

Living systems are often baffling when physical laws seem to be violated in their working. Such situations relate to (i) occurrence of processes or reactions which are not thermodynamically favourable, (ii) transport of matter against the gradient of chemical potential and (iii) tendency towards temporal and spatial organization. However, many paradoxes have been made clear during the last few decades due to developments in non-equilibrium thermodynamics[1-4]. The purpose of the present paper is to present a coherent and critical account of these developments along with the general non-equilibrium thermodynamics of the Living State.

2. Nature of Living Systems

Thermodynamics recognizes three types of systems (a) open, (b) closed and (c) isolated. Open systems are those which exchange matter and energy with the surrounding while closed systems are those which do not exchange matter but exchange only energy. Isolated systems do not exchange matter and energy either. The criteria for equilibrium for different types of systems are summarized in Table 1.

Table 1. Criteria for equilibrium for different types of systems

System	Exchange with the surroundings of		Criterion for equilibrium
	Matter	Energy	
Isolated	no	no	$(\Delta S)_{E,V} = 0$
Closed	no	yes	$(\Delta G)_{T,P} = 0$, $(\Delta F)_{T,V} = 0$
Open	yes	yes	no equilibrium possible

Further at equilibrium, Gibbs free energy G at constant temperature T and pressure P and Helmholtz free energy F at constant T and volume V tend to be minimum for closed systems while entropy S tends to attain a maximum value when internal energy E and V are kept constant. No equilibrium is possible for open system which can only attain a steady state.

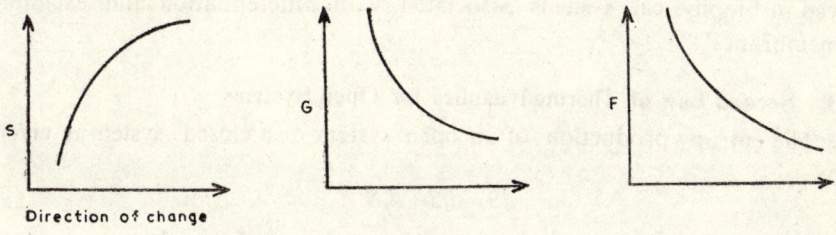

Fig. 1

The general equation of life may be written as $(Organism)_1$ + food $\rightarrow (Organism)_2$ + water + heat where the subscripts 1 and 2 denote the two states of the organism. The organism synthesises its essential metabolites, purine and pyrimidine bases and amino acids and organises them into specific macromoleucules. Energy is produced in this process and the living machine uses this energy for metabolic processes and other types of work at the cost of free energy which is degraded. Further, order increases and as a result entropy tends to decrease.

Steady state of the open systems can be visualized by imagining the system as follows:

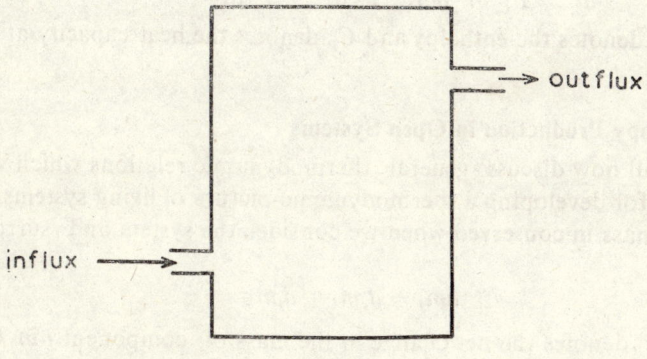

Fig. 2

where the influx and outflux of matter simultaneously occur. The above system would be in the steady state when influx is equal to outflux. The steady states can be of two types

(a) Stable
(b) Unstable

Stable steady states are common in physico-chemical systems such as in electro-kinetic phenomena. The living cell is another example of a system close to steady state. According to Norbert Weiner, "Living organisms are metastable. Maxwell demons whose stable state is to be dead". Unstable steady states are observed far from equilibrium and are associated with oscillatory phenomena. In addition multiple steady states are observed in biophysical systems associated with differentiation and excitable membranes[5].

3. Second Law of Thermodynamics for Open Systems

The entropy production of an open system or a closed system is given by

$$dS = d_iS + d_eS \tag{1}$$

where d_iS is entropy production in the system and d_eS is the entropy flow from the surroundings. The term d_eS can be positive, negative or zero but d_iS is positive definite. From second law of thermodynamics,

$$dS > 0 \tag{2}$$

so that

$$d_iS > -d_eS \tag{3}$$

It means that internal production of entropy has always to be greater than entropy outflux to the surroundings. For the special case when $d_eS = -d_iS$, $dS = 0$. Information about the entropy flow terms can be obtained from the experimental data on specific heat dissipation in animals[6] by noting that

$$\frac{d_eS}{dt} = \frac{1}{T}\left[\frac{dQ}{dt}\right]_{P,T} = \frac{1}{T}\cdot\frac{dH}{dt} = \frac{C_p}{T}\frac{dT}{dt} \tag{4}$$

where H denotes the enthalpy and C_p denotes the heat capacity at constant pressure.

4. Entropy Production in Open Systems

We shall now discuss general thermodynamic relations which would be required for developing a thermodynamic picture of living systems.

Since mass in conserved when we consider the system and surroundings together,

$$dm_j = d_em_j + d_im_j \tag{5}$$

where dm_j denotes the net change in the mass of component j in the open system. d_em_j denotes the influx of mass from the surroundings and d_im_j denotes the production of j inside the system itself due to chemical reactions. Dividing equation (5) by the molecular mass of j we have

$$dn_j = d_en_j + d_in_j \tag{6}$$

where n denotes the number of molecules and d_in_j is given by

$$d_in_j = \sum_{p=1}^{r} \nu_{jp}\, d\xi_p \tag{7}$$

where $d\xi_p$ denotes the degree of advancement of the pth reaction and ν_{jp} is the stoichoimetric coefficient of j in the pth reaction. ν_{jp} is counted positive when ν_j appears on the right hand side of the reaction equation and is counted negative when ν_j appears on the left hand side of the equation. It is supposed that r reactions are occurring in the system.

The first law of thermodynamics,

$$dQ = dU + PdV \tag{8}$$

where dQ is the heat absorbed by the system, dU is the change in internal energy and dV is the change in volume, would be interpreted for open systems as follows. In fact, for open system

$$dQ = d_e Q + hdm \tag{9}$$

where $d_e Q$ is the flow of heat due to exchange of energy with the surrounding and the term hdm represents the contribution of energy flow due to exchange of matter whereas h denotes the specific enthalpy per gram. Further, dU in equation (8) can also be split as follows for a sub-system I which is a part of the open system containing two sub-systems,

$$dU = d_e U^I + d_i U^I \tag{10}$$

where dU^I denotes the change in internal energy in sub-system I. $d_e U^I$ denotes the energy exchanged with the surroundings while $d_i U^I$ denotes the energy exchange due to internal transfer of energy from system II to system I.

Gibbs equation for entropy production for an open system containing n components is written as,

$$TdS = dU + PdV - \sum_{k=1}^{n} \mu_k d_e m_k - \sum_{k=1}^{n} \mu_k d_i m_k \tag{11}$$

where μ_k denotes the chemical potential of component k. When only chemical reactions are taking place in the system,

$$-\sum \mu_k d_i m_k = M \sum_{j=1}^{r} A_j d\xi_j \tag{12}$$

where A_j denotes the affinity of the reaction j i.e.

$$A_j = -\sum_{k=1}^{n} \nu_{kj} \mu_k \tag{13}$$

and M is the total mass of all the components i.e.

$$M = \sum m_k \tag{14}$$

With the considerations developed in the earlier section, it is easy to see that dS in equation (11) is made up of $d_e S$ and $d_i S$ in such a way that,

$$Td_e S = (dU + PdV - \sum_{k=1}^{n} \mu_k d_e m_k)$$

and
$$Td_iS = M \sum_{j=1}^{r} A_j d\xi_j \qquad (15)$$

We define the chemical reaction rate J_j by the relation

$$J_j = \frac{1}{M} \cdot \frac{d\xi_j}{dt} \qquad (16)$$

so that,

$$\sigma = \frac{d_iS}{dt} = (\sum_{j=1}^{r} A_j J_j)/T \qquad (17)$$

The living system although an open system is not an unitary system. It consists of numerous sub-systems which are divided into smaller sub-systems divided by barriers such as membranes through which transport of matter and energy can take place. For simplicity we may schematically represent the living system as follows,

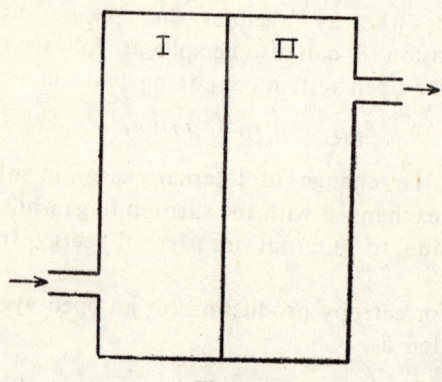

Fig. 3

where I and II are representative sub-systems separated by a membrane. In actual cases, one would have more of such sub-systems. When internal transfer of matter and energy between sub-system I to II is also considered, the entropy produced by irreversible processes within the system can be shown to be given by[2,4]

$$d_iS = \frac{\Delta T}{T^2} d_i U^I + \sum_{k=1}^{n} \Delta(\mu_k/T) d_e M_k^I + M^I (\sum_{j=1}^{r} A_j^I J_j^I)/T^I$$
$$+ M^{II} (\sum_{j=1}^{r} A_j^{II} J_j^{II})/T^{II} \qquad (18)$$

where T^I and T^{II} denote the temperature of sub-systems I and II respectively. Δ denotes the difference of quantities in sub-system II and the corresponding ones in sub-system I. Superscripts I and II denote the corresponding quantities in sub-systems I and II. In living systems, the concentration differences on the two sides of membranes for a number of species are quite large such as in extra cellular and inter-cellular fluid and hence the magnitude of second term on the left-hand side of equation (18) is often quite significant. The entropy production for such a general case is written as,

$$\sigma = \frac{d_iS}{dt} = J_U \Delta T/T^2 = - \sum_{k=1}^{n} J_k \Delta(\mu_k/T) + J_I(A^I/T^I) + J_{II}(A^{II}/T^{II}) \quad (19)$$

where

$$J_U + - \frac{d_i U^I}{dt} = \frac{d_i U^{II}}{dt} \quad (20)$$

$$J_k = - \frac{d_e M_k^I}{dt} = \frac{d_e M_k^{II}}{dt} \quad (k = 1, 2, \ldots n) \quad (21)$$

$$J_I = \left(\frac{1}{\nu_k}\right) d_i M_k^I / dt \quad (22)$$

$$J_{II} = \left(\frac{1}{\nu_k}\right) d_i M_k^{II} / dt \quad (23)$$

Both equations (17) and (19) show that entropy production can be written as the sum of the product of fluxes and conjugate forces so that

$$\sigma = \sum J_i X_i \quad (24)$$

In equation (17), the force $X_j = A_j/T$, whereas in equation (19) the forces corresponding to J_U, J_k, J_I and J_{II} are given by

$$X_u = - \Delta T/T^2 \quad (25)$$

$$X_k = - \Delta\left(\frac{\mu_k}{T}\right) \quad (26)$$

$$X_I = \frac{A^I}{T^I} \quad (27)$$

$$X_{II} = \frac{A^{II}}{T^{II}} \quad (28)$$

Equation (24) is based on Gibbs equation (11) which is supposed to be valid outside equilibrium. The validity of this assumption has been examined by kinetic theory by taking into account successive approximations to distribution function. The analysis shows that the assumption is valid so long as gradients of barycentric velocity are zero or small and local equilibrium is not disturbed. So long as convection does not occur the assumption would be valid for chemical reactions for an unlimited range. It is likely that the assumption would remain valid for a large number of sub-systems involving membranes in the living system so long as the gradients are not large enough to generate convective or turbulent flow.

4. Flux-force Relationship*

In irreversible processes, the vectorial fluxes are coupled with vectorial fluxes and the scalar fluxes are coupled with only scalar fluxes. The coupling of vectorial fluxes and scalar fluxes does not occur. In fact, the fluxes of a particular tensorial order couple with the fluxes of that very tensorial order. This is called Curie's principle. Thus a flux J of any tensorial order $0, 1, 2$ would depend on the forces of that particular tensorial order so that

$$\mathbf{J}_i = f(\mathbf{X}_i, \mathbf{X}_j, \ldots\ldots) \tag{29}$$

and

$$J_k = f(X_k, X_j, \ldots\ldots) \tag{30}$$

where $\mathbf{J}_i$ denotes the vectorial flux and J_k denotes the scalar flux. Close to equilibrium, the fluxes are linear in forces as in Ohm's law and Fourier's law for heat conduction and the transport equations are written as,

$$\mathbf{J}_i = L_{ik} \mathbf{X}_k \tag{31}$$

and

$$J_i = L'_{ik} X_k \tag{32}$$

where L_{ik} and L'_{ik} are called phenomenological coefficients and both are scalar. For a general case[7,8],

$$J_i = L_{ik} X_k + \tfrac{1}{2} L_{ijk} X_j X_k + \ldots \tag{33}$$

so long as entropy production can be represented as the sum of the product of fluxes and forces. In the linear range, matrix of the phenomenlogical coefficients is symmetric according to Onsager Reciprocity Relation. Lot of experimental data on cross-phenomenological coefficients for different phenomena are available which agree with Onsager relation,

$$L_{ik} = L_{kl} \qquad \ldots(34)$$

These are still obeyed in the non-linear regime. This can be proved[9] for the case of coupling of heat flux and mass flux by the kinetic theory formalism. Quite recently this has been proved experimentally from measurements of streaming current and electro-osmotic flux in the non-linear region[10].

5. Coupling of Chemical Reactions

Coupling is universal in the case of vectorial fluxes while this is not necessary in the case of scalar fluxes such as chemical reactions. Hence we shall examine the concept of coupling in greater detail since the living state is a factory where scores of chemical reactions are taking place in different subsystems.

Let us consider the entropy production in isothermal system in which two

---

*Philosophical implications of flux-force relationship in terms of causality principle have been recently examined (R.P. Rastogi, J. Sci. Ind. Res., 40, 565 (1981)).

chemical reactions occur at temperature T. We will have

$$\frac{d_iS}{dt} = \frac{J_1A_1 + J_2A_2}{T} > 0 \qquad (35)$$

where J_1, J_2 are the rates of reactions 1 and 2 and A_1 and A_2 are the corresponding affinities. If the reactions occur independently.

$$J_1A_1 > 0 \quad \text{and} \quad J_2A_2 > 0 \qquad (36)$$

and equation (35) is also satisfied.

On the other hand, if the reactions are coupled, the two reactions can occur simultaneously in such a way that,

$$J_1A_1 < 0 \,, \quad J_2A_2 \triangleright 0 \qquad (37)$$

provided equation (35) is obeyed. Relations (37) imply that a reaction which is not thermodynamically possible on account of negative entropy production can be made to occur by coupling with a reaction which has positive entropy production of larger magnitude. Numerous examples occur in biological systems, where thermodynamically unfavourable reactions are known to occur. In the normal liver, the synthesis of urea occurs but the reaction

$$2NH_3 + CO_2 \rightarrow (NH_2)_2CO + H_2O$$

is not thermodynamically favoured since the Gibbs free energy change for the reaction is $+11$ kcal. It is supposed that it is coupled with the following reaction which has a larger negative free energy change (-115 kcal),

$$\tfrac{1}{6}C_6H_{12}O_6 + O_2 \rightarrow CO_2 + H_2$$

Similarly, entropy is reduced during the synthesis of ribonucleic acids from constituent bases but this occurs due to coupling with other metabolic reactions which drive it on account of their capability to produce entropy of larger magnitude.

There are three ways of examining whether coupling is occurring in a reaction system or not[11,12]. First, if a reaction is driven by the affinity of another reaction, when its own affinity is zero, the two reactions are coupled, i.e. when

$$(J_k)A_{k=0} = \sum_{j=k} L_{kj}A_j \neq 0 \quad (k = 1, 2... \text{ but } k = j) \qquad (38)$$

Second, an inspection of the reaction scheme may itself suggest whether coupling is occurring or not. If the number of elementary reactions is equal to the number of independent reactions, coupling cannot occur. However, it does, if the former is larger than the latter. Thirdly, coupling can also be ascertained by a linear transformation of fluxes and forces keeping the entropy production constant. If by such transformation, we can get new type of fluxes, the number of which is less than the number of elementary reactions, coupling would occur between the new fluxes. These procedures have been examined in a recent communication[11].

Coupled reactions are not common in chemistry. Even complex reactions involving parallel, reversible and consecutive reaction do not show coupling. However, in case of cyclic reactions, the number of independent reactions is one less than the total number of elementary reactions and hence coupling can occur. Examples of such coupled reactions are the triangular isomerisation reaction of cymene[13] and of Δ^α-pentenoic acid[14].

Coupling of reactions although uncommon in chemistry is fairly common in biological systems. Biological reactions such as enzyme reactions involve a reaction net work having series of consecutive reactions. Cyclic reactions are often involved which hold the key to coupling. In the respiratory cycle[15] such a subset of reactions is as follows:

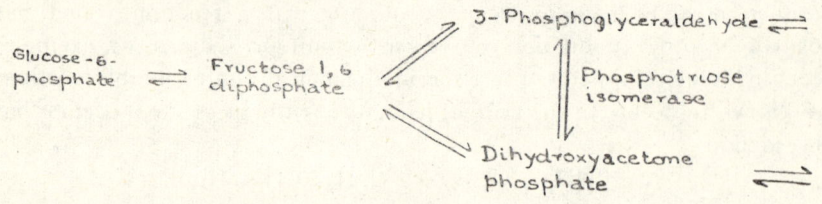

Fig. 4

such cyclic reactions have been postulated in general as catalysed reactions[16] such as non-competitive inhibition and reactions activated by metal ions.

We may make a reference to isoaffine reactions (having constant affinity)[17] whose potentiality in biological reactions has not been investigated fully. We can maintain the affinity A_j of any reaction j at constant value by counterbalancing the supply of matter from the surroundings and its consumption by chemical reaction as in a continuous flow reactor. The evolution in time of A_j at constant temperature and pressure would be given by,

$$\frac{dA_j}{dt} = \sum_{k=1}^{n-1} \left(\frac{\partial A_j}{\partial C_k}\right)_{T,P} \frac{dC_k}{dt} \tag{39}$$

where C_k is the mass fraction of component k. Further, $dC_k = d_e C_k + d_i C_k$ where $d_e C_k$ denotes the change due to influx from the surroundings and $d_i C_k$ denotes the change due to chemical reactions in the system. After suitable transformation, equation (39) can be written as,

$$\frac{dA_j}{dt} = \sum_{k=1}^{n-1} \left(\frac{\partial A_j}{\partial C_k}\right)_{T,P} \frac{d_e C_k}{dt} + \sum_{k=1}^{n-1} \sum_{l=1}^{r} \left(\frac{\partial A_j}{\partial C_k}\right) J_e^{\cdot} \tag{40}$$

It is obvious that when $\dfrac{dA_j}{dt} = 0$,

$$\sum_{k=1}^{n-1}\sum_{l=1}^{r} \nu_{lk}\left(\frac{\partial A_l}{\partial C_k}\right) J_l = -\sum_{k=1}^{n-1}\left(\frac{\partial A_j}{\partial C_k}\right)_{T,P} \frac{d_e C_k}{dt} \qquad (41)$$

6. Coupling in Transport Processes

By and large coupling is very common in vectorial processes. The intracellular fluid and extra cellular fluid contain number of electrolytes and if there are n fluxes and n forces the number of phenomenological coefficients in the transport equations would be n^2 out of which the number of Onsager reciprocity relations would be $n(n-1)$ so that number of unknown coefficients is $\frac{2(n+1)}{2}$. We have one additional relation on account of electronutrality $\sum Z_j J_j = 0$ where Z_j is the valency of the ion j. If the magnitude of the flux can be measured we have additional n relations that the number of unknown coefficients can still be reduced by $(n+1)$ so that effectively the number of unknown coefficients is $\frac{(n+1)(n-2)}{2}$. It may be possible to estimate these by suitable experiments.

Let us consider for the sake of simplicity, only the transport of three species across a cell membrane such as Na^+, K^+ and Cl^-. Denoting the fourth flux current as I, we can write,

$$J_{Na^+} = L_{11}X_1 + L_{12}X_2 + L_{13}X_3 + L_{14}X_4 \qquad (42)$$

$$J_{K^+} = L_{21}X_1 + L_{22}X_2 + L_{23}X_3 + L_{24}X_4 \qquad (43)$$

$$J_{Cl^-} = L_{31}X_1 + L_{32}X_2 + L_{33}X_3 + L_{34}X_4 \qquad (44)$$

$$I = L_{41}X_1 + L_{42}X_2 + L_{43}X_3 + L_{44}X_4 \qquad (45)$$

The forces X_1, X_2 and X_3 are related with the concentration differences of the species Na^+, K^+ and Cl^- respectively. X_4 is related to transmembrane potential. The phenomenological coefficients L_{ik} are given by

$$L_{ik} = f(T, P, G) \qquad (46)$$

where G denotes the structural factor which depends on the nature of the membrane. Further, $L_{ii} > 0$ but L_{ij} can be positive or negative and $L_{ii}L_{jj} > L_{ij}L_{ji}$ on account of positive definite character of σ. In voltage-clamp experiments on excitable nerve membranes, all the forces are constants, the variation of ionic current with time[18] would be due to change in the sign and magnitude of phenomenological coefficients due to change in the nature of the membrane. This is also expected to be true for similar time-dependent phenomena in nerve membranes.

7. Active Transport

Active transport is another common feature of biological systems. A process, where a flow of a substance occurs against an electrochemical potential gradient of the substance is called an active transport[19]. It has to be distinguished from "a diffusional flow against its conjugate gradient driven by dissipation of another diffusional process". It is implied in biology that active transport is based on the operation of internal metabolic processes coupled to external diffusional flows. In certain epithelial tissues flux is coupled to a metabolic reaction. Active transport may be schematically represented as follows[20],

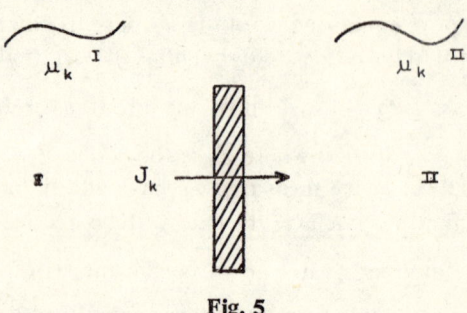

Fig. 5

where B is the membrane separating the fluid in two compartment I and II. The electrochemical potential $\widetilde{\mu}_k^{II}$ of the species k in II compartment is greater that $\widetilde{\mu}_k^{I}$ of the species k in the first compartment. Since, $\widetilde{\Delta\mu}_k = \widetilde{\mu}_k^{II} - \widetilde{\mu}_k^{I}$ is positive, flow of k should not normally be permitted from compartment I to II but it can be driven in that direction by metabolic energy. If we represent the velocity of the reaction per unit area of the membrane by J_r and its conjugate force being the affinity A_r, the entropy production for the system under consideration would be given by

$$\sigma = \mathbf{J}_k \mathbf{X}_k + \mathbf{J}_r(\mathbf{A}_r/T) \tag{47}$$

where $X_k = -\left(\dfrac{\widetilde{\Delta\mu}_k}{T}\right)$. $J_r A_r$ is $+$ve while $J_k X_k$ is not but σ is positive. Thus, the metabolic energy can drive a particular component against its gradient. In the linear domain which is quite wide in biological system, the phenomenological relations would be written as,

$$\mathbf{J}_k = L_{kk}(\widetilde{\Delta\mu}_k) + L_{kr}(\mathbf{A}_r) \tag{48}$$

$$\mathbf{J}_r = L_{rk}(\widetilde{\Delta\mu}_k) + L_{rr}(\mathbf{A}_r) \tag{49}$$

where $L_{kr} = L_{rk}$.

8. Steady States Close to Equilibrium

It has been shown by Prigogine that close to equilibrium, entropy production attains a minimum value in the steady state. This is true up to the range of validity of linear phenomenological equation.

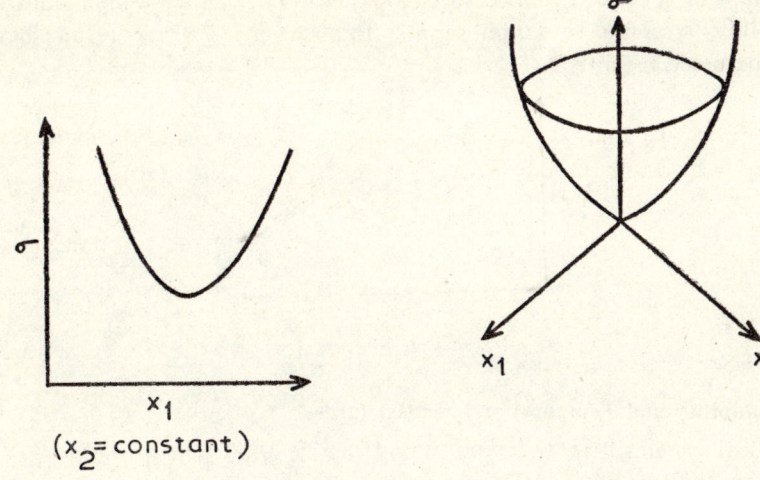

Fig. 6

If we consider entropy production in a system where two processes, X_1 and X_2 are operative corresponding to fluxes J_1 and J_2, the three-dimensional plot of against X_1 and X_2 would yield a paraboloid[21], a section of which at $X_1 = $ constant or $X_2 = $ constant would yield a parabolic curve displaying a minimum such that

$$\left(\frac{\partial \sigma}{\partial X_1}\right)_{X_2} = 0 \quad \text{and} \quad \left(\frac{\partial \sigma}{\partial X_2}\right)_{X_1} = 0 \tag{50}$$

It should be noted that equations (50) are obtained when linear phenomenological relations are assumed and when Onsager reciprocity relation is assumed to be valid.

When the only entropy producing processes are the chemical reactions, the entropy production can be shown to be equal to free energy dissipation at constant temperature and pressure. It means that for such a situation, the steady state would be characterized by minimum entropy production as well as minimum free energy dissipation. In other words, the non-equilibrium system tends to attain a state so that expenditure of free energy is minimum. This is partially true for living machine whose metabolic activity is primarily controlled by chemical reactions.

If $\frac{d_iS}{dt}$ is negligible as compared to $\frac{d_eS}{dt}$ which would be the case when the system is close to equilibrium, then

$$\frac{d_iS}{dt} \simeq -\frac{d_eS}{dt} \simeq -\frac{1}{T}\left[\frac{dQ}{dt}\right]_{P,T} = -\frac{\dot{Q}}{T} \tag{51}$$

where equation (4) has been used. Experimental data on $\dot{Q}$ for certain living organisms yields the following type of thermogram which shows a maximum in $\dot{Q}$. If σ is related to $\dot{Q}$ by equation (51), then a maximum in $\dot{Q}$ would correspond to minimum in σ. Implications of these results have been discussed recently[22].

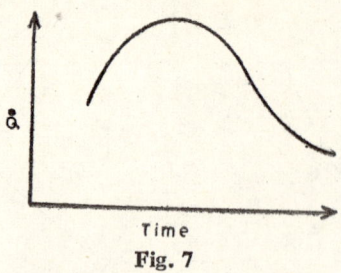

Fig. 7

9. Instability and Temporal and Spatial Order

Biological systems display temporal and spatial order which would appear to violate the second law of thermodynamics which predicts greater disorder in course of time. Professor Prigogine and his school in Brusselles has shown that such order is not in violation of second law of thermodynamics[23].

Temporal order and spatial order can occur in systems when they are far away from equilibrium and the system is unstable. Under these circumstances oscillatory phenomena appear. The most useful function which characterizes the non-equilibrium state is $\delta^2 S$ which is a homogenous function of degree. It can be shown that $\delta^2 S$ has a quadratic form, the associated constants of which have constant values. Further, it can be proved that Eulerian derivative of $\delta^2 S$ is just its time derivative. Hence $\delta^2 S$ is a Liapounov function*. Further since $\delta^2 S$ is negative, second method of Liapounov leads to the following condition of stability for non-equilibrium states[23,24],

$$\frac{\partial(\delta^2 S)}{\partial . t} > 0 \tag{52}$$

It is easy to show that $\delta^2 G$ and $\delta^2 F$ are also Liapounov functions where G and F denote the Gibbs free energy and Helmholtz free energy respectively. The corresponding stability conditions would be expressed as,

$$\left\{\frac{\partial(\delta^2 G)}{\partial t}\right\}_{T,P} \leqslant 0 \; ; \quad \left\{\frac{\delta(\delta^2 F)}{\delta t}\right\}_{T,V} \leqslant 0 \tag{53}$$

*Any function will be a Liapounov function if
  (i) the function is a quadratic form with constant values of the associated coefficients,
  (ii) it has either positive or negative values. If y^2 is a Liapounov function, the stability conditions are

$$\dot{y}^2 \leqslant 0 \text{ if } y^2 > 0 \text{ and } \dot{y}^2 > 0 \text{ if } y^2 < 0$$

Straight forward thermodynamic transformations show that

$$\frac{\partial}{\partial t}\{\tfrac{1}{2}\delta^2 S\} = \int \sum \delta J_k \delta X_k \, dV$$

and hence for stability, it is necessary

$$\int \sum \delta J_k \delta X_k \, dV > 0 \tag{54}$$

where the integration is over the whole volume of the system.

It should be noted that in equalities (52) to (54) would be valid so long as Gibbs equation remains valid outside equilibrium. In the case of chemical reactions occurring in isothermal-isobaric systems, this would always be true as long as the system is continuously stirred and concentration and temperature gradients are not generated.

Equation (54) shows that the perturbation δJ and the perturbation δX have the same sign if the system is stable. The stability condition is not violated by the common transport cross-phenomena or the common chemical reactions. However, autocatalytic reactions of the type,

$$X + Y \underset{k_{-1}}{\overset{k_1}{\rightleftharpoons}} 2X$$

can give rise to instability when

$$k_1 Y \rightleftharpoons 2k_{-1} X_0 \tag{52}$$

The concentration of X is perturbed from the steady state value $[X]_0$. This condition can be achieved in an open system. When the above criteria are applied to Lotka-Volterra Model, which shows periodic changes in X and Y with time,

$$A + X \rightleftharpoons 2X$$
$$X + Y \rightleftharpoons 2Y$$
$$Y + B \rightleftharpoons E$$

it is found that the reaction system obeys the condition of marginal stability[25]. The above scheme constitutes a set of three elementary reactions which are independent, hence the reactions are not coupled. Lotka-Volterra model predicts oscillations in X and Y and series of closed trajectories in X–Y planes but this does not correspond to limit cycle oscillations which are common in living systems. Prigogine and coworkers have shown that limit cycle oscillations beyond the critical value of a certain parameter would be obtained for the following reaction scheme,

$$A \rightleftharpoons X$$
$$B + X \rightleftharpoons Y + D$$
$$2X + Y \rightleftharpoons 3X$$
$$X \rightleftharpoons E$$

During biological organization from the molecular to super-cellular level, many oscillatory and other kinds of self-organization phenomena are observed. Amongst these, glycolytic oscillations[25] are the best known example of metabolic oscillations.

Biological structures in the living state are essentially non-equilibrium structures. The structures are maintained on account of the nature of living state which is an open system capable of exchanging energy and matter with the surroundings. A new structure is always the result of an instability governed by the principles enunciated in earlier sections. These structures are different from equilibrium structures which are maintained without any exchange of energy or matter with the surroundings.

10. Physico-chemical Analogues of Oscillators in Living Systems

Temporal oscillations of the limit cycle type have been observed in non-biological systems[26] and amongst these Belousov-Zhabotiuskii reaction, Bray's reaction and Briggs-Rauscher reaction have been intensively studied. Mechanistic studies show that each of these reactions involve at least one autocatalytic step which is primarily responsible for oscillations.

Saptio-temporal oscillations have also been observed in Belousov-Zhabotinskii reagent. Chemical vaves[27,28] are generated in the system which have been the subject of current research. It has been shown that autocatalytic reactions play a crucial role even in space oscillations which are associated with biological structures. Liesegang rings obtained in periodic precipitation aroused considerable interest sometimes ago but it now seems that these do not involve symmetry-breaking instability[29].

Mitochondrial oscillations have several unique features[30]. Damped oscillations of respiration and volume have been found under conditions of energized ion transport by EDTA treated mitochondria. Under appropriate conditions, oscillations of the respiratory rate is accompanied by oscillation in the oxidation-reduction state of respiratory carriers as judged by pyridine nucleotide flourescene studies. Further, it has been observed that the oscillations can be energized by ATP under conditions where electron-transport is inhibited. After a series of damped oscillations has faded, the response can be reinitiated by restoring the energy supply.

Nearest anologue to mitochondrial oscillator is the Teorell's membrance oscillator[31].

We shall review and make some comments on (i) Temporal oscillations (ii) spatio-temporal oscillations and (iii) membrane oscillations in non-biological systems.

(i) Temporal Oscillations

Temporal oscillations in the concentration of intermediates have been observed in a number of reaction systems. Among these, Belousov-Zhabotiuskii reaction driven by bromate has been more extensively studied[32].

Five types of B–Z reaction have been investigated. These are:

(a) Metal ion catalysed reactions in which the organic substrate e.g. malonic acid and acetylacetone etc. can be brominated by enolization mechanism.

(b) Reactions of organic substrates[33] such as all derivatives of phenol or of aniline in which no metal ion catalysts are needed. Oxidation products include quinones, quinone imines and polynuclear aromatic compounds.

(c) Cerous-catalysed oscillations with a mixed substrate[34] of tartaric acid + acetone mendelic acid + acetone and oxalic acid + acetone.

(d) Cerous-catalysed oscillations[35] with oxalic acid alone as substrate provided elementary bromine is scrubbed from the solution by a stream of inert gas.

(e) Reaction of H_2, with bromate, iodate and chlorite in aqueous sulphuric acid in the presence of a bright platinum plate[36].

Mechanistic studies show that all oscillatory reactions involves series of steps which include at least one autocatalytic step which is essentially responsible for creating instability under appropriate conditions. Following skeleton scheme which is called reversible Oregnator has been suggested to explain the mechanism,

$$A + Y \rightleftharpoons X + P \quad \text{(i)}$$
$$X + Y \rightleftharpoons 2P \quad \text{(ii)}$$
$$B + \bar{X} \rightleftharpoons 2X + Z \quad \text{(iii)}$$
$$2X \rightleftharpoons B + P \quad \text{(iv)}$$
$$Z \rightleftharpoons fY \quad \text{(v)}$$

where $A = B = BrO_3^-$, $X = HBrO_2$, $Y = Br^-$, $Z = 2Ce$ (iv) and $P = HOBr$. The factor f corresponds to the number of bromide ions produced per pair of Ce(iv) ions. Linear stability analysis shows that the reaction system can be unstable. Looking from thermodynamic angle, all the above five reactions are independent and uncoupled. Accordingly, the stability condition would be given by,

$$\delta A_i \, \delta J_i > 0 \quad (i=1, 2 \ldots 5)$$

under certain conditions as shown earlier step (iii) can destabilize the reaction system and hence it becomes exclusively the key step of the oscillatory reaction. Provided the sources of B and X and sinks for X and Z are properly controlled such that the rate of forward reaction is greater than twice the rate of back reaction, instability in the system can be generated. It seems that HOBr and $HBrO_2$ are rapidly converted into molecular bromine which is rapidly scavenged by (i) reactive methylene group, (ii) organic substrate or (iii) acetone.

The above concept is useful in the search of new oscillatory reaction. By scavenging I_2 by acetone in iodate + mandelic + acetone + H_2SO_4 reaction system, a new oscillatory reaction system has been reported by Rastogi and Varma[37].

It should be noted that thermodynamics by itself cannot predict specifically the conditions for generating the limit cycle type oscillations. For this purpose it is necessary to write the appropriate kinetic equations and then subject it to stability analysis.

(ii) Spatio-temporal oscillations

Periodicities in space are also observed in B-Z reaction system. These are related to morphogenesis in biological systems. Both one-dimensional and two-dimensional waves have been obtained[27,38-40]. These display symmetry breaking instability, the condition for which can be obtained by solving the appropriate reaction-diffusion equation.

Experimental data for one-dimensional chemical waves for malonic acid/Mn^{2+}/BrO_3^-/H_2SO_4 shows that the velocity v of wave propagation depends on concentration of sulphuric acid and bromate in the following manner,

$$v = k\,[H_2SO_4]^{-1}\,[KBrO_3]^{1/2}$$

where k is a constant. The velocity of wave propagation for citric acid/Mn^{2+}/BrO_3^-/H_2SO_4 also depends inversely on sulphuric acid concentration. The time-period of chemical waves is related to time-period of temporal oscillations. Waves usually start from a nucleus (a blue speck). In a typical experiment[39], the nucleation time in a steamed tube was 300 sec. whereas in unsteamed tube it was just 150 sec.

In the case of two-dimensional waves, the wave velocity depends on (i) the concentration of the reactants (ii) temperature (iii) depth of the solution (in order to ensure absence of connection) (iv) time of temporal oscillations and (v) rate of stirring. Here again, velocity of two-dimensional wave was found to depend directly on the square root of promate concentration and inversely on sulphuric acid concentration. However, Field and Noyes have observed square root dependence on [H^+]. This is probably due to difference in experimental conditions.

Both phase waves and trigger waves are obtained in B-Z medium. Trigger waves are controlled by the rate of chemical reaction and diffusion. The initial pulse occurs as a result of heterogenous catalysis at a physical imperfection such as a gas bubble, dust mote etc. It is supposed that decrease of Br^- concentration or increase in $HBrO_2$ concentration triggers a pulse of ferroin oxidation in the excitable reagent.

(iii) Teorell's membrane oscillator

Teorell showed that spontaneous periodic flux transitions occurred in a membrane system consisting of a highly porous ion-exchanger membrane which separates two compartments filled with sodium chloride solutions of different concentrations. Periodic oscillations of the transmembrane potential and hydrostatic pressure wave found to occur when the membrane was polarized by a constant current of appropriate strength. Such oscillations have also been observed with powders of quartz or aluminium oxide[41]. More refined experiments have been recently reported. The oscillations arise due to interplay of diffusive flux and electro-osmotic flux which gives rise to alternating changes in the concentration of the solution in the pores.

It should be noted that according to Mears and coworkers, the oscillations are unaffected by changes in stirring provided the speed of stirrers is greater than 100 rev/min and the fluid circulation rate exceeds 500 ml/h. The concentration of the electrolyte and its ratio on the two sides is important for regular periodicity. If the concentration ratio fell far below 10:1, the oscillations disappeared. Thirdly, a critical current density is essential for each experimental configuration.

In this context, following thermodynamic questions arise:

(i) Whether oscillations occur in non-linear regime?
(ii) Whether Onsager reciprocity is satisfied?
(iii) Is instability in conformity with Prigogine's stability criterion?
(iv) How close is the system to oscillations in biological membranes?

Recent experimental studies[44] on Teorell's oscillator with pyrex membrane show that the electro-osmotic flux J_v is non-linear in the region of oscillations and is found to depend on potential difference $\Delta\phi$ according to the relation:

$$J_v = ae^{b\Delta\phi}$$

where a and b are constants. The exponential increase is found to be associated with Ohmic polarization. Wherever possible, streaming current measurements showed that Onsager reciprocity is obeyed in the linear range. Apply the stability criterion due to Glandsdroff and Prigogine, it is found that

$$\delta J_v \delta X_v + \delta J_i \delta X_i > 0$$

where X_v and X_i are the corresponding fluxes. Thus, the system should tend to thermodynamic stability. The fact that the system is unstable simply shows that it is not allowed to reach a steady state around which perturbations could be generated. It is far from limit cycle behaviour.

In biological systems currents of the order of 15mA do not flow and electro-osmotic flux of the above magnitude in the region of Ohmic polarization do not occur. However, instabilities can be produced in biological membrane system where it is possible to have some autocatalytic vectorial reaction on the membrane surface. This is perhaps done by ATP based reactions on the mitochondrion membrane.

11. Acknowledgement

Thanks are due to Dr. Ram Shabd for helpful discussion and to University Grants Commission for financial support.

References

1. I. Prigogine, Ethde Thermodynamique des Phenomenes Irreversibles, Desoer, Liege, 1947.
2. S.R. de Groot, Thermodynamics of Irreversible Processes, North-Holland Publishing Company, Amsterdam, 1952.
3. I. Prigogine, Introduction to Thermodynamics of Irreversible Processes, Interscience Publishers, New York, 1961.
4. S.R. de Groot and P. Mazur, Non-equilibrium Thermodynamics, North-Holland Publishing Co., 1963.
5. G. Nicolis, Advances in Chemical Physics, **19** (1971), 209, Ed. I. Prigogine and S.A. Rice, John Wiley and Sons, Inc., 1971.
6. D. Lurie and J. Wagensberg, J. Theor. Biology, **78**, 241-250 (1979).
7. R.P. Rastogi, R.C. Srivastava and K. Singh, Trans. Faraday Soc., **61**, 854 (1965).
8. R.P. Rastogi, M.L. Srivastava and S.N. Singh, J. Phys. Chem., **74**, 2960 (1970); R.P. Rastogi, K. Singh and J. Singh, J. Phys. Chem., **79**, 2574 (1975).
9. R.P. Rastogi and B.P. Mishra, J. Phys. Chem., **74**, 112 (1970).
10. R.P. Rastogi, Ram Shabd and B.M. Upadhyaya, J. Non-Equil. Thermo., **6**, 273 (1981).
11. R.P. Rastogi and Ram Shabd, J. Non-Equil. Thermodynamics, **6**, 207 (1981).
12. R. Haase, Thermodynamics of Irreversible processes, Chapter 2, Addition Wesley Publishing Company, Massachusetts, 1969.
13. Allen, R.H., Alfray, T., Yato, L.D., J. Amer. Chem., Soc. **81**, 42 (1959).
14. Ives, D.J.G., Kerlogue, R.H., J. Chem. Soc., **33**, 1362 (1940).
15. Meyer, B.S., Anderson, D.B., Plant Physiology, p. 424, D. van Nostrand Comp., New Delhi, 1963.
16. Marshall, A.G., Biophysical Chemistry, Principles, Techniques and Applications, John Wiley & Sons, p. 227, 308, 330, New York, 1978.

17. Ref. 2, p. 174.
18. Mahendra K. Jain, The Bimolecular Lipid Membranes. A System, Van Nostrand, Rainhold Company, 1972, p. 342.
19. A. Katchalsky and P.F. Curran, Non-equilibrium Thermodynamics in Biophysics, Harvard University Press, Cambridge, Massachusetts, 1965, Chapter 14.
20. R. Lefever and A. Goldbeter, Molecular Movements and Chemical Reactivity, Advances in Chemical physics, Vol. 39 (Series Editor I. Prigogine and S.A. Rice) John Wiley and Sons, New York.
21. R.P. Rastogi and R.C. Srivastava, Physica, **27**, 265 (1961).
22. J. Hiernaux and A. Babloyantz, J. Non-Equilibrium Thermodynamics, **1**, 33 (1976).
23. P. Glansdorff and I. Prigogine, Thermodynamic Theory of Structure, Stability and Fluctuations, Wiley-Interscience, London, 1971.
24. R.P. Rastogi and Ram Shabd, J. Chem. Edu. (1981) in press.
25. G. Nicolis and I. Prigogine, Self-organization in non-equilibrium systems, John Wiley and Sons, 1977.
26. R.M. Noyes, Ber. Bunsenges, Phys. Chem. **84**, 295–303 (1980); J. Amer. Chem. Soc., **102**, 4644 (1980).
27. E.J. Reusser and R.J. Field, J. Amer. Chem. Soc., **101**, 1063 (1979).
28. R.P. Rastogi, K. Yadava and K. Prasad, Ind. J. Chem. **13**, 352 (1975).
29. R.P. Rastogi and S.N. Misra, unpublished results.
30. M.G. Mustafa, Kozo Utsumi and Lester Packer, Biochem. Biophys. Res. Comm. **24**, 381 (1966).
31. T. Teorell, J. Gen. Physiol. **42**, 831 (1959).
32. R.M. Noyes, J. Amer. Chem. Soc., **102**, 4644 (1980).
33. M. Orban and E. Koros, J. Phys. Chem. **82**, 1672 (1978).
34. R.P. Rastogi, H.J. Singh and A.K. Singh, "Kinetics of Physicochemical oscillations" Aachen Discussion meeting of Deutsche Bunsengesselchaft fur physikalische chemie preprint Vol I, 1979, p. 142.
35. Z. Nosticzius and J. Bodiss, J. Amer. Chem. Soc., **101**, 3177 (1979).
36. M. Orban and I. Epstein, J. Amer. Chem. Soc. **103**, 3723 (1981).
37. R.P. Rastogi and M.K. Varma, J. Amer. Chem. Soc., under publication.
38. R.P. Rastogi, K. Yadava and K. Prasad, Ind. J. Chem., **13**, 352 (1975).
39. R.P. Rastogi, K. Singh, P. Rastogi and R.B. Rai, **15A**, 295 (1977).
40. R.P. Rastogi and Jokhoo Ram (unpublished results).

41. U.F. Frank, Ber. Bussenges, Phys. Chem., **71**, 789 (1967).
42. P. Mears and K.R. Page, Phil. Trans. Roy. Soc. **272**, 1-46, (1972).
43. P. Mears and K.R. Page, Proc. Roy. Soc. London, A **339**, 531-532 (1974).
44. R.P. Rastogi, Ram Shabd and Ishwar Das (Unpublished results).

SELF ORGANIZATION
Some Aspects of Biological Phenomenology

Some Aspects of Synergetics

H. Haken

Institut für theoretische Physik
Universität Stuttgart

1. Definition of Synergetics

According to its definition, synergetics, an interdisciplinary field of research, is concerned with the cooperation of individual parts of a system that produces macroscopic spatial, temporal or functional structures. It deals with systems which acquire their ordered structures by *self-organization* and it focuses its attention on those situations where the macroscopic order of a system changes dramatically. Because its mathematical methods have been or will be presented elsewhere, I shall confine myself to a verbal description and shall be concerned with the question in how for synergetics can contribute to problems discussed in this volume, namely to problems of the living state [1]–[6].

Starting with the laser-problem more than 2 decades ago, we have studied numerous systems in physics, chemistry and other fields. These systems are open systems, driven far from equilibrium and exhibit pronounced transitions from disordered to ordered states. Whenever we found such a system we could show that the same basic principles operate.

Since living systems are open systems in which an energy input in form of light (photosynthesis) or chemicals is converted into specific actions or structures of a living organism it suggests itself to study life processes by means of concepts of synergetics. There are a number of questions of fundamental nature and we mention only a few of them. How does it come that energy is not only used up generating heat, but rather is upconverted so that only a few degrees of freedom are highly excited? Such degrees of freedom may manifest themselves in form of spatial organisations of tissues and organs, temporal correlations, for instance in brain function, locomotion, synergism of muscles (an old concept due to Sherrington) etc.

2. The Laser–A Prototype of Synergetic Systems

More than 20 years ago I was lucky enough to come across an example in the inanimate world which was very inspiring how nature may solve such problems. Before I briefly explain this example, I hasten to warn the reader not to take the conclusions which can be drawn from this example *too literally*. Indeed the conclusions and analogies are only valid if we adopt a sufficient degree of abstraction, which can be formulated mathematically rigorously, however [6]–[7].

Let us take a chemical laser in which chemical energy which usually would get lost (just as heat) is converted into coherent electromagnetic radiation. How can the individual complex molecules be subjected to a process in which a macroscopic coherent field is produced? Or, in other words, how does it come that the enormous number of individual degrees of freedom of the internal electronic motions give rise to a single macroscopic excitation which is selfsustained provided we feed enough new chemicals into the laser? The answer I found at that time for the incoherently pumped laser can be described as follows (by use of a more modern terminology which I have developed in the meantime).

Below a certain pump rate (input rate of chemicals) the individual molecules are excited only from time to time and emit incoherent tracks of light waves. When we increase the influx rate more and more, suddenly the old disordered incoherent state becomes unstable. At this point the system, through its fluctuations, tests all sorts of collective modes the total system, i.e. field and molecules, can acquire. At and somewhat above that instability point a competition between collective modes sets in and eventually one or a few collective modes win. In a way this is strongly reminiscent of Darwin's principle of the survival of the fittest but appearing now in the inanimate world. After such a process has set in, the whole system is governed by a few or a single degree of freedom, i.e. in our example the electromagnetic field of a single "mode". All other components of the system, in particular the individual molecules, are "slaved" by the field mode which acts as "order parameter". Close to such an instability point phenomena occur which are strongly reminiscent of phase transitions of systems in thermal equilibrium, namely enhanced fluctuations, an increase of time in which the system relaxes towards its new equilibrium state, symmetry breaking and other features. It soon turned out that these features make the laser an ideal prototype of a large class of phenomena in physics, chemistry, biology, mechanical and electrical engineering, ecology and even sociology and we have developed a mathematical apparatus to deal with the corresponding disorder-order or order-order transitions.

3. The Basic Principle of Selforganization

But inspite of the mathematical apparatus, which has become quite heavy, the basic principle has remained the same. A change of external conditions ("control parameters") can cause an instability. More precisely speaking one or several collective modes become unstable, and by means of competition or cooperation they eventually become stabilized, slaving all other degrees of freedom. Fluctuations decide in the case of equal "opportunities" which macroscopic order is realized. Since the change of external parameters can be achieved in different ways, different kinds of self-organization may occur.

(1) Self-organization through change of global parameters, such as temperature, energy flux etc.

(2) Self-organization through increase of number of components.

(3) Self-organization through transients. A sudden change of control parameters may cause the relaxation of the old state to a new one. On its way the system acquires a higher degree of order.

Order parameters govern the self-organization of temporal structures. In this way we can treat the onset of oscillations, quasiperiodic oscillations, subharmonic and harmonic oscillations and the approach allows us to have an overview over instability hierarchies. Surprisingly, these concepts apply to the formation and recognition of patterns also. They allow us to deal with the formation of biological patterns using Wolpert's concept of chemical prepatterns [8]. In particular our approach shows that quite different microscopic chemical processes can give rise to the same macroscopic patterns. More generally speaking, it seems that synergetics is bringing us a step further towards unification of science. For instance we now know that the formation of coherent radiation of the laser, the onset of chemical reactions, the formation of spatial or spatio-temporal patterns in the motion of liquids, models for the formation of patterns in biology, qualitative changes of quite different kinds of networks, even dramatic changes in the behavior of social groups or economic processes are governed by the same principle, of which we gave a very brief outline above. In this way, synergetics may provide us with a new approach to cope with complex systems. Rather than dealing with their individual parts it sorts out dominant collective modes of behavior which then allow us to study their interaction in further steps. In this way a new kind of hierarchical structure is revealed which intimately connects the micro- and macro-world. It provides us even with the possibility to understand how microscopic events may dramatically change or control macroscopic behavior close to instability points.

4. Micro- or Macroscopic Approach?

Over the past years some conflicts between different schools of thoughts have arisen on the issue which kind of approach is to be preferred. Of course, molecular biologists and others stress the importance of the study at the molecular levels. Indeed, molecular biology was extremely successful. But it may be allowed to contribute to the whole discussion something from the experience of a physicist. Undoubtedly the exploration of the microwarld of the atom has been one of the great achievements of physics of this century. But physicists soon have learned that knowing the properties of individual atoms is by no means sufficient to understand important collective phenomena such as ferromagnetism or superconductivity. Quite new concepts had to be developed to cope with these macroscopic phenomena, e.g. the Landau theory [9]. Therefore it seems that having understood the single cell does not guarantee that we understand the functioning of a whole organ or of a whole being. Rather it becomes more and more generally accepted that, for instance pattern recognition by the brain is not achieved by a "grandmothercell" but rather by the cooperation of very many neurons. From the point of view of synergetics it seems that the complex patterns of the outer world can be decoded only by a complex network with many collective states. In a way the complex macroscopic world is matched by the complex macroscopic network of neurons with its coherent collective excitations. The analogies which are advocated by synergetics have been very useful to give an interpretation to drug induced halluzinations. According to Jack Cowan, stripe-like parts of the visual cortex are collectively excited in analogy to roll patterns known in fluid dynamics [10]. Though these ideas may be still somewhat speculative, studies of EEG of normal brain functioning and, for instance, epileptic seizures clearly show how many neurons may cooperate coherently [11]. In this context, we must be very careful with the notion of "order". As is clearly exhibited by the "ordered" state of epileptic seizures, too much "order" in brain functioning may mean that less information is processed or produced. Clearly, it becomes more and more difficult to decode (or recognize) organization at higher levels as being ordered because the structures may become immensely complex. These studies may shed new light on the question of chaos which is now widely discussed in the natural and mathematical sciences. Namely, the seemingly irregular motion we observe may be just the outcome of an involved computational process which is prescribed by the interactions governing the evolution of these systems.

5. Links between Synergetics and Other Disciplines

In conclusion I wish to make a few comments on the relation of synergetics to other fields of science. In a way synergetics is like a chamelion. Depending on the background against which it is seen it acquires quite different shades. Physicists working on thermodynamics will recognize the analogies between synergetic systems and systems in thermal equilibrium which exhibit phase transitions. Specialists of statistical physics will notice the crucial role played by fluctuations especially at disorder-order transitions. On the other hand, synergetics may be looked at from the point of view of dynamical systems theory, since both fields deal with the temporal evolution of systems. Because synergetics focuses its attention on qualitative ("structural") changes, related problems of bifurcation theory come to a mathematician's mind. But while bifurcation phenomena may be richer than ordinary phase transitions, for instance thtough the occurrence of limit cycles, bifurcation theory is not dealing with highly important and decisive aspects of synergetic systems. For instance the crucial role of fluctuations, which become dominant at the transition point, is entirely missed. In addition, the methods of synergetics allow us to study stability and relaxation processes by means of probability theories. Synergetics contributes further to the theory of general systems, in that it has established general laws which apply to a large class of systems. But in a way synergetics has found an ecological niche. By confining its analysis to systems close to instability points it can cover a great many quite different systems. Scientists working in cybernetics will especially recognize the control processes taking place in self-organizing systems. But at the same time synergetics created new general principles and concepts which are alien to cybernetics. An example is provided by the concept of the order parameter. In general, it is **not** of any material nature (in contrast to the laser field of our initial example), but it can be a mere construction of our thinking (which attempts at matching "reality" by macroscopic views), (of course, the order parameter is a well defined mathematical quantity).

While undoubtedly synergetics has links to many fields and each field contributes important methods and ideas to synergetics, at the same time and quite obviously each of the traditional disciplines entirely misses other aspects of equal importance. Therefore it is worth to study the complex processes of self-organization within a new interdisciplinary approach. Whether this field carries the name "synergetics" or any other name is, of course, of no importance at all. But I think that quite independently of its name this new field of research has come into existence.

One final remark : The generality of synergetics gives rise to a fascinating aspect : Precisely the same coherent, well ordered processes may be supported by quite different substrates. Thus even logical procedures may be exerted on quite different substrates.

References

The number of papers related to synergetics has become enormous and it is impossible to list here any adequate number of them. Therefore I list here only those papers which are most closely linked up with the development of synergetics. Quite obviously there are numerous important cross-correlations with many of the other contributions to this volume and the reader is referred to the corresponding articles.

1. The word "synergetics" was introduced in my 1970 lectures at Stuttgart University.

see also H. Haken and R. Graham: "synergetik-Die Lehre vom Zusammenwirken", Umschau **6**, 191 (1971).
2. H. Haken: Rev. Mod. Phys. **47**, 67 (1975).
3. H. Haken: Synergetics. An Introduction. Nonequilibrium Phase Transitions and Self-Origanizing in Physics, Chemistry and Biology, 2nd enlarged edition, Springer, Berlin 1978 H. Haken: Advanced Synergetics, to be published.
4. H. Haken ed. : Synergetics. Cooperative Phenomena in Multi-component Systems, Teubner, Stuttgart 1973.
H. Haken ed. : Cooperative Effects. Progress in Synergetics, North Holland, Amsterdam 1974.
5. See also Vols. 2–12 of the Springer Series in Synergetics.
6. H. Haken, talk given at the International Conference on Optical Pumping, Heidelberg 1962.
H. Haken and H. Sauermann, Z. Physik **173**, 261 (1963); **176**, 47 (1963) (semiclassical approach).
7. H. Haken, Z. Physik **181**, 96 (1964) (quantum theoretical approach).
8. L. Wolpert, Positional Information and Spatial Pattern of Cellular Differentiation, J. Theoret. Biol. **25**, 1–47 (1969).
9. Landau-Lifshitz, Course of Theoretical Physics, Vol. 6: Fluid Mechanics, Pergamon Press, London-New York-Paris-Los Angeles, 1959.
10. J. Cown and G.B. Ermentrout in: Pattern Formation by Dynamic Systems and Pattern Recognition, Springer Series in Synergetics, Vol. 5, ed. H. Haken, 1979.
11. See e.g. A. Babloyantz in: Dynamics of Synergetic Systems, Vol. 6 Springer Series in Synergetics, ed. H. Haken, 1979.

Towards a Theory of Biological Morphogenesis

R. Thom

1. Philosophical Generalities on the Living State

Although the question "What is life" is known not to have any precise formal answer, I think it is fairly clear that when dealing with higher organisms, the property of life can be quite clearly ascertained. One of the most obvious criteria for the living state is the *organizational* one: organisms are divided into tissues and organs, and their configurational pattern is—for a given species—relatively fixly determined. Moreover reproduction of these organisms takes place also according to a relatively well defined spatio-temporal evolution: for sexual reproduction it involves the passage through the one-cell stage, the egg cell. The egg, once fertilized, will undergo a sequence of transformations (its embryology) leading back to the adult form. Moreover, specific organs can sometimes be given specific functions, having regulatory value for the whole organism. Although the problem of morphology has since a long time ceased to be one of major interest to present day Biologists, I think it is fair to say that it is among the most fundamental ones. Historically, it was the subject of a celebrated discussion at the Paris Academy in 1830 between G. Cuvier and Geoffroy Saint-Hilaire. Geoffroy had stated the general principle of the "uniqueness" of the "Bauplan" for all organisms. Cuvier retorted that no such unique "plan" could be detected, that no uniqueness existed for Invertebrates nor even for Vertebrates, but that correlation between orgnns had to be founded on their functional synergy. Although this dispute is thought to have only historical interest, I believe it still has considerable impact on our way of understanding biological morphology. Most biologists tended to praise Cuvier—as being more rigorous in his approach than Geoffroy, who was more prone to propose far-fetched generalities. There is little doubt, that when we restrict ourselves the Vertebrates, for instance, the notion of a general Bauplan is firmly grounded, especially when considering terrestrial animals (tetrapods). And the number of general symmetries for an animal organism is severely limited (bilateral, pentagonal, longitudinal metamery). What I am proposing here is a kind of general formalism allowing to describe (if not to explain) the morphogenesis of the embryo. Here again the general sequence of primitive events (gastrulation, neurulation) is fairly well known and shows a relative constancy in its beginning stages (as exemplified by the so-called Haeckel's recapitulation law).

To understand the way I intend to attack these questions, may I call upon a personal remembrance. In 1966, at a congress I met Francis Crick (of the DNA molecule). He told us that in a few years it should be possible to write down all the chemical reactions taking place inside a Bacterium as Escherichia Coli, such that a complete knowledge of its metabolism could be reached. Now, in 1981, I do not know to which extent this promise has been fulfilled. But even if it were so, as the number of chemical species involved in this mechanism approximates 3000, the determination of the kinetic of the global metabolism would imply the solution of a differential system

$$\frac{dX}{dt} = F(X; \tau)$$

where X is a vector of $\mathbf{IR}^{3000}$, τ external parameters, long range time.

The instantaneous state of the global metabolism is then described (for τ fixed) by an asymptotic set in $\mathbf{R}^{300}$ (an attractor). Now this attractor will modify itself when τ varies; its shape depends on the phase of the mitotic cycle—as well as of other external parameters. Such changes of attractors are called, mathematically, bifurcations. Now defining a set in a space of dimension 3000 exceeds by far the capabilities of the most intuitive geometer. As a result, a complete description of the global variation of the attractor eludes us, and we won't be able to really describe the variation of the metabolism; especially if some reactions depend on the catalytic effects of enzymatic proteins, and if this effect depend on the tertiary structure of such proteins, which may themselves depend on some external parameters. This is why the pure chemical knowledge—without taking into account spatio-temporal localization—is fundamentally illusory. Hence the need to take space-time as basis if we want to end finally with a spatio-temporal morphology. This is what happens with the now classical formalism of reaction diffusion formalism:

$$\frac{\partial X}{\partial t} = k\Delta X + F(X) \tag{1}$$

As such equations are invariant under dilatation, they are by themselves unable to create an "specific" morphology. (As all homothetic morphologies are also solutions). Such an equation requires the homogeneity of the ambiant space, whereas the effect of the reaction is in general to destroy such an homogeneity. Perhaps the main interest of such equations is to propagate constraints given by boundary data inside the domain of the solution. A more refined type of equations would be

$$\frac{\partial X}{\partial t} = P_X(X) + F(X) \tag{2}$$

where $P_X(X)$ is a P.D.E. operator depending locally on x, where this local operator could describe the local inhomogeneities of the medium.

In that respect, one should perhaps call attention to the fact that even determining the spatio-temporal growth of a phase inside a medium (for instance, crystallisation of a sursaturated solution) leads to open problems, as the theory of defects of ordered structures obviously enters into play in such a case. The formalism I am going to propose does not take such local effects into account, and gives up any attempt to determine quantitatively the growth of what plays the role of a "phase", namely cellular differentiation.

The underlying assumption is that the local metabolism in a given tissue can be described by a differential system

$$\frac{dX}{dt} = F(X, v) \tag{3}$$

where v denotes "external parameters" of general nature
,, X ,, "chemical concentration" (internal parameters)
(external parameters)

We admit that v varies very slowly with time, in such a way that the representative state point rapidly reaches an asymptotic regime, characterized by an attractor of the metabolism dynamics described by (3). We suppose here that this attractor is sufficiently well-behaved to admit a local Lyapunov function denoted by $V(x; v)$, which we may consider as a local "entropy" around this attractor[1]. In fact we shall admit that from the point of view of stability of these regimes, we may replace the flow dX/dt by the gradient of $V(x, v)$ with respect to some Riemann metric. The basic idea here is the suggestion—made by Max Delbruck in 1949, that a particular type of cellular differentiation could be associated to an asymptotic regime of the metabolism. And to that we add the "ad doc" assumption that such a regime can be described by an attractor of the local metabolism symbolized by a differential system

$$\frac{dX}{dt} = F(x; \tau)$$

In most cases we shall deal with punctual minimizing points of the potential $V(x, v)$. But in some constructions (as for instance for mesoderm in Vertebrate Embryology) we shall accept variations of the potential and of the metric. The variations of the local potential have to satisfy genericity assumptions (structural stability requirement). In some cases we shall use an hyperbolic metric instead of the usual Riemann metric. (The underlying motivation is to allow transferring some kind of potential energy from one degree of freedom to an other).

These local potentials $V(x, u)$ have to be considered as "local negentropies" which always tend to decrease along trajectories. At least in the beginning phases of Embryology, they have some biological interpretation which can be phrased in terms of the global regulative functions of animal life. "Predation" in particular plays an essential role in these interpretations.

It may happen that when $F(x; u) = \frac{dx}{dt}$ varies as a function of the parameters (u), an attractor becomes unstable and ceases to be an attractor (A_u). As a result, it is changed into one (or several attractors) arising from (A_u) either by bifurcation, if the transformation takes place continuously with u, or by "catastrophe" (if $A_{(u)}$ jumps rapidly into a new attractor $A_u^\sim$) These internal transformations manifest themselves, generally after some lapse of time, by new ordering or pattering of tissues leading later to the formation of organs.

2. Mathematical Formalism

The basic mathematical tool to study such transformation is the notion of "unfolding" of a singularity. It is not the place here to describe this theory in its full mathematical generality. Let me give only a few hints about the intuitive meaning of such a notion.

Consider first a potential well defined for instance by a parabola $V = \frac{x^2}{2}$. The origin $0 n = 0$ is a critical point of the V function, as $dV = xdx$ vanishes at 0. Then it is easy to prove that any function W near V in the topology C^2 (defined by the diffeomorphism $|V - W_1|$, $|V'_x - W'_x|, |V''_{x^2} - V''_{x^2}|$) has the same structure of a parabolic potential well.

Proof. Take an interval $-\alpha, \alpha$ around 0 with

$$V'(-\alpha) < 0 \; V'(\alpha) > 0, \text{ Inf } |V'| - \alpha), |V(\alpha)| = b \gg 0.$$

Then we have (Cf. Fig. 1).

$$|V''(x)| = 1 - m | -\alpha, \alpha |.$$

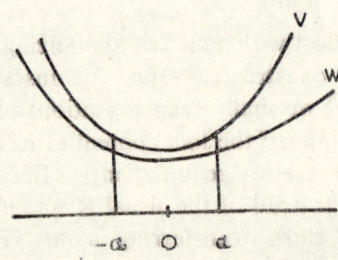

Fig. 1

Suppose $W(x)$ satisfies

$$|V'_{(x)} - W'_{(x)}| < \frac{b}{2} \text{ for } (x) < \alpha$$

and
$$|V''_{x^2} - W''_{x^2}| < \tfrac{1}{2} \text{ for } (x) < \alpha.$$
Then
$$W'(-\alpha) < -\frac{b}{2} \quad W'(\alpha) \blacktriangleright \frac{b}{2}$$

hence W' vanishes $]-\alpha, \alpha[$; but as

$$W''_{x^2}(x) < \frac{1}{2} \text{ in }]-\alpha, \alpha[,$$

W' vanishes at a single point $c(W)$. This point is a single minimum of the W function in $]-\alpha, \alpha[$. One may state this result by saying that the quadratic minimum $V = \frac{x^2}{2}$ is "rigid", or *structurally stable*.

On the contrary, the function $V = \frac{x^3}{3}$ is not structurally stable, as it may be deformed either in $V = \frac{x^3}{3} - x$ (curve with a bump (Fig. 2a) or $x^3 + x$ (Fig. 2b) curve without a bump.

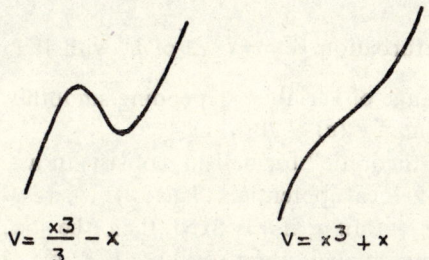

$V = \frac{x^3}{3} - x$ \qquad $V = x^3 + x$

Fig. 2

One can show that any function W sufficiently near V in the C^3 topology is equivalent, by a change of variables, either to $\frac{x^3}{3} - x$, or $\frac{x^3}{3} + x$, or $\frac{x^3}{3}$ (this last case being exceptional). All perturbed functions near V are equivalent by a change of variables to a polynomial $\frac{x^3}{3}$.

The function $V = \frac{x^4}{4}$ is very important in that respect (as it is positive) its unfolding is

$$V = \frac{x^4}{4} + u\frac{x^2}{2} + vx$$

The set of maxima and minima is defined by

$$V_x = x^3 + ux + v. \text{ The surface } V_x = 0 \text{ in } \mathbf{R}^3 \ (x; u, v)$$

which projects on $\mathbf{R}^2$ (u, v) according to the critical curve
$$V''_{x^2} = 3x^2 + u \quad u = -3x^2$$
whose projection is $\quad u = -3x^3$
$$v = x^3 - 3x^3 + v = -2x^3$$
or $\quad 4x^3 + 27v^2 = 0$ (discriminent of equation $V_x = 0$).

This is the so-called cusp singularity. Inside the cusp we have a curve with two minima (Fig. 3a), outside a curve with a single minimum (Fig. 3b).

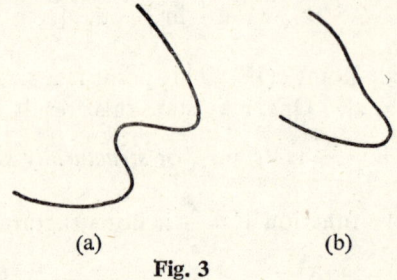

(a)   (b)
Fig. 3

The plane Ovu is called the unfolding space of the singularity $V = \frac{x^4}{4}$. Any multiparameter perturbation $W = (x; c)$ of V, with $W(x, 0) = V = \frac{x^4}{4}$ is equivalent by a change of variables depending smoothly on c to a polynomial induced by a map $g:(c) \to R^2(u, v)$.

The "catastophe-theoretic" formalism consists in associating to a living tissue a family of local potentials $V(x; u)$ where the spatiotemporal coordinates belong within u; if u is fixed, then the local metabolic regime is described by a minimizing point $\mu(u)$ of $V(x; u)$. If one follows the so-called Maxwell's rule, the dominating regime $\mu(u)$ is the lowest minimum. In some other cases, when we have in u a natural dynamics, than we may follow the rule of *perfect delay*: an attractor stays along a trajectory as long as it is not destroyed by bifurcation.

For instance, in the cusp singularity $V = \frac{x^4}{4} + u\frac{x^2}{2} + vx$, "catastrophes" ruled by Maxwell's rule occur when the representative point crosses the "Maxwell set" $u < 0$, $v = 0$, locus of potentials like $x^4 - x^2$ which reach their minimum at two distinct minima. Perfect delay catastrophe occurs along the semi-cubic parabola (c) $4u^3 + 27v^2 = 0$, projection of the fold of the critical set $V'_x = V''_{x^2} = 0$.

3. Schemes for Embryology (Vertebrates)

In the steady of biological morphogenesis, we have to point out that our methods are unable to give a "quantitative" modelization. Our only aim is to describe qualitatively the succession of the great accidents of Vertebrate Embryology: gastrulation, neurulation, formation of the neural axis. We shall stop at somites formation—associated to a new phenomenon: metamery, i.e. longitudinal symmetry breaking, which is not in itself describable to elementary catastrophe formalism.

Let us first describe our basic object, the "predation loop". In the cusp model let us construct a circle $u^2 + v^2 = 1$ of radius one around the center 0; let J, K be the intersection of this circle (Γ) with the semi cubic parabola (P). Let us run around this circle in the positive sense. Above the

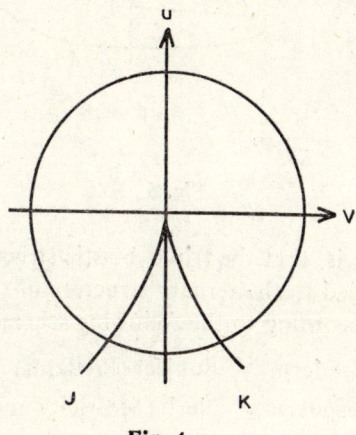

Fig. 4

arc JK there are two minima for the potential V. This will correspond to two basic actions "predator" and "prey". At K the lower lying minimum, the predator, captures the upper lying minimum, the "prey". (Fig. 5)

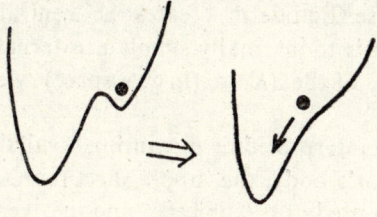

Fig. 5

In the upper arc KsJ, we have only one minimum, which we interpret as due to coincidence of predator and prey. (I interpret this upper state as "sleep", fusion of subject and subject). On the left of J, the predator awakes on the sheet of the prey: this means that the hungry predator *is* (symbolically) his prey. At J, the predator perceives the shape of a prey in the outside world, and recognizes it as such: this is the "pesception catastophe": the predator is created in J, whereas the true prey is put on its upper sheet. Then the predator chases his prey, and finally captures it at K. (Cf. Fig. 6).

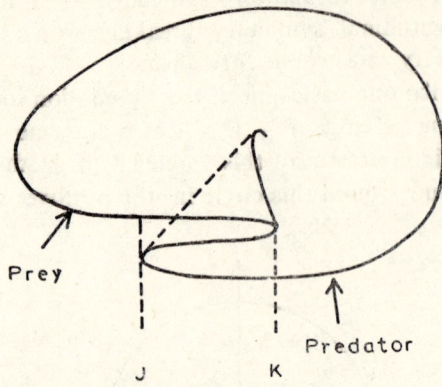

Fig. 6

Our basic postulate is that the triplo blastic structure of the Vertebrate Embryo can be identified to the ternary structure of the transitive sentence Subject-Verb-Object according to the following scheme:

    Entoderm   Subject (Predator)
    Mesoderm   Verb (Motricity and capturing)
    Ectoderm   Object (Prey)

Entoderm is the "subject" because it is there, at the intestinal mucosa, that final assimilation of the prey's flesh into the predator takes place.

Mesoderm identifies with the Verb, because mesodermal tissues have movement and spatial relation as their basic function Ectoderm identifies with "object", because Ectoderm creates the central nervous system, an organ whose basic aim is to internally simulate external objects.

In the counter image of the JK arc (in x, v space) we have the following diagram.

As the v axis can be interpreted as measuring (available) chemical energy content in the predator's body, the upper sheet represent prey ($=$ object). The lower sheet, the predator (subject), and the vertical arrows at JK perception, and capture of prey respectively. The mesoderm, globally, can be considered as a tissue describing the hystéresis loop γ between the two stable sheets: it pumps energy grom the lower sheet (entodermal liver) to bring it to organs which may pull the prey along the arc jk of the upper sheet. (As specialists of Van der Pol Equations well know, such an hysteresis loop may occur as a limit of a smooth limit cycle in the equation

    $\varepsilon V_x = x^3 - x + v$ for ε tending to zero
    $V_v = x$ when applying singular perturbation theory.

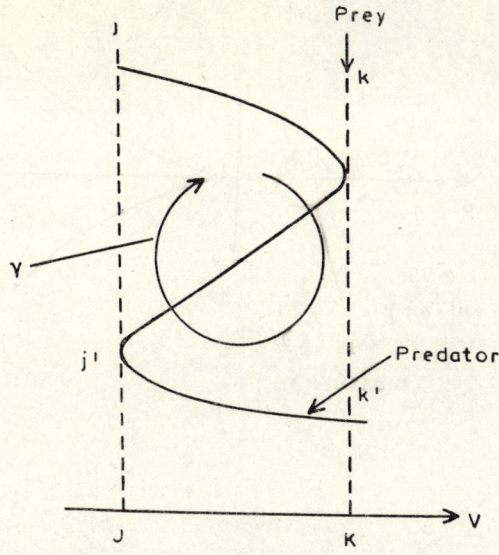

Fig. 7

Let us now start with the egg (unfertilized ovum); it is "sleeping", and can be considered as a small *cell* (in biological on topological sense) D^3 mapped into (Ouv) around the s point (Fig. 8a). As soon as it gets fertilized, it is activated and starts moving downward along the Ou axis. In case of Amphibians it becomes a blastula, a 2-dimensional sphere mapped onto Ouv as a linear projection parallel to the poles' axis.

The apparent contour in Ouv is then a small circle E. When u decreases, the circle E will formally touch the origni O. Then, applying originally Maxwell's rule, we shall get a cellular differentiator starting at a small circular arc on one side of the blastula. This appears as beginning of a groove, the blastopore on the dorsal side of the egg (Fig. 8b). When E moves further toward the negative axis Ou, this groove extends along a circular section of the blastula, till finally it closes, thus separating mesectoderm ($v > 0$) from entoderm ($v < 0$). (Fig. 8c). Later on, the mesectoderm develops a velocity of its own, directed towards $v < 0$, such that, spatially, the mesectoderm wraps around the entoderm (epiboly movement) (whereas the entoderm remains relatively fixed). (Fig. 8d). The unstable branch of the S curve $x^3 - x + v = 0$ becomes stabilized by an auxiliary dynamic process which we may interpret as "internalizing" the external coordinate v. A part of the energy derived from gliding down the $\text{grad}_x V$ gradient is diverted towards climbing up the $\text{grad}_v V$ gradient. As a result, we have to consider the dynamics $X = -\text{grad}_\mu V$ where μ is the hyperbolic metric $ds^2 = dx^2 - \lambda dv^2$. Such a flow X has for components.

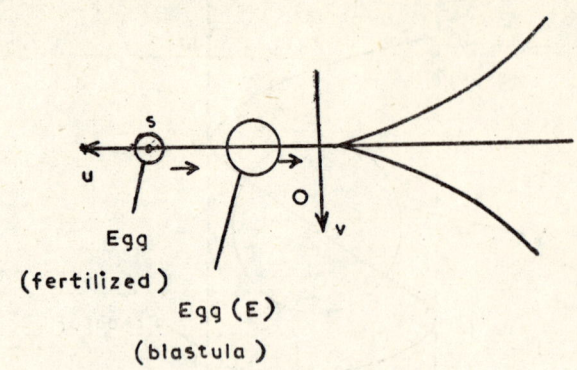
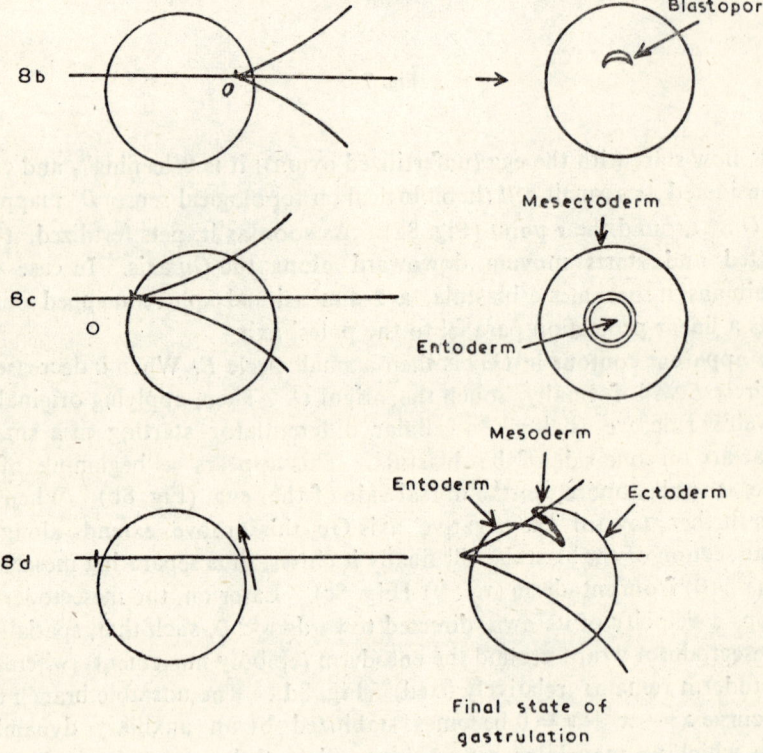

Fig. 8

$$X_x = x^3 - x + v$$
$$X_v = -\frac{x}{\lambda}$$

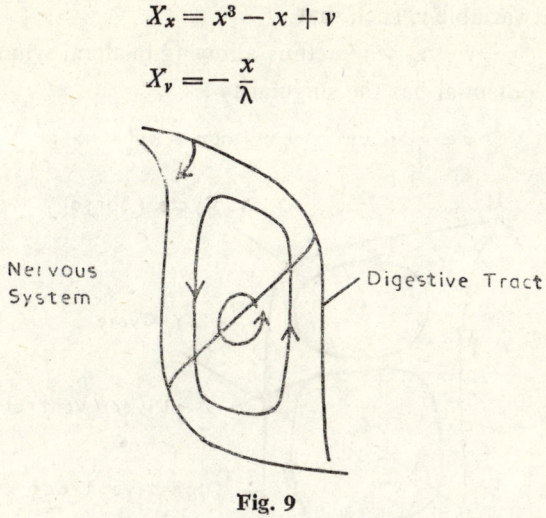

Fig. 9

Hence $(-X)$ has a repeller f_0 point (a focus) at 0, and it has also an attracting limit cycle γ_1 turning around in the opposite sense to the standard hysteresis loop. We postulate that his part of the mesectoderm which has become "ectodermally unstable" when nearing or crossing the half parabola (P) (OK arc) is captured by this new attractor. Hence it moves back both internally- and spatially towards the ectodermal layer, thus forming the intermediate sheet of the mesoderm. It glues itself first against the ectodermal layer, thus provoking neural induction; and the effect of neural induction is symbolized by the descending vertical arrow, which captures a part of the upper lying ectoderm which becomes transferred into the neural axis (process of neurulation). The sequence of events affecting later the mesoderm (scission between axial mesoderm and lateral mesoderm, which becomes dorsal somatapleure+ventral splanchnopleure can be symbolized as follows: the upper arc of the hysteresis loop γ_1 separates itself (internally from the "prey" upper layer, creating that way an auxiliary loop γ_2 acting as a roll rotating contrary to γ_1 : so we get two cycles in reverse directions. This upper cycle γ_2 has namely a functional significance: it describes the process of capturing the prey. On the left side, it describes by the cusp in v the elaboration of information given by the sensory apparatus (arc j_1v) transferred into motor activity bringing the prey to the mouth. This is the "dorsal cycle". The old loop γ_1 becomes the "ventral cycle", describing essentially blood circulation (blood is charged with chemical energy in the liver (middle layer λ) to distribute it to the external motor organs. In fact this loop γ_1 is realized in a K kinematic way in the beating of the heart. A more complete morphological description of the Embryology may be obtained if one admits that "internalization" of the v variables arises through the appearance of

a new internal variable y, such that
$$v = v_0 + y^2 \quad \text{(thus allowing bilateral symmetry)}$$
then the local potential has the singularity
$$v = \frac{x^4}{4} + xy^2 + v_0 x + wy + sy_x^2$$

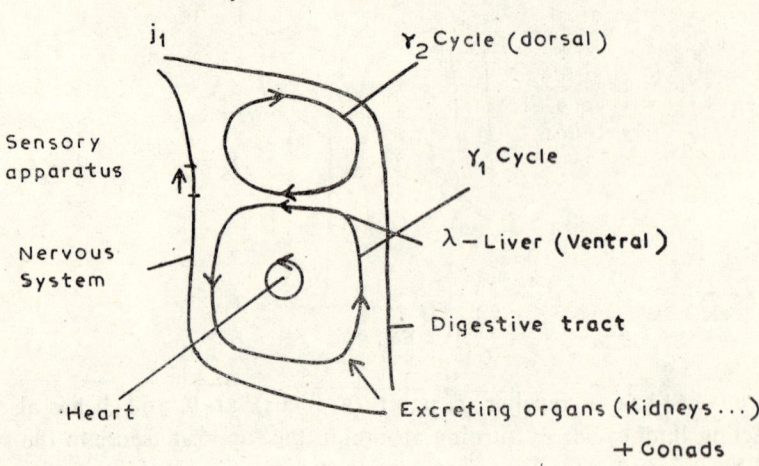

Fig. 10

this singularity known as the parabolicumbilic. The unfolding of this singularity contains four parameters which locally have a specific spatio-temporal interpretation; u is time, v_0 is external internal depth, w medio lateral gradient, s is cephalo-caudal gradient, which is later broken by metameres formation.

For instance, in the unfolding, one gets the successive sequence of sections describing the closing of the neural tube.

In this sequence for neurulation we even have a subfigure 11e describing neural *crest dissociation*. The axial mesoderm will correspond to the formation of an elliptic umbilic symbolizing the notochord. It is noteworthy that even in the global structure of a vertebra one finds as a rigidified scheme of the diagram Fig. 10, the upper half surrounds the neural axis, whereas the lower half surrounds the dorsal artery (Fig. 12).

Of course this is no more than a preliminary attempt to throw some light into some of the most mysterious processes offered to the contemplation of the human mind. Many questions remain unanswered, as for instance the fundamental differences between Insect Embryology (where entoderm practically does not exist) and Vertebrate Embryology. In the first case, the main differences are to be found in the indentification :

Ectoderm — Subject
Mesoderm — Verb
Vitellus — Object

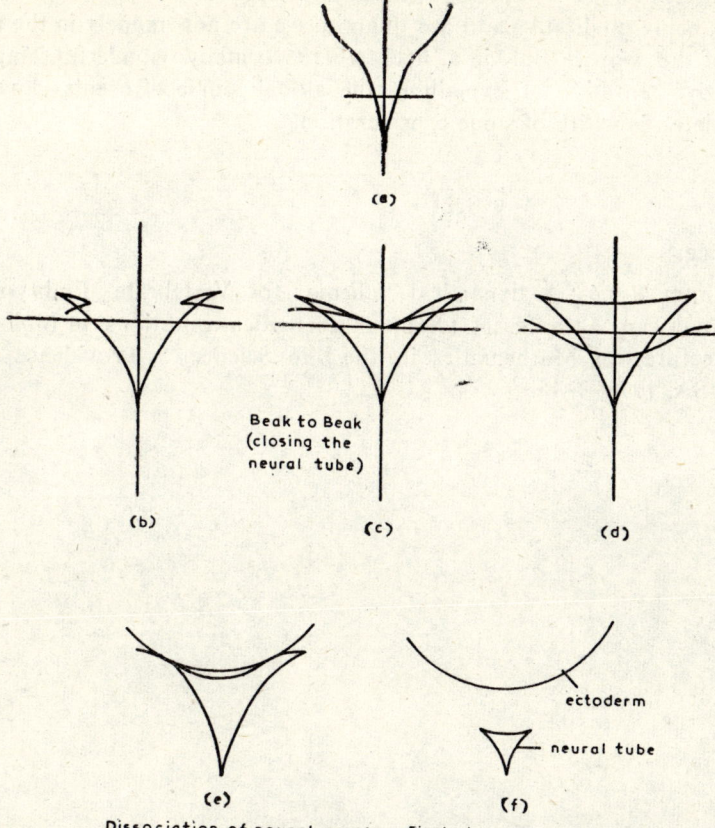

Fig. 11

Dissociation of neural crests — Final stage of neural tube

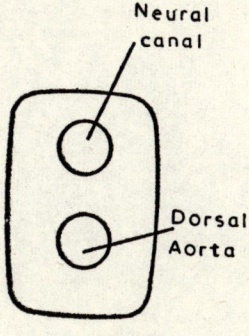

Fig. 12

allowing for a very different strategy of regulation. Of course these models, being qualitative and not quantitative are not models in the usual sense of the word. But in a field where so many wonderful things do occur, any tentative of explaining the global course of events—however incomplete—is worth of some consideration.

Reference

1. Thom René : A dynamical Scheme for Vertebrate Embryology, American Math. Series : Some mathematical questions in Biology 4, Lectures on Mathematics in the Life Sciences 5, Providence, R.I., USA, 1973, 3–45.

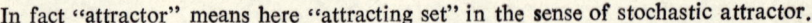

In fact "attractor" means here "attracting set" in the sense of stochastic attractor.

Emergence of Self-organizing Systems at the Molecular and Cellular Levels Viewed in a Darwinian Framework

U. N. Singh

Molecular Biology Unit
Tata Institute of Fundamental Research
Bombay 400 005, India

1. Introduction

The nature of life, its origin and purpose have occupied the attention of mankind since antiquity. The problem of its origin may be essentially viewed as the emergence of a self-organizing system capable of revolutionary adaptation to the changing environment. It was only at the turn of this century following a provocative postulate by A. I. Oparin and J. B. S. Haldane regarding the reducing nature of the primitive terrestrial atmosphere that many of these basic questions became amenable to a truly scientific investigation. That the Oparin-Haldane postulate was indeed a radical suggestion may be judged from the fact that the astronomers at that time were yet to recognize the abundance of hydrogen in the universe. The presence of ammonia in the jovian atmosphere was demonstrated several years later in 1934 (Shklovskii and Sagan, 1966).

The contemporary activities in theoretical biology may be broadly classified as *idealistic* or *instrumental* (the terms borrowed from a recent article by G. S. Stent, 1980). While the former underscores the inadequacies of the available logico-mathematical structure to cope with the enormous complexities of the living system, the latter is content to make use of the existing deterministic or stochastic formulations in the analysis of its varied manifestations at different levels of organization. In this regard I wish to emphasize that a theory (or theories) of biological processes can only be set in a contextual framework. For example, the role of a protein molecule as an enzyme or a receptor can be viewed in reference to a substrate or a ligend, respectively. Among the nucleic acids, DNA as a repertoir of genetic information, mRNA's as intermediate transcripts and tRNA's as adaptor molecules are of relevance only in relation to a formal description of the code and the nature of the translational machinery.

The universal appeal of the principles underlying the Darwinian evolution lies in the fact that they are set in the broadest contextual frame-the organism and its environment. The two basic tenets of neo-Darwinianism—*conservative multiplication* and *mutational diversification*—have been recently extended to molecular level to understand the nature of pre-biotic self-organising systems. At the molecular level, I suggest that an early integration of the informational (nucleic acids) and catalytic (proteins) molecules into a functional nucleoprotein complex may provide a plausible basis for a coherent evolution of a spatio-temporal order leading to the formation of a primeval cell. At the cellular level, metabolic cooperation afforded by specialized structures like gap junctions is viewed as an essential factor in the evolution of multicellular organisms.

2. Structural versus Functional Organization: Proteinoids and Hypercycles

Until recently much of the speculations on the nature of a primitive prototype of contemporary cells leaned heavily towards the structural organization implicit in the formation of coacervates, microaggregates etc. These were discussed at length by Oparin (1953) in his classical monograph. The works of chemical evolutionists over the past 2-3 decades (Fox and Dose, 1977), particularly those of Fox and his group on the formation of "proteinoids" under simulated pre-biotic conditions, provided some credibility to these notions. However, with an overmuch emphasis on the structural organization, the proponents of this viewpoint paid little attention to subtle interactions between nucleic acids and proteins, which may be rightly considered as paradigmatic of the living system on the earth.

Formulations of irreversible thermodynamics by Prigogine and his group (Glansdorff and Prigogine, 1971) and their applications to chemical reactions gave a new insight into this problem. It has become amply clear that an order (dissipative structures) can arise from interactions between molecules in a seemingly homogeneous population, particularly in the region far away from the equilibrium. The concept of hypercycles developed by Eigen and Schuster (1979) with reference to nucleic acids and proteins provided a plausible basis for the emergence of a functional order relevant to the evolution of the living system.

Although the old paradox of 'the chick and the egg' with reference to nucleic acids and proteins was amicably settled by the hypercycle concept, we are still confronted with it in a different context, i.e., structural versus functional organization. The emphasis on the question as to which came first in the current literature seems unwarranted as the proponents of both the viewpoints recognize a simultaneous existence of the basic elements of the structural and functional organization in their speculations on the nature of primeval cells. The real problem is to define a plausible basis for a meaningful structural organization of the functional units, which could ensure a coherent evolution of a spatio-temporal order.

Chemical evolutionists have provided a wealth of information on the synthesis of diverse types of biologically relevant molecules under simulated pre-biotic environment. Current activities in this area are largely directed towards an evaluation of various physico-chemical factors which may lead to the formation polymeric compounds like nucleic acids and proteins, with particular emphasis on template-directed synthesis. Although much of these studies do not envisage any functional constraint leading to a selective advantage in favour of a sub-set of these molecules, they do provide a rationale for the generation of a large repertoire of such polymeric molecules. Indeed, this is considered to be an essential prerequisite in the Eigen-Schuster hypercycles model.

The conditions required for the synthesis of RNA and proteins from their activated monomers are not significantly different. This ensures the presence of both the compounds in the pre-biotic soup, as tacitly assumed in the hypercycle model. Apart from transient interactions between nucleic acids and proteins envisaged in the formulation of hypercycles, the possibility of formation of stable complexes cannot be excluded. Such structural elements devoid of any functional attributes may be viewed as superfluous, or even counter-productive. However, if we consider them as integral components of the hypercycles then they acquire a meaningful role in the structural organization of the functional units. A closer look into the principles underlying the organization of subcellular organelles in contemporary cells has provided some support to this viewpoint.

3. Basic Principles Underlying Structural Organization of the Functional Units in Contemporary Cells

Apart from the outer envelope (plasma membrane) which defines the cell as a unit of all living systems, the organization of organelles like chloroplasts, mitochondria, lysosomes, golgi, apparatus etc. in eukaryotes is essentially based on the compartmentation afforded by the membranes. Prokaryotes, on the other hand, devoid of such intracellular structures supposedly represent a more primitive organization, and hence closer to the primeval cells. It is instructive to look into the principles underlying the organization of their transcriptional and translational machinery.

(A) *Ribosomal structure—non-informational RNA as a skeletal framework*: Crick (1968) considered the role of non-informational RNA (rRNA) in ribosomes. Having made a rather casual observation— "...RNA is 'cheap' to make than protein", he left the question open with the remark—"Even though this may be true we cannot help feeling that the more significant reason for (involvement of) rRNA) and tRNA is that they were part of the primitive machinery of protein synthesis".

A striking feature of ribosomal structure is the multitude of protein components ($\simeq 54$ in prokaryotes). This indeed puts these organelles in a class by themselves distinct from that of multimeric enzymes containing a relatively modest number of subunits. Since all the three types of rRNA (23s, 16s and 5s) are known to have a high degree of secondary structure due to intra-molecular hybridization, it seems reasonable to think that such double stranded regions possess a wide range of specificities to function as anchoring sites for proteins. Thus the non-informational rRNA in ribosomes may be viewed as a skeletal framework for the spatial organization of a host of ribosomal proteins. This indicates to a novel organizational principle which may be of relevance in understanding the nature of primitive self-organizing systems.

(B) *Transcription-translational complex*: The transcription translational apparatus in-prokaryotes (Miller, Jr. *et al.* 1970) presents a still higher level of organization designed to coordinate the synthesis of mRNA – an intermediate transcript, and its translation by ribosomes. Recently I suggested that the stringent coupling between the transcription and traslation may not be incidental, and that it is ensured by the elongation factor Tu-Ts (at one time mistaken to be a new factor ψ in view of its affinity towards RNA polymerase) functioning as a link between the polymerase attached to DNA and the ribosome (Singh, 1976).

(C) *Host ribosomal proteins as integral constituents of $Q\beta$ replicase*: $Q\beta$ replicase has provided a versatile model system for an evaluation of the evolutionary implications of the concept embodied in hypercycles. It is interesting to note that out of four subunits in a replicase molecule three are derived from host ribosomes—i.e., 30s protein S_1 and elongation factors Tu and Ts (Küppers, 1979). Remarkably, the elongation factors Tu and Ts mentioned above (3B) in the organization of the transcription-translational complex appear again as a part of an RNA-synthesizing enzyme.

4. Organization of Functional Units into a Nucleo-protein Complex during Pre-biotic Evolution

With a minimum of intracellular compartmentalization and the absence of skeletal networks which decorate eukaryotic cells, prokaryotes come closest to our simplistic notions of a primeval cell. However, it is also apparent from the above considerations (section 3) that these cells find it expeditious to organize their RNA– and protein-synthesizing machinery into a structural unit for an effective coordination of their synthetic activities. The role of Tu and Ts-peptide elongation factors—in the organization of transcription-translational complex in prokaryotes, and as an integral component of a viral RNA-synthesizing enzyme ($Q\beta$ replicase) may not be just coincidental. Indeed this suggests an early integration of the two basic processes during cellular evolution.

It is reasonable to assume that the presence of both nucleic acids and proteins in the same *milieu* (pre-bioti soup) will necessarily lead to the formation of a multitude of complexes. While the formulations of hypercycles recognize a putative existence of such complexes in reference to the catalytic activities of proteins, they take no cognizance of the structural elements inherent in them.

In the model presented here, shown schematically in Fig. 1, I suggest that stable nucleoprotein complexes constitute the functional elements of hypercycles. It is based on two basic premises: (a) the catalytic activities of proteins with reference to quasi-replicational and quasi-translational processes are manifested at the level of nucleoprotein complex and (b) nucleic acid (RNA) functions as a skeletal frame in the organization of relevant proteins into a composite structural unit (see section 3). Some of the essential features of the model may be summarized as follows:

(i) The proteins P_1 and P_2 (Fig. 1 (b)) possessing rudimentary RNA- and protein-synthesizing activities are organized around a core RNA molecule.

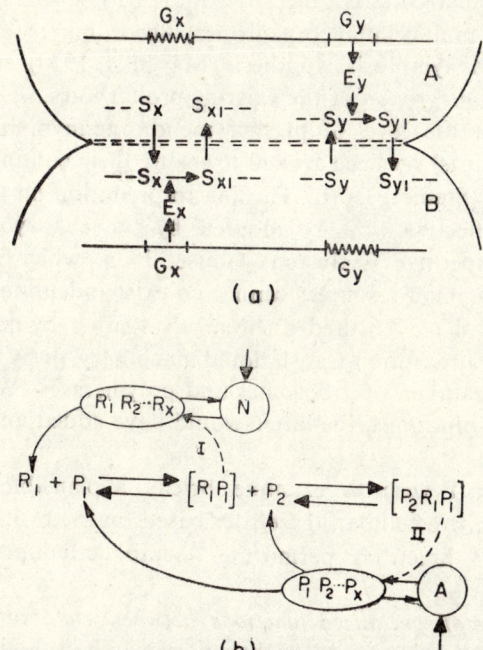

Fig. 1. (a) A schematic representation of organization of the functional units comprising quasi-replication (I) and quasi-translation (II) into a nucleoprotein complex.

(b) A minimal model based on the scheme in (a) for a primitive organization of RNA- and protein-synthesizing proteins P_1 and P_2 into a composite structural unit around a core RNA molecule.

(ii) RNA in the nucleoprotein complex may actively participate in the synthetic processes by holding back the growing RNA and peptide chains preventing thereby their premature release.

(iii) There is now a long list of peptides synthesized under simulated prebiotic conditions, which possess rudimentary catalytic activities. It is conceivable that appearance of peptides with ribonuclease-like activity, and capable of benign coexistence in association with functional nucleoproteins may have added additional selection pressures.

(iv) The stabilizing influence of proteins may lead to an increase in size of RNA. The latter in turn may facilitate the synthesis of longer peptide chains necessary for a progressive increase in their catalytic efficiency.

In summary, I have emphasized the need of incorporating elements of structural organization in the formulations of hypercycles. Interactions between the nucleic acids and proteins leading to the formation of stable nucleoprotein complexes provide a logical basis for such a formalism.

5. Evolution of Multicellular Organisms: Role of Intercellular junctions

Emergence of eukaryotes is generally believed to be a major landmark in the evolution of multicellular organisms. According to a popular view, often referred to as symbiosis hypothesis (Margulis, 1970), eukaryotes arose by *ingestion* and *integration* of the existing prokaryotes. A similar facultative expression at the level of plasma membrane involving *ingestion* and *digestion* (endocytosis) has been evoked to ensure their continued coexistence with prokaryotes (Stanier, 1970). Turning to predation on the prokaryotes was indeed an effective strategy adopted by the eukaryotes against the Principle of Competitive Exclusion (Gause, 1934) which states that two species sharing common resources cannot co-exist indefinitely. Eukaryotes, of course, could have ensured their survival either by improving on the effeciency of the transcription-translational machinery, or by a proportionate increase in the number of ribosomes and polymerases. While the former had its inherent limitations, the latter would have added an excessive load on their genomes.

In this section I wish to examine briefly a plausible model for the emergence of primitive cellular aggregates based on the evolution of specialized intercellular junctions permitting metabolic cooperation between constituent cells.

(A) *The nature of specialized junctions between eukaryotic cells*: While the evolution of cell wall around prokaryotes may have added significantly to their survival value by providing them a rigid envelope, it also severely restrained their further evolution beyond unicellular existence. In metazon animals coordination of the activities of various functional units (organs) and their spatial organization depend to a large extend on an unhindered access to cellular plasma membranes. Ultrastructural studies have led to the recognition of a variety of intercellular junctions such as tight and gap junctions, desmosomes, synaptosomes, neuromuscular junctions etc. responsible for short-range communications (Gilula, 1974).

I wish to restrict these discussions to gap junctions—specialized structure formed by apposition of plasma membranes of competent cells. They are widely distributed both in vertebrates and invertebrates. In electron micrographs they are easily recognized by their septilaminar organization with a distinctive 20–40 Å thick electron-lucent region between opposing membranes (hence the name 'gap' junctions). The techniques based on electronopaque tracers like lanthanum hydroxide, negative staining with heavy metal salts and freeze-fracture have revealed an arrangement of 70–80 Å subunits in a two-dimensional polygonal lattice (Gillula, 1974). These units constitute two-way transmembranal channels for unobstructed flow of ions and metabolites. A protein of a molecular weight of about 26,000 daltons has been identified as an essential constituent of gap junctions from a variety of sources (Benedetti, 1981).

(B) *Complementation at cellular level—metabolic cooperation through gap junctions*: Although the implications of gap junctions in cellular differentiation, or in the sustenance of the differentiated states of the cells are yet to be fully appreciated, studies on cells growing in tissue culture have convincingly demonstrated the efficacy of these structures in metabolic cooperation (Cox *et al.*, 1981).

An ensemble of cells, sustained by mutual interdependence of the constituent cells, may be rightly considered as a primitive precursor of multicellular organisms. It is suggested that evolution of gap junctions-like structures may have been an important factor in bringing about this transition. A minimal formalism based on this contention is schematically shown in Fig. 2. The diagram in Fig. 2(a) depicts two cells A and B, presumably derived from a common ancestor, having defective structural genes G_x and G_y, respectively. These cells will not survive in isolation due to metabolic blocks. However, if they are competent to form gap junctions, the metabolic block due to the defective gene in one cell can be effectively circumvented by the flow of the missing metabolite from the other. In essence, the formalism represents complementation at the cellular level, in which the cells retain their identities, as opposed to complementation at the genetic level brought about by the fusion of two genomes in diploids or in somatic hybrids.

We have examined the behaviour of a simple model system shown in Fig. 2(b). The corresponding reaction scheme may be described as follows:

$$2A \underset{k_2}{\overset{k_1}{\rightleftarrows}} B \quad \overset{k_3}{\rightarrow} N \quad \downarrow k_4$$

For the sake of simplicity we assume the cells (A) to be identical which form a doublet B. Assuming unlimited supply of nutrients (N), the rate equations can be written as:

$$\frac{dB}{dt} = k_1 A^2 - k_2 B \tag{1}$$

$$\frac{dA}{dt} = (2k_2 + k_3) B - k_1 A^2 - k_4 A \tag{2}$$

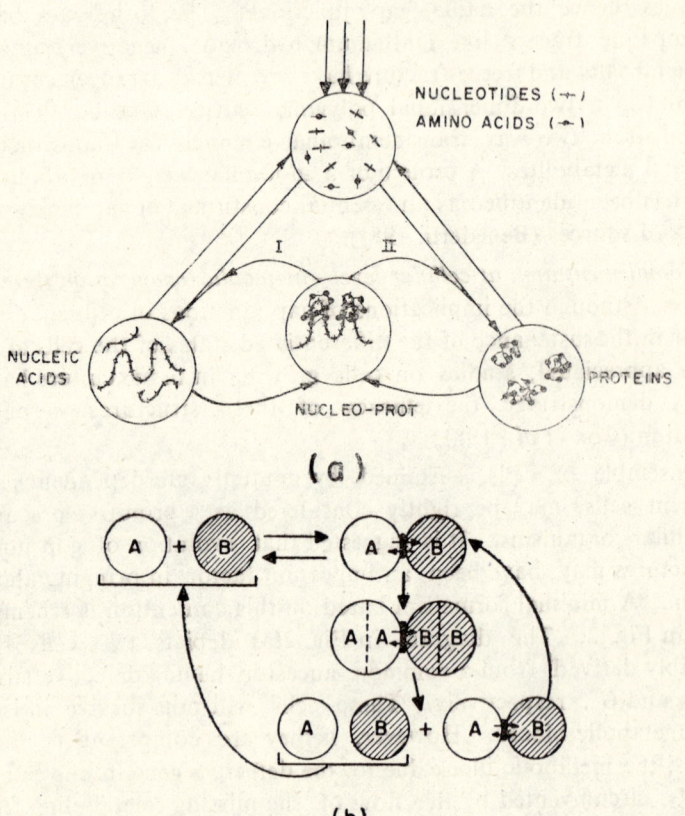

Fig. 2. (a) Complementation at the cellular level involving metabolic cooperation through gap junctions. Cells A and B have defective genes Gx and Gy for enzymes Ex and Ey, respectively. The metabolites Sx and Sy accumulating in respective cells can diffuse into adjoining cells. The products Sx_1 and Sy_1 diffusing back ensure their multiplication.
(b) A simplified scheme used in computer-simulation (Fig. 3).

The behaviour of the system for $k_1 = 1.0$, $k_2 = 0.5$, $k_3 = 0.5$, $k_4 = 1.0$, and for different initial values of A and B is shown in Fig. 3. It is noticed that the system is characterized by a metastable equilibrium state corresponding to $A = 0.5$ and $B = 0.5$ marked by a cross. Depending upon the initial values the curves either converge to the origin, i.e. the species

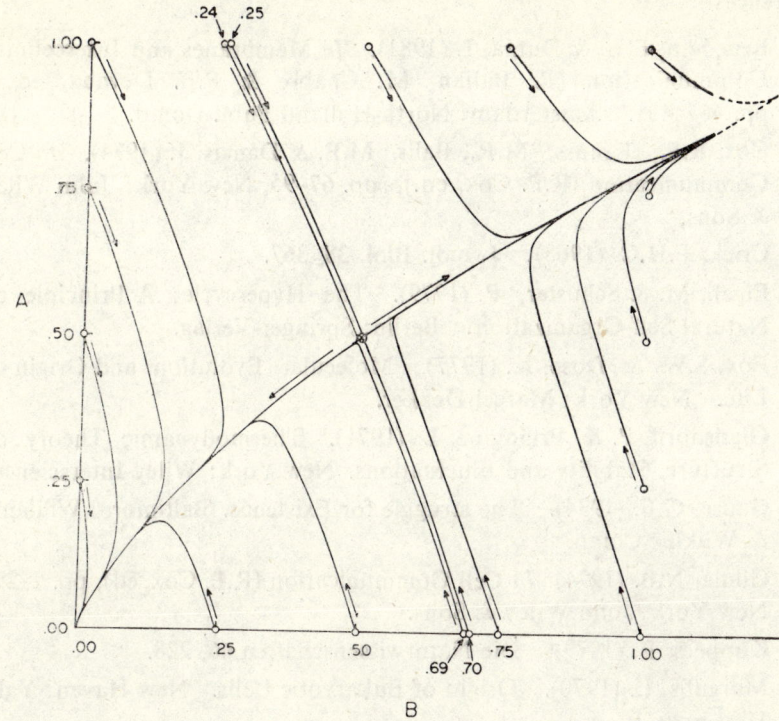

Fig. 3. A computer-simulation of the model shown in Fig. 2(b), and described by equations (1) and (2) for $k_1=1.0$, $k_2=0.5$, $k_3=0.5$ and $k_4=1.0$. The cross at the center represents metastable equilibrium state. Note that a diagonal passing through equilibrium point, and lying between set of curves in its neighbourhood defines two distinct domains. For all initial values in the region bounded by the diagonal and the axes the curves converge to '0', and for those outside this region merge into a straight line.

goes to extinction, or merge into a straight line representing a state of dynamic equilibrium with a constant ratio between A and B.

Although a wide spread occurrence of gap junctions in the animal kingdom has provided some support to the above viewpoints, a critical evaluation of the hypothesis presented here has to await informations on the molecular evolution of their protein constituents.

Acknowledgement

Assistance by Mr. Chetan Premani in the computational work is gratefully acknowledged.

References

1. Benedetti, E.L. & Dunia, I. (1981). *In* Membranes and Intercellular Communication (R. Balian, M. Chabre & P.F. Devaux, eds.), pp. 407–431. Amsterdam: North-Holland Publ. Comp.
2. Cox, R.P., Krauss, M.R., Balis, M.E. & Dancis, J. (1974). *In* Cell Communication (R.P. Cox, ed.), pp. 67–95, New York: John Wiley & Sons.
3. Crick, F.H.C. (1968). J. mol. Biol. **38**, 367.
4. Eigen, M. & Schuster, P. (1979). The Hypercycle: A Principle of Natural Self-Organization. Berlin: Springer-Verlag.
5. Fox, S.W. & Dose, K. (1977). Molecular Evolution and Origin of Life. New York: Marcel Dekker.
6. Glansdorff, P. & Prigogine, I. (1971). Thermodynamic Theory of Structure, Stability and Fluctuations. New York: Wiley-Interscience.
7. Gause, G.F. (1934). The struggle for Existence. Baltimore: Williams & Wilkins Comp.
8. Gilula, N.B. (1974). *In* Cell Communication (R.P. Cox, ed), pp. 1–29. New York: John Wiley & Sons.
9. Kuppers, B. (1979). Die Naturwissenschaften **66**, 228.
10. Margulis, L. (1970). Origin of Eukaryotic Cells. New Haven: Yale University Press.
11. Miller, Jr., O.L., Beaty, B.R., Hamkalo, B.A. & Thomas, Jr. C.A. (1970). Cold Spring Harb. Symp. quant. Biol. **35**, 505.
12. Oparin, A.I. (1953). The Origin of Life. New York: Dover.
13. Shklovskii, I.S. & Sagan, C. (1966). Intelligent Life in the Universe. Holden-Day, Inc.
14. Singh, U.N. (1976). J. theor. Biol. **59**, 107.
15. Stanier, R.Y. (1970). *In* Organization and Control in Prokaryotic and Eukaryotic Cells (H.P. Charles & B.C.J.G. Knight, eds), pp. 1–38. Cambridge: Cambridge University Press.
16. Stent, G.S. (1980). Trends in Neurosciences, February, p. 49.

Coherent Excitations in Biological Systems

H. Fröhlich

Department of Physics,
The University, Liverpool

1. General

One of the most characteristic properties of biological systems is their use of supplied energy to develop and maintain a characteristic organisation. Many ordinary physical systems use energy supply to become hot. There are notable exceptions to this rule, however, such as lasers, or many machines that use energy to do work.

The question is often asked whether macrophysics, and biology in particular, follows from micro (i.e. particle) physics. We take the view here, that in biology the laws of microphysics are never violated. Beyond this, however, the general question is ill defined; for macro-physics, and biology in particular, use concepts that do not exist in microphysics. Usually it is relatively easy to describe ground states of macro-systems in terms of microphysics. Low excitations then usually follow in a systematic way in terms of small displacements leading to theories of linear response. Larger excitations, however, can only rarely be treated in such a systematic way. It can be shown, in fact, that the number of such excited states is so large that it is impossible, in principle, to calculate them all if one is beyond the range in which a systematic treatment is possible [1].

In these circumstances it is necessary to choose certain types of excitation for theoretical treatment, to evaluate their consequences, and then in collaboration with experiment to verify their existence or adandon them. Usually such theories are based on the introduction of general concepts; these then will lead to more specific consequences in a way which is not always unique and whose particular formulation should be arrived at with the help of appropriate experiments. In this approach it is the task of the theory to introduce appropriate concepts, and to propose experiments which may verify their validity and will then lead to specification.

In the present article the existence of coherent excitations will be proposed, their consequences will be presented and experimental evidence will be discussed.

2. General Physical Properties

The possibility of the introduction of general physical concepts for the dynamic description of biological systems rests on the existence of common material properties. Particularly notable amongst these are the dielectric properties, and here the establishment of fields of the order of 10^5 volt/cm in active biological membranes is most prominent. As a consequence, material within a membrane will be very strongly polarised, and in fact non linear features will be relevant at such strong fields. This in turn indicates the possibility of changes of certain properties of molecules like proteins when they are transferred from regions outside strong fields into regions of strong fields. Non linear properties, in fact, appear to be a general concept, required in order order to describe activation and deactivation of biological systems.

It has been shown some time ago ([2], [3]) that under very general conditions, highly polarisable systems posses a metastable excited state which is very strongly polar i.e. they possess a quasiferroelectric metastable state. This arises because the free energy of a polarised system depends on its shape so that the polarisation field interacts with the elastic field. This would ultimately lead to a break up of the system, but stabilisation is obtained from non linear terms. In the model that has been considered in detail, sufficiently strong excitation of polarisation waves leads in the first place to mode softening, but at sufficiently strong excitation the system may jump into the metastable state, thus storing the excitation energy. By slight extension of the model, energy transport of the soliton type may arise [4]. In this way contact is made with the specific introduction of solitons by Davydov [5].

In the above model excitation of a particular mode of polarisation waves (zero wavenumber) has been considered. Very strong, coherent, excitation of this mode is required in order to transfer the system into the metastable state. The possibility of coherent excitation of a single mode has been treated in its own right, with the help of a simplified model in which mode softening was disregarded; introduction of this latter should not lead to undue difficulties.

In this model, all permissible wave numbers of polarisation waves are considered, and their excitation is calculated when this system is strongly coupled to a heat bath of temperature T, and when it is supplied with energy from metabolic processes, or otherwise (first introduced in [6]; of [7] for a review). Linear interaction then leads to exchange of energy with the heat bath in terms of single quanta. Essential non linear terms are introduced, however, which can lead to exchange of energy between excitation of polarisation waves waves under the influence of the heat bath. It is then found that a single mode, say with frequency ω_j, becomes coherently excited if the rate of total energy supply, s, exceeds a critical value s_0,

$$s > s_0 \qquad (1)$$

When this is achieved, thus, a drastic change in the behaviour of the material will take place.

Relevant frequencies can range over a large region. Thus, the membrane, in regions where there are no proteins should have frequencies of dipolar oscillations of the order $10^{10}-10^{11}$ Hz. This follows from its thickness of about 10^{-6} cm assuming elastic constants corresponding to a sound velocity of 10^5 cm/sec, or less, perpendicular to the surface. Proteins have well known frequencies mostly in a higher region. Lower frequencies have been predicted [8] for double helical molecules.

3. Consequences

In the previous section it has been shown that if the rate of energy supply s to certain materials exceeds a critical value s_0, a single mode will be coherently excited and the behaviour of this material should, therefore, change in a drastic way. Furthermore, if the energy of excitation is sufficient, then the system may transfer into the metastable ferro electric state, and thus become stabilised. The possibility of activation—deactivation of particular materials thus arises.

The excitation of coherent vibrations of given frequency leads (a) to a long range frequency selective interaction between systems of nearly equal frequencies and (b) to the possibility of energy storage and transport via the metastable state or the related solitons. Both these possibilities act in addition to (not in contrast with) the well known chemical processes i.e. (a) short range chemical interaction and (b) transport through energy storing chemicals (ATP).

The range of long range interaction depends on the dielectric properties of the medium in which two or more vibrating units interact. If this medium is water then the range of the electric fields carrying the interaction is much larger than the diameter of a cell in frequency regions of membrane vibrations or even beyond i.e. up to about 4×10^{13} Hz. Near 5×10^{13} Hz, however, a very strong water absorption band occurs, thus restricting the range of interaction to much smaller distances, probably below cell diameter. Above this frequency, strong water absorption occurs at selective frequencies thus restricting interactions to distances below cell diameter, but still large compared with the size of proteins. Such interactions, clearly, may have important influences on the dynamic properties of biological systems. Thus, when the range is larger than a cell diameter, it may lead to selective interaction between cells, or to the attraction of molecules of the nutrient medium to cells. In the short range region it may yield a selective attraction between enzymes and substrates, or between other materials within a cell.

The metastable polar state may represent the active state of enzymes and act not only as energy storage and transduction, but also by reducing activation energies, thus accounting for the high catalytic power of enzymes (cf [1], [2], [9]). In its connection with solitons it may play an important role in the transduction of free energy required in a whole range of processes [10].

The treatment of the polar metastable state as activated enzyme state in conjunction with the selective enzyme—substrate attraction through excitation of coherent modes will lead to periodic enzyme reactions, as first shown in connection with a model for brain waves (cf [7], IV E, with literature). For periodic activation—deactivation leads to local periodic polarisation, and, if the enzymes are spatially organised, to periodic electric vibrations. Their frequencies can be very low, and at any rate lower than the frequencies of most of the coherent vibrations considered earlier. Under special circumstances they can take the form of limit cycles which are oscillations that stabilize at particular amplitudes.

The properties described above provide the possibility of causing large biological effects through the supply of energy at very low rate in the form of electric waves of a correct frequency. In the case of an established limit cycle external application of appropriate fields may destroy the limit cycle, and thus liberate the energy stored in it.

In the case of coherent excitations, dominated by inequality (1), the situation may arise that the biological system is preparing for the excitation by pumping energy at a rate s_B close to, but below the critical s_0. External application of electro-magnetic waves at a rate $s_m > s_0 - s_B$ will thus lead to a total supply at a rate $s = s_m + s_B > s_0$ and hence cause excitation of the coherent mode.

4. Experimental Evidence

The basic excitations proposed above are properties of active biological systems. They are activated-deactivated according to the stage of development, and to other circumstances which cannot always be controlled. As a consequence, reproducibility poses a considerable experimental problem. These difficulties have been overcome in a number of cases that will be discussed below.

A. Quite recently it has been possible to find the highly polar metastable state outside active biological systems through excitation of large molecules in external electric fields. For this purpose, the electric characteristics of multilayer samples of haemoglobin (Langmuir-Blodgett films) were investigated [11]. When the applied field strength exceeded 10^5 vols/cm, a dramatic change in the response took place. The capacitance measured at frequencies above 100 Hz increased a hundredfold and the change persisted for a considerable period during which the electric characteristics gradually reverted towards the initial values. Clearly the protein was, therefore, not denatured (as has also been shown in other ways) but had been lifted into the metastable state.

Other proteins should also show this effect, and in fact this has been found [12] in serum albumin. Others are being investigated.

B. Excitation of coherent modes may lead to strong long range interaction between equal cells provided the frequency is in the region below 4×10^{13} Hz. They may, in particular, depend on the membrane field and hence be in the $10^{10} - 10^{11}$ Hz range if one deals with membrane oscillations, or in a higher frequency region if they refer to proteins, dissolved in membranes and polarised through the membrane field. If the membrane potential is removed then the electric fields carrying the interaction will disappear, as will the interaction between these cells. Also, only cells of the same kind are expected to have equal frequencies.

Evidence for the above has been obtained from an investigation of the interaction between human erythrocytes—red blood cells ([13], [14]). It is found that these cells, suspended in plasma, carry out Brownian motion when at sufficient distance from each other; at closer distance, however, they attract each other, and form so called rouleaux (aggregations) much faster than predicted by Brownian motion. If the membrane potential is brought to zero, however, or if the cells are metabolically depleted, the rate of aggregation agrees with that required by the theory of Brownian motion. Furthermore earlier results on rouleaux formation of a mixture of mammalian cells indicate that equal cells aggregate principally in contrast to the prediction based on random Brownian motion.

It follows that a selective attractive force between active cells occurs, and that it satisfies a number of requirements of the theory. It should be mentioned in this context that mammalian erythrocytes are not motile.

An unresolved question refers to the presence of certain long molecules in the nutrient, without which the attraction is absent. As a possible explanation it might be assumed that through their dielectric properties, these molecules carry the electric vibrations in a more efficient way than water with its strong absorption. Such an interpretation would also link with the suggestion that the laws of non-linear optics should hold for our coherent vibrations; this includes self-focusing and hence propagation through filaments (cf. [15] although we need not necessarily accept the author's complicated composition of certain frequencies).

C. Excitation of coherent vibrations can in principle be measured by laser—Raman effect provided their frequencies are above about 20 cm^{-1}.

For an excitation above the thermal one will show itself on an above thermal anti Stokes: Stokes ratio. Appropriate experiments have been carried out on suspensions of bacterial cells ([16]). Very considerable experimental difficulties arise in these experiments, and they have to be overeome to ensure reproducibility. This has been achieved only quite recently ([17]). Two types of main difficulties are relevant: (a) the laser beam has adverse effects on the cell acitivity, and (b) the density of bacterial cells has to be very low in order to assure that they are active. An estimate then shows that an amplification of at least a factor 10^5 is required to make Raman lines visible ([18]). Even strong excitation cannot achieve this in a single cell, and it is required, therefore, that cells are carefully synchronised so that they can oscillate coherently.

In the recent experiments ([17]) a flow system was used such that each cell is in the laser beam for a few milliseconds only. Furthermore, a special method was employed which synchronises DNA synthesis. The principal result is a well reproducible line near 990 cm^{-1} whose intensity decreases with the age of the cells and which also appears on the anti-Stokes side with an estimated augmentation of a factor seven above thermal. This line also occurs in the pure nutrient—with thermal Stokes—Anti-Stokes ratio. The interpretation suggests that in the cells, the corresponding vibrations are very strongly excited, and by long range interaction attract the molecules of the nutrient which are in resonance. It is hoped that these can be specified in the near future.

It has been pointed out in section 3 that the range of forces in this frequency region is larger than the cell diameter. Around 1680 cm^{-1}, however, a very strong broad water band arises; excitation in this region leads to interaction of much shorter range. It is of importance, therefore, that the Stokes spectrum indicates excitations above this water band though exact separation from it has so far not been achieved. Most remarkably, however, the anti-Stokes spectrum shows a definite line at 1632 cm^{-1} while the only thermally excited water band is completely absent. It follows that the 1632 cm^{-1} line must be very strongly excited. It is noticeable, in this context, that the amide I line at 1660 cm^{-1} has been considered as a strong candidate for collective excitation [19] and has also been used as basis for calculations on soliton excitation [5]. The shift from 1660 to 1632 cm^{-1} must then be considered as the beginning of mode softening, required by the theory in the case of strong excitations.

D. It has been pointed out in section 3 that irradiation of biological materials with low intensity microwaves might achieve large biological responses provided the frequency overlaps with a band of internal vibrations, one of which, on occasions, is coherently excited by biological pumping.

It has been mentioned that the microwave intensity s_m must be larger than a lower limit, $s_m > s_0 - s_B$, which might be very small. The system has the ability of storing energy, and some time is usually required to channel this energy into the mode that ultimately will be excited coherently. This time should be much smaller when the imposed microwave has already this frequency than at other frequencies. Consequently strong frequency dependence of the observed effect is to be expected.

The biological effects that usually are measured in the relevant investigations usually consist of a number of elementary steps, only one of whom may be sensitive to microwaves. Under these circumstances it has been shown that the response is steplike ([7]), equ (63) i.e. there is no response below $s_m = s_0 - s_B$, and the response becomes independent of the supplied microwave intensity at sufficiently larger values of s_m. Other responses may, however, interfere with these findings, especially at high values of s_m.

The theory thus has predicted three main properties of response: (a) strong frequency dependence, (b) steplike intensity dependence at very low intensities, (c) requirement of sufficient time of irradiation. A series of experiments with mm waves on a wide range of biological objects has confirmed all three (cf. [20] and [21]). The frequencies used in these experiments are of the order of 50 GHz; their origin thus is likely to be in the membrane and the arising interactions have a range larger than the cell diameter.

It must be mentioned now that an enormous literature on the action of various microwaves on biological systems exists. Only in the case of mm waves have investigations on frequency and intensity dependence been carried out, without which an appreciation of the results may be very difficult although they might demonstrate the influence of vibrations on biological properties if effects are found at sufficiently low intensities.

It should be pointed out then that biological effects may arise either from an increase in temperature due to the absorption of microwaves which follows in the region of linear response; or they may arise from changes in the properties of materials i.e. from non-linear response. Temperature effects can relatively easily be recognised, and eliminated by going to sufficiently low intensity. In the case of non-linear effects (cf [22]) one must distinguish the cases of involvement of storage from the cases where it is absent. The latter is often suggested to give the possibility of rearrangement—e.g. change in the direction of a protein—by the electric field of the microwave. One can estimate, however, that this would require intensities above a million Watt/cm^2 whereas effects are achieved at intensities below 10 mW/cm^2.

Oscillating systems have a certain storing possibility. The amount of energy that can be stored depends on the frequency in a resonance like manner, on the friction, and it is proportional to the intensity of the exciting wave. Usually the presence of water causes large friction i.e. the just mentioned resonance curve will be broad and low. Sharp frequency response can be achieved with non-linear excitations only such as our coherent excitations (maser like) or the limit cycles mentioned in sections 2 and 3.

Repetition of the experiments published in [20] has in some cases led to further refinements in the results (cf [7]). In other cases, however, reproduction was possible only after long experimentation; it appears that the requirements on both biological and physical parameters are not yet fully understood in these cases.

Such difficulties have arisen in investigations on mono cellular materials such as bacteria. In multi cellular materials a possible long range interaction between cells as described earlier might lead to greater stabilisation. The basic problem then arises to identify the particular region and activity that is influenced by the weak intensity microwaves. For this should then permit identification of the activities which biologically make use of coherent vibrations, on which the externally applied microwave acts as a trigger.

In this context the impressive experiments of effect of mm waves on the fertility of Drosophila (fruit fly) should be mentioned. It has been found [23] that after irradiation the fertility in the second generation is reduced by nearly 40%, the effect depending in a resonance way on the frequency with the largest effect at 40 GHz.

It has been possible to reproduce this large effect [24] by application of radiation with the extremely low intensity of $10\,\mu W/cm^2$. This, together with further precautions, clearly excludes temperature effects, and the frequency—resonance like response supports the interpretation that the system itself is making use of vibrations in the relevant frequency region and thus very sensitive, in a manner described earlier, to externally applied microwaves.

Our interpretation, thus, asserts that the sensitivity of certain biological systems to microwaves in a narrow frequency region, at extremely low intensities, can be understood only if the system itself makes use of vibrations in this frequency region and, therefore, has developed a particular sensitivity in this respect. Clearly it is of prime importance, therefore, to identify the regions on which the microwave acts. The above experiments seem particularly suited for this purpose, and investigations in this direction have been planned.

References

1. H. Fröhlich Rev. Nuovo Cimento 3, 490, 1973.
2. H. Fröhlich J. Collect. Phenom. 1, 101, 1973.
3. H. Fröhlich Bio Systems 8, 193, 1977.

4. H. Bilz, H. Büttner, H. Fröhlich, Z. Naturforsch. **36b**, 208, 1981.
5. A.S. Davydov, Int. J. Quantum Chem. **16**, 5, 1979.
6. H. Fröhlich, Int. J. Quantum Chem. **2**, 641, 1968.
7. H. Fröhlich, Adv. Electronics and Electron Physics, Academic Press, **53**, 85, 1980.
8. E.W. Prohofsky, J.M. Eyster, Phys. Lett. **50A**, 329, 1974.
9. H. Fröhlich, Proc. Natl. Acad. Sci. U.S.A., **72**, 4211, 1975.
10. D.B. Kell, J.J. Clarke, J.G. Morris, FEMS Microbiol. Lett. **11**, 1, 1981.
11. J.B. Hasted, H.M. Millang, D. Rosen, Trans. Farad. Soc. in print.
12. J.B. Hasted, personal communication.
13. S. Rowlands, L.S. Sewchand, R.E. Lovlin, J.S. Beck, E.G. Enns, Phys. Lett. **82A**, 436, 1981.
14. S. Rowland, L.S. Sewchand, E.G. Enns, Can. Journ. Physiol. Pharmacol. in print.
15. E. Del Giudice, S. Doglia, M. Milans Phys. Lett. **85A**, 402, 1981.
16. S.J. Webb, Phys. Rep. **60**, 201, 1980.
17. F. Dissler, personal communication.
18. H. Frohlich, Tisza Festschrift, MIT Press, in print.
19. G. Careri, Cooperative Phenomena (ed. Haken and Wagner) p. 391, Springer Verlag 1973.
20. N.D. Devyatkov, Sov. Phys. Usp. (Engl. Transl.) **16**, 568, 1974, and following papers.
21. H. Fröhlich, Phys. Lett. **15A**, 21, 1975.
22. H. Fröhlich, Proc. Symp. Biomedical Thermology 1981, in print.
23. N.P. Zalyubovskaya, Sov. Phys. Usp. (Engl. Transl.) **16**, 574 1974.
24. G. Nimtz, personal communication.

Co-operativity and Long-ranged Correlations in Biosystems

Debajyoti Bhaumik*, Kamales Bhaumik** and Binayak Dutta-Roy***

1. Introduction

Even the smallest autonomous unit of biology—the cell-whether looked upon from the point of view of morphogenesis or in the context of processes involved in its functioning, indicates, through the organic unity manifest in its form and activity, the imperative necessity for the existence of co-operativity, of long-ranged correlations and of communication between molecules over distances many orders of magnitude larger than the range of inter-molecular interactions. Similarly, in the temporal domain the need for 'memory' and 'history' in implied, in a time scale much larger than typical molecular collision time. Both these features of the living state, in space and time, are concommitant with non-linear and open systems in metastable states far from thermal equilibrium maintained and developing through sequences of ordered states in a milieu characterised by disorder and thermal chaos. Indeed the ideas of Professor Fröhlich [(1)] involving the process of Bose condensation in low-frequency polar modes and long-ranged electric forces [based on certain extra-ordinary dielectric properties of biological molecules], as well as the point of view emphasised by Professor Prigogine and his group [(2)] implicating the ubiquitous role of the reaction-diffusion process and attendant chemical waves, provide mechanisms for long-ranged and selective communication which apppear to be a logical necessity in so far as the understanding of the underlying physical principles of the living state is concerned. In the contributions of Professor Davydov [(3)] on solitons. as also the work of Professor Knox and his group [(4)] on excitation transfer, the same dominant theme may be discerned.

The possible role of phase transitions in biological processes [(5)], and

*Bose Institute, 93/1 APC Rd., Calcutta 700009.
**All India Institute of Medical Science, New Delhi.
***Saha Institute of Nuclear Physics, 92 APC Rd., Calcutta 9.

the consequent long-ranged correlations associated with critical phenomena, represent yet another approach directed to serve the same purpose. Indeed, it is very likely that the living state which is, perhaps, the highest and most subtle stage of aggregation of atoms and molecules, makes use of all these mechanisms, and others not yet formulated, in its exquisite and intricate functioning and genesis. The present authors have been mainly inspired by the works of Professor Fröhlich to whose ideas and attitude of encouragement this contribution is dedicated.

To concretise, to a certain extent, the general philosophy delineated above, it is necessary, for the present purpose, to concentrate on three particularly promising aspects of biosystems:

(i) Energy storage and transfer
(ii) Long-ranged selective interaction (recognition) between macromolecules
(iii) Transport of ions and molecules.

These aspects will generally be inter-related cf. energy storage (in an ordered form with maximum free energy) can take place in a collective giant dipolar mode of a macromolecule leading in turn to long-ranged selective interactions.

2. 'Bose Condensation' and Coherent Excitation of a Mode

In a series of papers [(1)] Fröhlich has pointed out that a system possessing low frequency oscillatory modes [with energies $\{\epsilon_i\}$] in interaction with a bath [characterised by a temperature T (with $\beta = 1/k_B T$, where k_B is the Boltzmann constant] may, when considered as an open system with energy supplied to these modes at a rate S greater than a certain critical amount, undergo Bose Condensation in the lowest mode. This result follows from a contemplation of the rate equations for the occupancy n_i of the ith mode, namely,

$$\dot{n}_i = S_i + \phi(T)[(n_i + 1) - n_i \exp(\beta\epsilon_i)] + \chi(T)[\sum_j \{(n_i + 1)n_j \exp(\beta\epsilon_j) - (n_j + 1)n_i \exp(\beta\epsilon_i)\}], \quad (1)$$

where the terms successively represent: feeding by the external pump, gain and loss from the bath and non-linear processes involving transfer from one mode to another. At stationarity ($\dot{n}_i = 0$) the distribution of quanta in the different levels is given by,

$$n_i = [1 + (\phi/\chi)S_i\{1 + \exp(-\beta\mu)\}/(\sum S_i)]/[\exp\{\beta(\epsilon_i - \mu)\} - 1], \quad (2)$$

wherein, it is to be noted, a chemical potential μ, given by

$$\exp(-\beta\mu) \equiv [\phi + \chi\sum_i(n_i + 1)]/[\phi + \chi\sum_i n_i \exp(\beta\epsilon_i)], \quad (3)$$

has appeared. Furthermore, in the absence of non-linearity ($\chi = 0$) and/or external pump [$S_i = 0 \Rightarrow$ closed system $\Rightarrow$ regression to equiliurium $\Rightarrow n_i = 1/[\exp(\beta\epsilon_i) - 1] \Rightarrow n_i \exp(\beta\epsilon_i) = n_i + 1$] the chemical potential μ vanishes, as the implementation of these conditions in eqn. (3) reveals. Due to the existence of the chemical potential in the open non-linear system, with the pumping rate above a certain critical amount the lowest level achieves a macroscopic population and an ordered state of energy storage results. This transition bears a superficial resemblance to what obtains in systems such as liquid Helium in thermal equilibrium where lowering of temperature (removal of energy) results in condensation. Here, however, in an essentially non-equilibrium, albeit steady state, situation (typical of biosystems considered as open systems), it is the supply of energy which elevates the system into a state of order. These ideas have also been framed [(6)] in a microscopic Hamiltonian formalism and certain objections [(7)] to the procedure outlined above effectively countered (through an analogy with the laser) and it has been contended that this should provide a natural basis for the understanding of several remarkable experiments on the effects of electromagnetic radiation on the biological activity of various organisms [(8)] and on the rate of enzyme catalysed reactions [(9)].

3. Polar Modes with Elastic Restoring Forces and Meta-Stable 'Ferroelectric' State

With the strong excitation of a polar mode the system would tend to be deformed which in turn would call into play elastic restoring forces [(10)]. This may lead to the softening of the polarisation mode, which in the presence of quartic self-interaction of the polarisation field (invoked to prevent the disruption of the system) may derive the system into a ferroelectric metastable state (with permanent electric dipole moment). The Bose condensed and ferro-electric state evokes co-operativity and long-ranged forces constituting the central issue to which we are addressing ourselves. Thus consider a system consisting of a polarisation density $P(\mathbf{r}, t)$ and elastic field $A(\mathbf{r}, t)$ at a space-time point $(\mathbf{r}, t)$. For simplicity, considering a macro-molecule, we may adopt the direction along the chain as the only relevant dimension and write the Hamiltonian density as

$$H = \tfrac{1}{2}\dot{P}^2 + \frac{\lambda^2}{2}\left(\frac{\partial P}{\partial x}\right)^2 + \tfrac{1}{2}\omega_0^2 P^2 + \tfrac{1}{4}d^2 P^4 \\ + \tfrac{1}{2}\dot{A}^2 + \frac{\sigma^2}{2}\left(\frac{\partial A}{\partial x}\right)^2 + CP^2\left(\frac{\partial A}{\partial x}\right), \quad (4)$$

where ω_0 is the frequency of the polarisation mode, σ the velocity of the elastic wave, c the strength of the coupling of the elastic and polarisation fields and d^2 the strength of the quartic self coupling of the polarisation field inserted for stabilisation. The elastic field has a time scale much slower than that associated with the polarisation field and thus the elastic forces will play a rather passive role in counteracting the deformation of the system arising from polar excitation and will thus adjust itself (acting like a parameter) to a value minimising the potential energy to give

$$\left(\frac{\partial A}{\partial x}\right)_{\text{effective}} = -\frac{c}{\sigma^2}\langle P^2\rangle, \qquad (5)$$

where the braces designate time averages taken over a period large compared to the time scale associated with the polar mode. Implementing this condition, the potential term in the Hamiltonian density corresponding to the restoring force in the potential becomes $\frac{1}{2}\left(\omega_0^2 - \frac{2c^2}{\sigma^2}\langle P^2\rangle\right)P^2$. When $\langle P^2\rangle$, which is proportional to the energy stored in the polarisation mode, exceeds $\sigma^2\omega_0^2/2c^2$ the mode softens, the system tends to disrupt which is in turn prevented from doing so by the quartic self-interaction and accordingly the polarisation mode oscillates about a new minimum of the potential located at a non-zero value of the polarisation, namely

$$P_0 = \pm\frac{1}{d}\sqrt{\left\{\frac{2c^2\langle P^2\rangle}{\sigma^2} - \omega_0^2\right\}}. \qquad (6)$$

Thus, the system, provided sufficient energy resides in the polarisation mode, is driven into a metastable ferro-electric state with a permanent giant dipole moment, bringing into play long-ranged electrical forces.

Furthermore, the model described above has the virtue of providing a mechanism for the propagation of localised excitations [(11)] along macromolecules and their aggregates, which is often considered to play an important role in transport and transduction processes occurring in biology. At the same time a mechanism of control and variability is provided. The antagonists of the view-point, however, draw attention to the smallness of the lifetimes of these excitations, which would indeed be the case if these were associated with local chemical groups; on the other hand, it is our contention, that the existence of collective modes in macromolecules (or their aggregates), with concomitant non-linearities. makes possible relatively stable localised excitations, in the form of solitary wave solutions for the polarisation density of the type [$a\,\text{sech}\,b(x-vt)$] to the equations of motion of the system governed by the Hamiltonian density [eq. (4)], to wit

$$\frac{\partial^2 P}{\partial t^2} - \lambda^2\frac{\partial^2 P}{\partial x^2} + \omega_0^2 P + c\left(\frac{\partial A}{\partial x}\right)P + d^2 P^3 = 0, \qquad (7a)$$

$$\left(\frac{\partial^2}{\partial t^2} - \sigma^2\frac{\partial^2}{\partial x^2}\right)\frac{\partial A}{\partial x} - \frac{c}{2}\frac{\partial^2 P}{\partial x^2} = 0, \qquad (7b)$$

after due recognition is accorded to the slower time scale of the elastic mode.

The potentiality of solitary waves (in adopting the role of carriers of excitation) implicating amide vibrations in proteins has been beautifully emphasised by Professor Davydov [(3)] in the context of muscle action involving the actin-myosin complex and the contraction of the sarcomere of muscular fibres.

4. Role of Polar Modes in Macro-molecular Interactions

It is often conjectured that long-range attraction and recognition may be essential to the understanding of problems relating to goal-directedness, ordering and morphogenesis underlying biological processes, even if one takes resort to the reaction-diffusion mechanism for pattern formation. Yet it is known that, at the microscopic level, the usual Van der Waals interaction between molecules falls off rapidly with distance (inversely as the sixth power of the separation and at larger distances, due to retardation effects as the seventh power). Again, while specificity and other details of the ubiquitous enzyme reactions require, no doubt, detailed stereo-chemical and structural considerations, nevertheless, the as yet unexplained "essential mystery of the enormous catalytic power of enzymes", in the words of Koshland [(12)], calls for an understanding in terms of new physical concepts involving the interaction of large molecules. Within the framework reported above, it has been suggested [(13)] that the softening of vibrational modes of macromolecules when they approach each other, the consequent large amplitude dipolar vibrations and the possibility of excitation of highly polar metastable states, may qualitatively alter the nature of the interaction between such molecules and that too in a frequency selective manner and thereby suggesting a possible clue to these mysteries. It may be remarked that low frequency modes would be particularly amenable to softening due to interactions and their existence indicates a 'looseness' or lability in the structures. As the separation between two molecules decreases the strength of the interaction between them increases, the looseness of the structures (existence of low frequency modes) would make possible a situation where the interaction energy is comparable to the energy of elastic binding. This would lead to the softening (vanishing frequency) of one of the eigenmodes of the system and consequent instability, were it not for the stabilising inharmonic forces in each molecule which prevent them from being torn apart; and each molecule driven into a strained configuration (metastable states) with non-vanishing static dipole moments as a result of the interaction which in turn modifies the interaction itself. To illustrate the underlying idea consider two molecules with polarisation vibrations P_1 and P_2 (with natural frequencies ω_1 and ω_2, respectively) corresponding to the correlated motion of their constituent charged groups. The energy of interaction between the two molecules separated by a distance R is taken to be $-g(R)P_1P_2$ where the vector nature of the polarisation has been suppressed to simplify and to emphasise the basic principle involved. $g(R)$ the coupling constant between the two modes is of the form α/R^3. Accordingly, the equations of motion for the two interacting polar modes read,

$$\ddot{P}_1 = -\omega_1^2 P_1 + g(R)P_2 - d^2 P_1^3, \qquad (8a)$$

$$\ddot{P}_2 = -\omega_2^2 P_2 + g(R)P_1 - d^2 P_2^3. \qquad (8b)$$

Contemplation of these equations reveal some interesting possibilities. The effective interaction between the molecules would of course involve a thermal averaging which would result in the vanishing of the first order contribution in the coupling constant and normally the first non-vanishing term would in the second order leading to the $1/R^6$ dependence of the Van der Waals interaction. However, if the effect of the medium is incorporated into the equations of motion through a frequency dependent dielectric constant, it is observed [(1), (13)] that the interaction between the molecules (defined as the difference in free energy at a distance of separation R from that at infinite separation) behaves as $1/R^3$ (and not as the usual $1/R^6$) for $\omega_1 = \omega_2$, provided the medium possesses appropriate dispersion properties, to wit, the normal mode frequencies straddle a frequency at which the medium shows anomalous dispersion. Such a frequency selective long-ranged attraction may be invoked in problems such as the long-ranged attraction and recognition of homologous chromosomes [(14)]. Apart from this effect of the medium it is easily seen that when the molecules approach each other and the magnitude of $g(R)$ increases, a stage is reached when $|g(R)| \gg \omega_0^2$ (where for emphasising the basic mechanism we have taken $\omega_1 = \omega_2 = \omega_0$), where the branch corresponding to oscillations about $P_1 = 0 = P_2$ becomes unstable and the molecules vibrate around non-zero polarisation values namely, $\sqrt{[\{g(R) - \omega_0^2\}/d^2]}$. As a consequence, the Van der Waals interaction which has a $1/R^6$ dependence at large distances undergoes a modification at distances smaller than a critical separation R_c [where $g(R_c) = \omega_0^2$] to a $1/R^3$ form (typical of interaction between permanent dipoles). The effect of the intervening medium as well as the feeding of energy into the polar modes on macromolecular interactions has also been studied [(15)]. Thus, macromolecules possessing pliable structures could be raised through interactions into metastable states with the excitation of giant dipolar modes, which would, in turn, qualitatively alter the nature of the Van der Waals interaction giving rise to long-ranged effects ($1/R^3$) and a mechanism for long ranged attraction and recognition.

5. Possible Role of Phase Transition and 'Memory' in Biology

A mechanism for long-ranged correlation in biosystems could lie through the intervention of phase transitions with attendant fluctuations and correlations at the transition point, enabling the system (or part of the system) to act as a whole. Thus for example the helix-coil transition in proteins and the unwinding of the DNA helix may be cited as possible illustrations of the mechanism which has important functional implications. In this connection we shall focuss our attention on the role of phase transitions in bilayer lipid membranes as a prototype system to concretise the underlying ideas.

The cell membrane is mainly composed of lipid molecules, arranged in the form of a bilayer, with polar head groups lying on either side of a hydrophobic hydrocarbon chain region, typically 30°A thick, which, for all practical purposes, provides an impermeable barrier to the flow of inorganic ions (such as those of sodium and potassium), because the electrostatic energy of such an ion in an aqueous environment is lower than that in lipid surroundings by about an electron volt (due to the considerable disparity in the dielectric constants ~ 80 for water and ~ 2 for lipid) which is large compared to thermal energies. This remarkable property helps in the maintenance of large gradients in ionic concentrations across the membranes by specialized enzymes that act as ion pumps. Nevertheless, the application of a stimulus (provided it exceeds a certain threshold value) results in a flow of ions across the excitable membrane, a phenomenon that constitutes the basic response of the system. This flow occurs, it is believed, through certain ion-channels, composed, presumably, of protein bundles present in the membrane. It is being increasingly appreciated that the lipid in the membrane is not merely a passive matrix, in which various active elements (such as ion pumps and channels) are positioned, but has a crucial dynamical and functional role to play and could very well be an instrument of control. Indeed, anomalies in the temperature dependence of several membrane-associated functions have been related to transitions in the physical state (liquid to gel) of the lipid constituent. In particular, transport properties have been found to be correlated with the phase of the lipid, the corresponding Arrhenius plots possessing biphasic shapes with the characteristic temperature, defining the change in slope, exquisitely reflecting the melting point of the fatty acid supplement in the membrane, as well as changes thereof brought about by growing the cells in media wherein different fatty acids have been added. A further corroboration of this correlation is provided by the interesting proposal that the action of local anaesthetics on nerve membranes is actuated by the increase in fluidity of the lipid. Accordingly, it behoves us to adopt the ansatz that the sodium channel is open when the annulus of boundary lipid is in the crystalline or gel state and is closed when this surrounding lipid is in the disordered or fluid (albeit, liquid crystalline) phase.

Normally, phase transitions are effected by changing the temperature, but in vivo, the temperature is, however, fixed at the ambient physiological value. Hence, in the context of the position adopted by us regarding the mechanism of response in excitable membranes to stimuli, it is the appropriate dependence of the 'melting' temperature on the stimulus (external electric field) which should be held responsible for the change in state of the boundary lipid and consequent opening of the sodium channel.

The mechanism of response in excitable membranes described above illustrates the possible role of co-operative phenomena (phase transitions) in biology. Furthermore, it may be conjectured that the movement of excitation along the membrane is actuated by the phase propagation (growth from region of nucleation in gel phase). This enables the system to act as a whole and correlates the state of the different ion channels. It may be remarked that the active role of the lipid component of the membrane in its functional behaviour has also been emphasised in several [(16)] critiques of Mitchell's theory of proton translocation (chemiosmosis) across the membrane. These developments, in the view adopted by the present authors, point to the possible importance of the mechanism of phase transitions in establishing co-operativity and correlations in biosystems.

Another mechanism for co-operativity and correlation in biosystems is revealed from the studies in excitation transfer in the chloroplasts of green plants involved in the process of photosynthesis [(4)]. With the absorption of photons by antenna chlorophyll molecules in the chloroplast and consequent excitation, the question of transfer and propagation of this excitation from molecule to molecule to the active site where the dark reactions occur, may be posed through a master equation, namely;

$$\frac{dP_i}{dt} = \sum_j [F_{ij} P_j(t) - F_{ji} P_i(t)], \tag{9a}$$

where $P_i(t)$ is the probability of finding the excitation at the ith molecule and F_{ij} is the rate of transfer of excitation from the jth to the ith site. With the introduction of "memory" (dependence on past history) it is necessary to write down a generalised master equation [(4)]

$$\frac{dP_i(t)}{dt} = \int_0^t ds \sum_j [W_{ij}(t-s) P_j(s) - W_{ji}(t-s) P_i(s)]. \tag{9b}$$

The memory-less Markoffian case $[W_{ij}(t-s) = F_{ij}\delta(t-s)]$ leads, in the continuum limit to the diffusion equation $\left[\frac{\partial P}{\partial t} = D \frac{\partial^2 P}{\partial x^2}\right]$, while perfect memory corresponding to maximal correlation $[W_{ij} = \lambda_{ij}\theta(t-s)]$ give rise to the wave-equation $\left[\frac{\partial^2 P}{\partial t^2} = C^2 \frac{\partial^2 P}{\partial x^2}\right]$. Exponentially decaying memory, however, yields the telegrapher's equation. The velocity and diffusion parameters are related to the constants in the memory function as well as to the lattice spacing. These studies emphasise the role of correlations in the time domain on important processes such as energy transfer in biology.

6. Concluding Remarks

In conclusion the object of our presentation is mainly one of emphasis. The necessity for long-ranged correlations and co-operativity in biology is almost self-evident. Some of the approaches have been outlined above. The need for further studies along these lines and investigations of other possibilities is strongly felt by the present authors.

References

1. H. Frohlich: Nature **228**, 1093 (1970); Int. J. Q. Chem. **2**, 641 (1968); Phys. Letts. **39A**, 153 (1972); Phys. Lett. **51A**, 21 (1973);
   Proc. Nat. Acad. Sci (USA) **72**, 4211 (1975).
2. G. Nicholis and I. Prigogine: Self Organisation in non-equilibrium systems, John Wiley, New York 1977.
3. A.S. Davydov: J. Theo. Biol. **66**, 379 (1977).
   A.S. Davydov: Solitons in Quasi-one-dimensional Molecular-structures. Preprint ITP–80–133E.
4. V.M. Kenkre and R.S. Knox. Phys. Rev. **9B**, 5279 (1974).
   K. Colbow and R.P. Dunyluk, Int. J. Quant. Biol. Symp. –No. 3, 151 (1976).
5. H.W. Wu and H.M. McConnell. Biochem. Biophys. Res. Commun. **55**, 484 (1973).
   C.D. Linden, K.L. Wright, H.M. McConnell and C.F. Fox. Proc. Nat. Acad. Sci. USA, **70**, 2271 (1973).
   R.M.J. Cotterill: Physica Scripta **18**, 191 (1978); **22**, 188 (1980).
   D. Bhaumik, B. Dutta Roy, A. Lahiri and T.K. Chaki: Bull. Math. Biol. (to be published).
6. D. Bhaumik, K. Bhaumik and B. Dutta Roy: Phys. Lett., **59A**, 77 (1976).
7. M.A. Livshitz: Biophysics (translation of Biophysika), **17**, 726 (1972).
8. S.J. Webb and M.E. Stonehane: Phys. Lett. **60A**, 267 (1977).
9. N. Kollias and W.R. Mellander; Phys. Lett, **57A**, 102 (1976).
10. H. Frohlich; J. Collect. Phenomena **1**, 101 (1973).
    D. Bhaumik, K. Bhaumik, B. Dutta Roy and M.H. Engineer; Phys. Lett. **62A**, 197 (1977).
11. D. Bhaumik, B. Dutta Roy, A. Lahiri; Bull. Math. Biol. 1982 (in Press).
12. D.E. Koshland and K.E. Neet, Ann. Rev. Biochem., **37**, 359 (1968).

13. D. Bhaumik, B. Dutta Roy and A. Lahiri; Phys. Lett. **68A**, 131 (1978).
14. B.W. Holland: J. Theor. Biol. **35**, 395 (1972).
    H. Jehle: J. Chem. Phys. **18**, 1150 (1950).
15. D. Bhaumik, B. Dutta Roy and A. Lahiri: Phys. Lett. **69A**, 68 (1978).
16. R.J.P. Williams, FEBS Lett. **102**, 126 (1979).
    D.E. Green, Proc. Nat. Acad. Sci. USA **78**, 2240 (1981).

Variations of the Micro-Dielectrophoresis of Cells During Their Life Cycle

Herbert A. Pohl, Tim Braden and Douglas G. Pohl

Department of Physics, Oklahoma State University, Stillwater, Oklahoma, USA

Introduction

In 1968, H. Fröhlich[1-4] predicted that the chemical energy of cells could evoke high frequency electrical oscillations. We have now observed such outputs from living cells of various types. We see them to be associated with cell division.

To observe such rf (radio frequency) oscillations we have used two very different techniques, each needing only relatively simple apparatus. The first technique is called micro-dielectrophoresis (μ-DEP) and requires essentially only a microscope[5-7]. Here, one observes the collection of various highly polarizable particles by a cell so as to examine the rf field emitted from it. In the second techniques, direct observation is made of the spinning or tumbling of cells as evoked by external rf fields. Both methods yield similar conclusions as to the nature, frequency, strength, and occurrence of the natural rf oscillations of cells. We believe, therefore, that the presence of the postulated rf oscillations has been established beyond reasonable doubt. It now remains to study their meaning. Are they cause or effect, necessity or frill, in the life of cells? Where and how do they operate? What causes them? What controls do they evoke or reflect...intra-cellular or inter-cellular? With what processes are they associated in the cellular life cycle? In the present paper we shall describe experiments aimed at answering some of these questions, especially as to where in the cell life cycle these natural rf oscillations occur. We have also presented studies of the dependence of the μ-DEP of various cell types upon time and upon the compositions of the cellular suspensions and and test powders so as to provide firmer understandings of the phenomenon of μ-DEP. The experiments here deal mainly with the first named technique, micro-dielectrophoresis.

The phenomenon of *dielectrophoresis*[8] (DEP) is the motion induced by the action of a nonuniform electric field on a *natural* body. It was observed back in antiquity (Thales of Miletus, ca. 600 B.C.) and is the phenomenon by which electricity was discovered when it was observed that rubbed or "electrified" amber attracted bits of fluff and paper, (although it was not understood as DEP until recently). The action of the nonuniform electric field can be considered to act in two steps. First, the applied field induces a charge separation (polarization) of the neutral body. Second those shifted charges lying in the more intense region of the nonuniform field are pulled upon more strongly than their oppositely charged couterparts are repelled towards the weaker field region. The result is a net force on the neutral body, one pulling it towards the region of higher field intensity. The effect does not depend upon the sign (+ or −) of the field direction. It, therefore, acts in the same direction, i.e. towards the region of most intense field, even though the applied field might be one of alternating character. This induced motion of neutral objects in nonuniform is called DEP. In contrast, the motion of a *charged* body in any field (uniform or nonuniform) is known as *electrophoresis*. It works in static fields, but unlike DEP, is repressed in ac fields.

Because the polarization of matter varies substantially with the frequency of the applied field, the DEP force upon an object suspended in a real medium can give rise to a possibly complex (difference) spectrum of the DEP response in ac fields as the frequency is altered. This feature supplies useful characterization information not available in electrophoresis. This fact can be used, for example, to analyze and sort cells in a delicate and sensitive manner[8,9]. In the present study, we have turned the matter about, and use the DEP motion of tiny and highly polarizable test particles to examine the nature and extent of the minescule electric fields provided by the cells themselves. If the cells generate electromagnetic fields, they may collect highly polarizable particles more readily than they do ones of low polarizability.[5-7] Such a study can be done with simple means; a good microscope and a knowledge of what to look for. Because the experiments are done on a microscopic scale, we refer to it as "micro-dielectrophoresis" (μ-DEP).

Experimental
In this section we describe the preparation of the high and low dielectric constant powder suspensions and the methods for determining the micro-dielectrophoresis of various cells.

Powder Suspensions
The pure powders (e.g., $BaTiO_3$, $NaNbO_3$, $BaSO_4$, SiO_2, and the organic polymer of high dielectric constant, DPIA) were ground in an agate mortar and pestle under ca. 10 ml of water containing a drop of liquid detergent (Joy). The product was then diluted with 70 ml of a 0.1% solution of

soluble potato starch (0.1% S) and poured into a Petri dish to form a layer 11 mm deep. This was let stand, in the case of the BaSO$_4$, for example, for 5 min, and the supernate then carefully decanted into a second Petri dish and let stand for 10 min before pouring off and discarding the supernatant liquid containing the fines. The residual powder in the second dish was now of rather narrow size range. This was now rinsed four times with the 0.1% strach solution, collected and provisionally labeled "2 um BaSO$_4$", or as appropriate for each powder. The settling times for each case were chosen in accordance with the settling times calculated from the densities and Stokes law for spherical particles.

The powder materials used were: BaTiO$_3$, Alpha Products Co. 99.99% "2 micron grade"; BaSO$_4$, "certified" grade 2 micron nominal particle size, Fisher Scientific Co.; NaNbO$_3$, kindly supplied by Dr. P.C. Held, Ceramics Dept., University of Illinois; SiO$_2$, amorphous powder, Imsil A-15, Illinois Minerals Co., Cairo, Illinois; DPIA, a synthetic copolymer of high dielectric constant prepared by D.G. Pohl from an equimolar mixture of anthraquinone (recrystallized) and pyromellitic dianhydride (re-sublimed).[10] The relative dielectric constant of this polymer varies with frequency, being 5000 at 300 Hz, and 200 at 300 kHz, for example. The conductivity also varies with frequency, being about 1.4×10^{-5} mho cm^{-1} at 300 Hz, and about 2.8×10^{-5} mho cm^{-1} at 300 kHz, and at 1 kbar pressure, but is lower at one atmosphere pressure.

The concentrated suspensions of these purified and size-graded powders in the 0.1% soluble strach were each shaken together with a pair of 4 mm and 2 mm diameter pyrex glass balls for 1 min in a "Wig-L-Bug" shaker (Crescent Dental Mfg. Co.) immediately before use with cell preparations. The suspensions were 3 to 20% powder by volume.

Cells

Murine ascites tumor fibroblasts (Sarcoma 180) were obtained from the peritoneal fluid of Swiss mice grown by E.M. Hodnett. The cells were separated from the fluid by centrifugation first at ca. 200 g for 30 sec to rid of detritus, drawn off in the supernate and re-centrifuged at 2000 g for 1.5 min. The cells were then suspended in M/4 sucrose containing 0 1% soluble potato strach (S/S) and centrifuged at 2000 g for 1.5 min whereupon the supernate was discarded. This was repeated 4 times to obtain a cellular suspension of low conductivity. Typically, a portion of this suspension was taken (e.g. 200 μl) and mixed gently with 2 to 5 μl of the desired powder suspension and let stand for a short time, usually 3 min, then sampled by dipping in it a flat capillary "microslide" of 100 μm path length (#5010, 50 mm length, from Vitro Dynamics, Inc.). The microslide filled in seconds and was then laid flat on a microscope slide for the settling and subsequent count of the cells and particles and of their distribution on and about the cells.

The cells were examined at 430 X magnification and the particles counted with the aid of a grid reticle having (projected view) sides of 12.5 μm. The number of powder particles seen to be associated with a cell was termed 'n' and the number of particles lying in the immediately adjacent 8 grid squares was termed 'p'. From a count of some 40 cells and their surrounding particles, the ratio n/p was determined in each run.

Results and Discussion

The technique of dielectrophoresis (DEP) is the motion of neutral particles as induced by nonuniform electric fields. Cells and other particles suspended in a liquid may readily be moved about by the application of external nonuniform fields. Such forces depend upon the difference in the polarizability (relative effective dielectric constant) of the particle and its surrounding medium.

The micro-dielectrophoresis (μ-DEP) experiments use this phenomenon, but use tiny polarizable particles to explore the electric field about cells. The μ-DEP experiments are relatively simple and direct. A typical experiment consists of mixing a suspension of cells with a suspension of smaller powder particles, then observing the number of particles associated with the cells after a short time. This rate (collection factor, n/p) is then compared for similarly sized particles of various polarizable characters (DK). The particles used are typically in the size range of 2 μm and are size-selected by appropriate settling procedures. In contrast the mouse cells are rather larger (ca. 2 μm ×5 μm for erythrocytes, 10–15 μm for murine "L" fibroblasts or ascites tumor cells). The bacterial cells are typically about 2 μm by 5 μm, (*B. cereus*), while the yeast (*S. Cerevisiae*) are about 8 μm in diameter.

For particles of high polarizability, one uses materials such as $BaTiO_3$, $SrTiO_3$, $NaNbO_3$, or a polymer such as DPIA, an anthraquinone, pyromellitic dianhydride condensation polymer exhibiting strong nomadic polarization and, hence having a very high dielectric constant, about 2000; 400; 650; and 300 respectively. For particles of relatively low polarizability (i.e. less than that of the suspending medium H_2O, (static DK ~ 80) numerous materials would serve. We have used $BaSO_4$, and SiO_2 (static DK 11.5 and 3.8 respectively). In prior studies we have shown that cells prefer to attract and hold on their exteriors those particles which are the more polarizable, especially if the cells are in the reproductive phase. Here at this time we investigate the kinetics of cellular μ-DEP, its dependence upon the conductivity of the medium, and upon the life cycle.

The frequency of the natural rf oscillations of cells can be estimated from the conductivity effect. An estimate of the field intensity at the cell surface can be made from the fact that one of the highly polarizable particles, $BaTiO_3$, is a ferroelectric solid. It has a hysteresis of polarization and requires a certain minimum electrical field, E_{crit}, to override its internal intrinsic polarization. That $BaTiO_3$ is attracted to and held by fields in in excess of that amount found for ordinary particles of similar size, shows that the field is greater than E_{crit}, about 50 V cm^{-1}. From this one can also estimate that the natural rf dipole, assumed to be a point dipole situated at the cellular center for this estimate, is about 10^{-21} Coulomb-meter, corresponding to a swinging of some 10^8 electrons a distance of ca. 1 Å in the cell.

It is natural at this point to ask why we can be sure that the natural cellular dipole is an oscillating one. Might the natural dipole be a static one and still cause the observed μ-DEP effects to be observed? No, because if the dipole were simply a static one, the presence of the conducting medium H_2O would quickly release ions so as to mask the static dipole.[7] It can be shown easily from dielectric theory that the natural cellular dipole must be one which oscillates at a rate faster than the masking rate (dielectric relaxation rate) of the aqueous medium in order for the cellular dipole to be appreciably sensed by the test particles (e.g., $BaTiO_3$ or $NaNbO_3$). As shown earlier[5-7] this frequency must be at least about 5000 Hz. There cannot be a simple static charge effect remaining effective and due just, for example, to the differing DK's of the cell and the various test particles. For example, it has been suggested that it was well known that simple 'triboelectric' charging (Coehn effect)[11,12] of particles occurred upon separating materials of differing DK in a vacuum. Might not charges so generated result in the mutual attraction of the cells for high DK particles? The answer is no, because of the presence here of the conductive medium, H_2O, which would rapidly provide ions to destroy such a charge imbalance. To show this explicitly and experimentally, a simple experiment was done using suspensions of particles of widely different polarizability, to see if co-clustering by such means would result as predicted by this (erroneous) model.

Dielectric Constant and "Co-Clustering" of Colloids:

To see if suspended particles of very different dielectric constant would interact in the presence of water such as to evoke mutual coagulation, suspensions of such materials were mixed so as to study the possible increase of settling rate caused by larger particle clumps produced by such a mechanism. Using this extension of the Coehn model, a material of high dielectric constant would become positive and the lower dielectric constant material would become negative and, thus, mutually attract each other, despite the presence of a conductive medium. This was tested by comparing the settling times of three aqueous suspensions: (a) $BaTiO_3$, (b) $BaSO_4$, and (c) a mixture of these. Dilute suspensions (ca. 0.1% by weight of the 2 μm diameter size range) of the particles were prepared in aqueous 0.1% soluble potato starch which served as a protective colloid agent. The conductivity of the suspensions was purposely kept low (5×10^{-6} mho cm^{-1}). The settling rates were measured by comparing the change in optical transmission in a Lumetron colorimeter, Model 401, operated at 540 nm. The change in optical density with time was observed to be linear for the duration of the experiment and was observed to be d $(OD)/dt = 8.0$ for the $BaTiO_3$ suspension, 1.0 for the $BaSO_4$ suspension, and to be an intermediate value, 3.0, for the 1:1 mixture. If mutual clumping had occurred, one would have expected the 1:1 mixture to have settled faster than either pure suspension. We, therefore, conclude that mutual clumping did *not* occur. This simple result dictates against the correctness of the notion of simple charge transfer due Coehn-type triboelectric charging in aqueous media as the cause of the observed preferred attraction of cells for particles of high DK over that for ones of low DK. Recall that the static dielectric constants (DK) of $BaTiO_3$ and $BaSO_4$ are about 2000 and 11.5, respectively, while that for water is 78. We conclude that as argued earlier[5-7] the evidence for the observed μ–DEP of cells showing preference for high DK materials rather than for low DK materials is very strongly in favour of its being based upon oscillating (ac) fields and not upon static (dc) effects. No static effect is plausible.

At this point it is essential to apply the concept of the "effective dielectric constant" as developed in both theoretical and experimental studies of DEP[8]. By this is meant the idea that one must consider the overall effect of *all* mobile charges in the system during ac measurements. Thus, the conductivity *and* the dielectric constant contribute to the effective dielectric constant at any particular frequency. Failure to carefully include both factors can lead to unwanted consequences and misunderstandings. Without going into great detail (see however ref. 8) we can put the point that for many cases of a body suspended in a real fluid medium, the DEP force depends upon the difference of the effective dielectric constant of the body from that of the surrounding medium. The DEP force in a specific case depends, of course, upon the volume of the body, and upon the gradient of the square of the field. For the moment, we need only concern ourselves with the dielectric factors. For simple cases the effective dielectric constant $|K_{\text{eff}}|$, depends upon the relative dielectric constant, K_i, the specific conductivity, σ_i, and the frequency, f, in the following way:

$$K_{\text{eff}} = [K_i^2 + (\sigma_i/2\pi\varepsilon_0 f)^2]^{1/2}$$

where ε_0 is the permittivity of free space.

The theoretical problems in real cases become quite complex, but these details need not concern us further at this point. Suffice it to say that for the moment the effective dielectric constant, K_{eff}, of a liquid such as H_2O can be changed at will over wide ranges by the addition of conductive salts. In this manner the relative polarizabilities and hence the DEP forces upon bodies such as cells or crystals can be widely varied. Thus, at a given high frequency (say, 1 MHz) although in pure water ($DK \approx 80$) a pure (insulating) crystal of $BaTiO_3$ has a higher effective static relative dielectric constant, $DK (\approx 2000)$, yet adding a trace of a salt (e.g. 1 mM) increases the effective DK (K_{eff}) on the aqueous medium to over 13,000. Said another way, the erstwhile high DK material, $BaTiO_3$ is attracted to the high field region by DEP while in pure water, but is repelled even when in very dilute aqueous salt solutions at this frequency. It experiences positive DEP in the pure H_2O and negative DEP in the very dilute salt solutions. This is a most useful fact.

We have used this fact to examine the natural rf oscillations from cells.[5,7] When murine ascites fibroblasts, for example, are in de-ionized M/4 sucrose, they show an extra gathering rate for the high DK materials ($BaTiO_3$, $NaNbO_3$) over that for the low DK materials ($BaSO_4$, SiO_2). The addition of 0.1 mM KCl or NaCl clearly suppress this preference. The preference disappears completely in 1 mM solution, supporting the ideas given above. Kinetic studies of the rate of accumulation of the 2 μm diameter particles shows that only a few minutes mixing time suffices for comparative rate studies.

The Time Dependence of μ-DEP:

Upon sampling freshly mixed suspensions of cells and test particles at various times, one observes that the μ-DEP reaches a maximum in only two or three minutes. The results for ascites mouse fibroblasts as they accumulate either $BaTiO_3$ or $BaSO_4$ particles are shown in Fig. 1. That for $NaNbO_3$ or SiO_2 particles is shown in Fig. 2. The difference in n/p, the gathering factor, the ratio of the number of test particles associated with a cell to the number of particles lying free in a fixed volume (8 grid squares $\times 12.5$ μm $\times 12.5$ μm $\times 100$ μm $= 1.25 \times 10^5$ μm³) is seen to rise to a maximum and then fall. The peak height at the maximum is about 1.5 fold higher for the high DK materials than for the low DK materials. The difference then disappears after a rather longer time. We can understand this as being due to the eventual failure of the energetic processes causing the natural rf oscillations as these mammalian cells sit in the very restricted confines of the small glass microslide. The changes in n/p with time of yeast cells (*Saccharomyces cerevisiae*) as they accumulate either DPIA polymer particles or SiO_2 particles, in contrast (Fig. 3) do not show a marked fall off with time indicating that these cells perhaps more stable under these conditions. The n/p values for the high DK polymer are about twice those for the SiO_2 in this case. We can draw several conclusions from these data on mouse and yeast cells. First, both cell types, the ascites tumor cells and the yeast cells are actively propagating and also show a preference for the high DK material over that for the low DK materials. This confirms earlier work[5,7] on this aspect. Second, we can choose an experimental sampling time of say 3 minutes as a reasonable interval following the mixing of the cells the their test powders in further comparison tests of the μ-DEP on cells.

The Effects of Ionic Conductivity upon μ-DEP:

As noted earlier, one would expect the conductivity and hence the

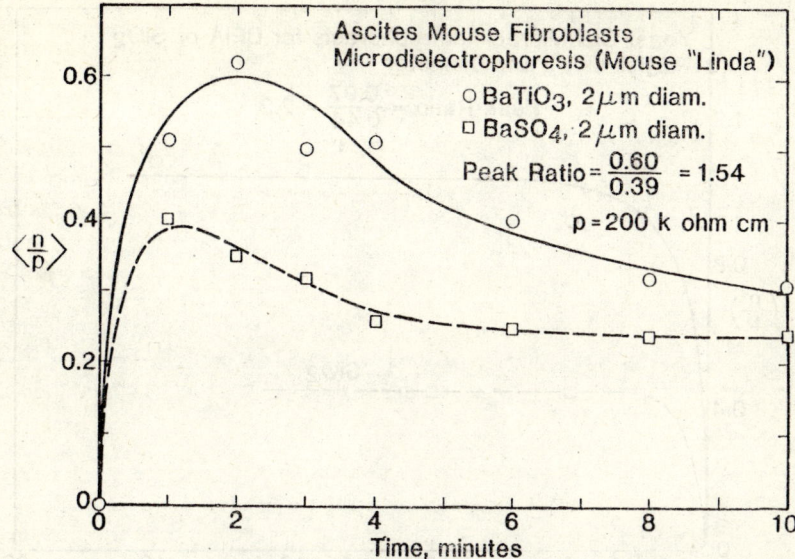

Fig. 1. The time dependence of micro-dielectrophoresis. The ratio, n/p, of the number of test particles associated with each cell compared to the concentration of free particles immediately surrounding the cell is shown for $BaTiO_3$ (circles) and for $BaSO_4$ (squares) as a function of time. Ths cells are ascites mouse fibroblasts. The peak ratio of the n/p values is 1.54.

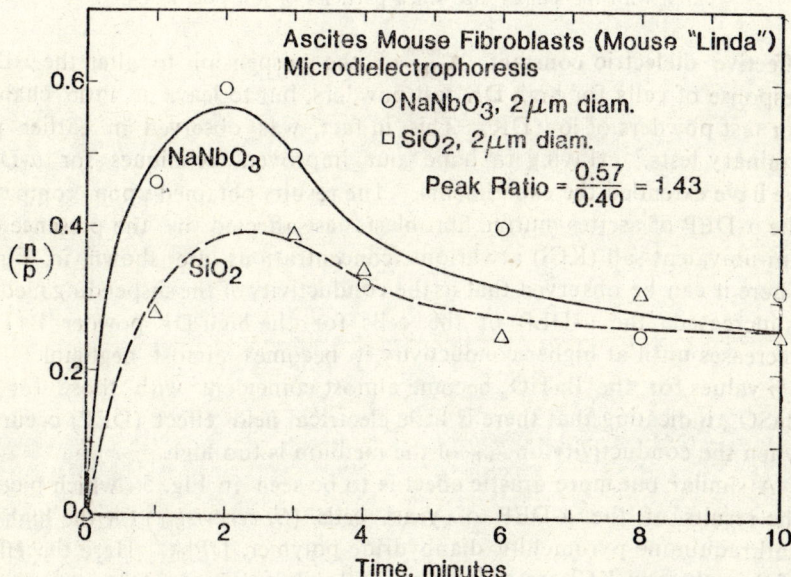

Fig 2. The time dependence of µ-DEP. Ascites mouse fibroblasts. Test particles: $NaNbO_3$ (circles) and SiO_2 (squares). The peak ratio is 1.43.

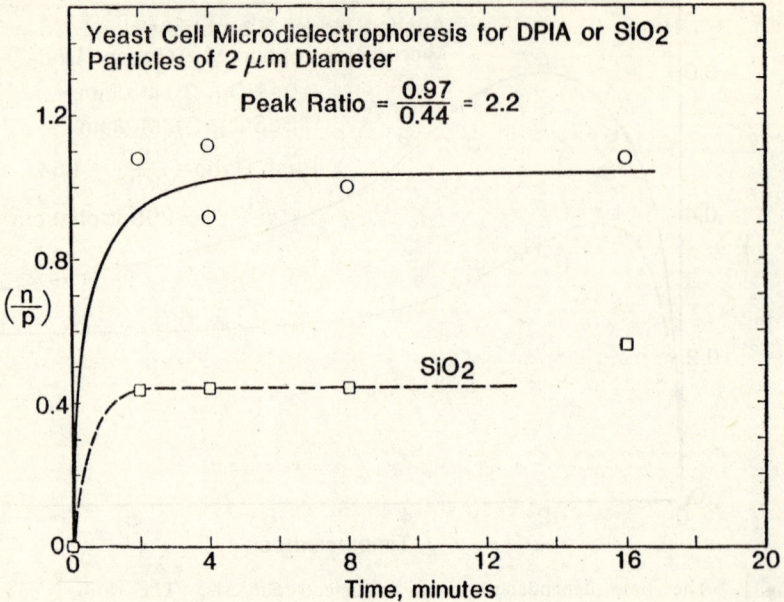

Fig. 3. The micro-DEP of yeast cells (*S. cerevisiae*) for 2 micrometer test particles of the high DK organic polymer, DPIA (a copolymer of anthraquinone and pyromellitic dianhydride) (circles) and for SiO_2, squares shown as a function of time. The peak ratio of the n/p values is 2.2, showing that the highly polarizable polymer particles are pulled preferentially to the cells when compared with the similar size silica particles of low polarizability.

effective dielectric constant, K_{eff}, of the suspension to alter the μ-DEP response of cells for high DK test powders, but to leave it little changed for test powders of low DK. This, in fact, was observed in earlier preliminary tests.[7] Having in hand our improved techniques for μ-DEP, we have extended the experiments. The results obtained upon comparing the μ-DEP of ascites murine fibroblasts as affected by the presence of a uni-univalent salt (KCl) at various concentrations are shown in Fig. 4. There it can be observed that as the conductivity of the suspending medium is increased, the μ-DEP of the cells for the high DK powder, $BaTiO_3$, decreases until at higher conductivity it becomes almost negligible. The n/p values for the $BaTiO_3$ become almost coincident with those for the $BaSO_4$, indicating that there is little electrical field effect (DEP) occurring when the conductivity or K_{eff} of the medium is too high.

A similar but more drastic effect is to be seen in Fig. 5, which pictures the results of the μ-DEP of yeast cells (*S. cerevisiae*) for the high DK anthraquinone pyromellitic dianhydride polymer, DPIA. Here the effects of the salts and KCl are seen to be similar but to cause the suppression of μ-DEP even at very low conductivities. This can be understood in terms of the lower dielectric constant of the polymer compared to that of

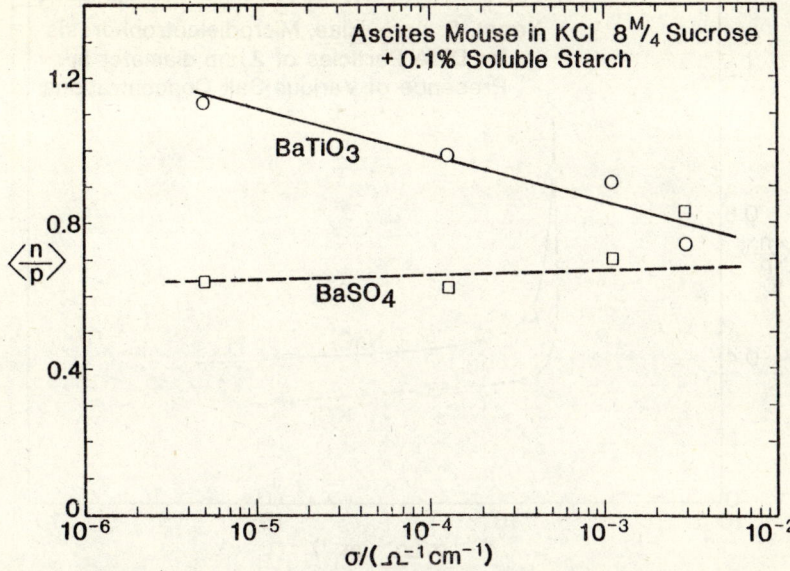

Fig. 4. The micro-DEP of ascites mouse fibroblasts for $BaTiO_3$ and $BaSO_4$ as affected by the conductivity of the suspending medium. Here the presence of added KCl is used to increase the conductivity. The mouse cells are suspended in M/4 sucrose containing 0.1% soluble starch. Note that at the higher conductivity range the preference of the cells for the high polarizability particles is lost.

the $BaTiO_3$, especially the probable high frequency of the cellular rf oscillation.

The Effect of a dc Pulse on the μ-DEP of Cells:

Having in mind a model for the cellular rf oscillations based upon the presence of coherent and collimated oscillating charge waves in the cell interior, it seemed worth asking the question. Were these oscillating charge waves completely coherent or only incompletely coherent? Were they completely collimated or only partly so? If the charge waves were only partially collimated or partially coherent among the internal oscillating regions, then possibly the application of a strong external dc field pulse might serve, at least temporarily, to increase either the coherence or collimation of the charge wave regions within the cell. With this in mind, we examined the μ-DEP of murine ascites fibroblasts before and after the application of a strong dc pulse. Ascites cells were rinsed in $\overline{M}/4$ sucrose until the suspension resistivity was 200 to 230 kOhm-cm with the aid of centrifugal separation. To this suspension was added the test powder suspension. Three min. thereafter, the suspension was then drawn into a micro-slide 15 mm long, with an internal height of 400 μm within which two parallel Pt wires, 75 μm in diameter and spaced

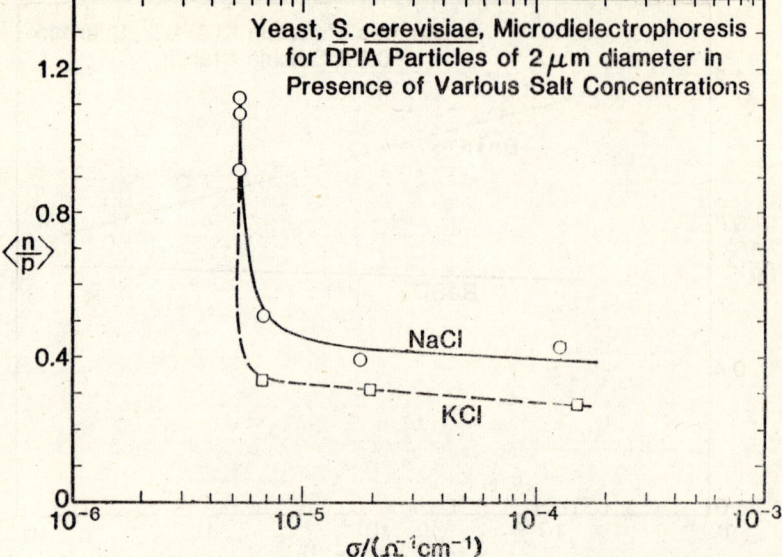

Fig. 5. The micro-DEP of yeast (*S. cerevisiae*) for the high DK organic polymer, DPIA in the presence of various simple salts; NaCl (circles); KCl (squares). The micro-Dep, as shown by the n/p values drops radically in the presence of even low suspension conductivities to values typical of those of the particles of low polarizability (cf. Fig. 3).

400 μm appart were affixed. The Pt wires served as electrodes. A capacitor (5000 μμf) previously charged at 300 V dc, was then discharged through the electrodes. This provided the dc pulse to treat the cells. A count was then made of n/p, the gathering rate, for the test particles. The results are shown in Table I. Recalling that in the phenomenon of dielectrophoresis (DEP) the more polar material tends to collect in the region of the strongest electric electric field, we would then expect to have high DK materials such as $BaTiO_3$ or $NaNbO_3$ be attracted (in preference to water) and low DK materials such as $BaSO_4$ or SiO_2 to be repelled from the region of strong field. We would expect, in other words, positive and negative dielectrophoresis in these respective cases. The ratios calculated from the data and shown below Table II tend to confirm this expectation. It would appear then that a strong dc pulse can increase the μ-DEP of these ascites cells, and thus indicate that the internal charge waves of these cells can have their collimation or coherence increased, at least temporarily.

The μ-DEP of Living and Dead Cells:

In the following experiments the yeast cells from a 7 hr. old culture (Difco/Sabaurad broth) were exposed to various chemicals for 30 min, then rinsed free with deionized water with the aid of four successive centrifugations and dilutions. The final suspension resistivity was 200 to

Table I. The effect of pulsing with a dc field upon the μ-DEP of murine ascites fibroblasts

Experiment	Pulse Applied	n/p	σ, std. dev'n
BaTiO$_3$	yes	1.22	.35
BaTiO$_3$	no	0.90	0.13
BaSO$_4$	yes	0.60	0.11
BaSO$_4$	no	0.60	0.02
NaNbO$_3$	yes	1.08	≈ 0.2
NaNbO$_3$	no	1.07	≈ 0.2
SiO$_2$	yes	0.41	≈ 0.06
SiO$_2$	no	0.63	≈ 0.06

Calc. of Corrected ratios:

$$\frac{(n/p)_{\text{Ti. pulsed}} - (n/p)_{\text{S. pulsed}}}{(n/p)_{\text{Ti. unpulsed}} - (n/p)_{\text{S. unpulsed}}} = \frac{1.22 - 0.60}{0.90 - 0.60} = 2.07$$

$$\frac{(n/p)_{\text{Nb. pulsed}} - (n/p)_{\text{Si. pulsed}}}{(n/p)_{\text{Nb. unpulsed}} - (n/p)_{\text{Si. unpulsed}}} = \frac{1.08 - .41}{1.07 - .63} = 1.52$$

250 kOhm-cm in each case. To it was then added a small quantity of the test powders (2 μm BaTiO$_3$ or BaSO$_4$) which were mixed, let stand, and then sampled with a 100 μm path-lenth microslide, which was then placed flat on a microscope slide and examined at 430 X. The n/p values observed are shown in Table II, as are the percent of viable cells determined using the following rapid assay.

A Rapid Vital Stain Procedure:

In this procedure, as in earlier dry film techniques[13] using neutral red, the suspension of cells is contacted with a dry film of dye. We have preferred to use methylene blue because of its usually low toxicity to microorganisms, and because it undergoes a very strong color change with change in state of oxidation. Methylene blue (Basic Blue 9, Matheson Coleman, and BE 11, F.W. 373.92, U.S.P.) was used. A solution of 60 mg Me Blue in 60 ml neutral 95% ethanol (pH 6–7 on test paper) was flooded onto scupulously clean microscope slides using a carefully and previously cleaned camel's hair brush, and let dry. This formed a thin film (ca. 2.5 cm × 2.5 cm) of dry Me Blue on the central portion of each slide. The test procedure consists simply of placing a drop of the cellular suspension to be tested onto the dry dye region, and rapidly (within 3 to 5 sec) covering with a cover slip. The "dead" cells rapidly develop a deep blue to blue-black color in 3.6 min. More functionally able cells remain leuco in color. The test is rapid and test slides can be prepared in large number

Table II. The effect of heat and various toxic materials upon the µ-DEP of yeast (*S. cerevisiae*)

Test particle	Yeast Exposure	n/p	% live by Me Blue
BaTiO$_3$	control	.092	98
BaSO$_4$	control	0.024	98
BaTiO$_3$	Heat: 18 min. at 65°C	0.024	0
BaSO$_4$	Heat: 18 min. at 65°C	0.024	0
BaTiO$_3$	0.1% phenol, 30 min.	0.025	15
BaSO$_4$	0.1% phenol, 30 min.	0.022	15
BaTiO$_3$	84% i-propanol, 30 min.	0.024	9
BaSO$_4$	84% i-propanol, 30 min.	0.020	
BaTiO$_3$	1:1 volume Gram's iodine with suspension	0.028	2
BaTiO$_3$	7.6 m$\bar{\text{M}}$ NaN$_3$	0.029	25
BaTiO$_3$	7.6 m$\bar{\text{M}}$ NaCN	0.10	80

well in advance of the vital stain test. In the case of yeast especially, the Me Blue test gave more visible staining than the neutral red.

The Variation of µ-DEP during the Cell Life Cycle:

To help answer the question, "What is the function of the natural *rf* oscillations of cells?" Or "Are these *rf* oscillations frill or necessity, effect or cause?", we have sought to find out where in the cell cycle they are most prominent. We have had a clue from prior studies that they are associated with the process of cell division.[5-7] We have observed, using µ-DEP, a preference of cells for high DK particles in the following cases: (a) synchronized early phase *Bacillus cereus*,[14] (b) yeast (*Saccharomyces cerevisiae*),[2,3] (c) rapidly dividing fetal mouse fibroblasts,[7,14] (d) rapidly dividing mouse "L" fibroblasts, and (e) mouse ascites tumor cells.[5-7,14] On the other hand, normal non-preferential pickup of cells for either high or low DK particles was observed for non-dividing cells, e.g., confluent mouse "L" fibroblasts, or the (enucleate) mouse erythrocytes.[5]

We have extended and confirmed earlier results on *B. cerus*. Cell preparation was similar to that for *B. megatherium* developed by Hunter-Szybalski *et al*.[15,16] Synchrony of the growth was obtained by chilling the logarithmically growing cells for 30 min. to 15°C, then returning them to a constant 34°C, which is an optimum temperature for growth. A 40 min lag, in the case of *B. megatherium* is followed by several cycles of usual length corresponding to the normal generation time, about 30 to 32 min. Each cycle, according to the optical density curves we obtain with *B. cereus*, appears to be comprised of a sudden duplication in numbers followed by a relatively stationary period. Cold appears to arrest division or perhaps to accumulate the nuclei or nucleoli in a condensed state, which in the phased population seems to appear only immediately following division. Pure stock culture was made from streak plates grown on nutrient agar slants. Broth cultures were grown in nutrient broth (Difco). Growth curves obtained on the latter showed the expected growth synchrony population curves as judged by optical density. The results for the bacterial μ-DEP tests are shown in Table III.

Table III. The micro-dielectrophoresis of *B. cereus* for $BaTiO_3$ or $BaSO_4$ particles of 2 μm diameter, on comparing the doubling and stationary phases in synchronous culture

Test Powder	n/p, cells in stationary mitotic phase	n/p, cells in cytokinesis, post-mitotic phase	Ratio, $(n/p)_{mit} : (n/p)_{cytokin}$
$BaTiO_3$	0.37±0.07	0.16±0.01	2.3
$BaSO_4$	0.18±0.06	0.17±0.03	1.1

As can be seen from these results, the cells in the stationary phase, presumably those containing a high fraction of their population in the mitotic phase, do show a preference for the high DK particles. This indicates that an electrical field is present about the mitotically active bacterial cells.

Perhaps the most informative system to study the relation between μ-DEP and life cycle phase is the budding yeast, *S. cerevisiae*. In this organism the morphology is distinctive at each phase[17,18,19] of the cell cycle. We have determined the μ-DEP of the yeast cells for high and low DK powders at four phases of the cells cycle; viz.: "Single cell", diameter 5 to 9 μm; "early-budded", with a bud less than 1/3 cell width; "late-budded", with bud diameter greater than 1/3 and less than 2/3 cell diameter; "doubles", with two cells joined. For graphical purposes, these were somewhat arbitrarily assigned a numerical life cycle "fraction" of 0.15, 0.4, 0.6, and 0.8. The results shown in Figs. 6 and 7 are based upon usually quadruplet runs, and 40 counts at each cell phase type in each run.

In Fig. 6 is shown a plot of the data obtained directly for $(n/p)_{average}$ for the two high DK powders, $BaTiO_3$ and $NaNbO_3$ and for the low DK powder, $BaSO_4$. The results for the two highly polarizable test powders are very similar. This demonstrates the strong preference of cells for high DK materials, especially during the cell phase in which the mitotic spindle is present. It will be noticed that there is a low peak in the n/p curve for

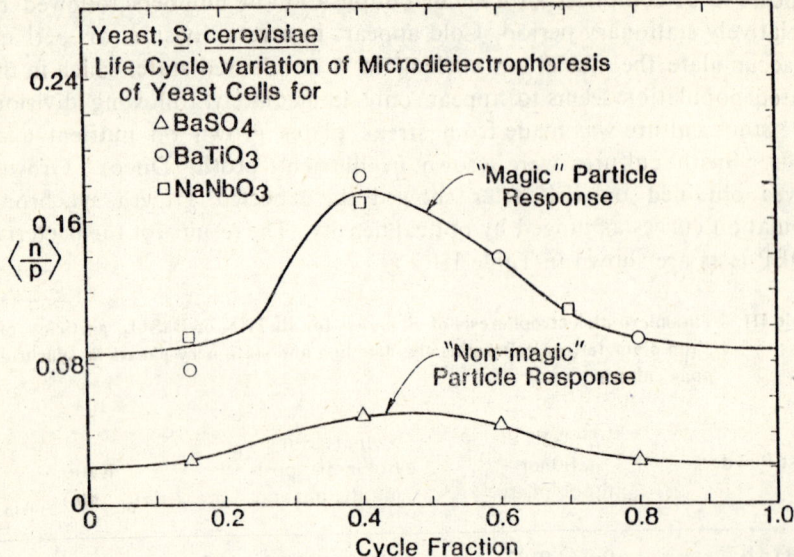

Fig. 6. Life cycle variations of the micro-DEP of yeast cells (*S. cerevisiae*) for BaTiO$_3$ (circles); NaNbO$_3$ (squares), both powders of very high polarizability; and for BaSO$_4$ (triangles); a material of low polarizability. The peak in the micro-DEP is seen to occur at the cell cycle phase corresponding to the appearance of the small bud. This places the maximum in n/p, the micro-DEP, near the occurrence of mitosis.

the low *DK* BaSO$_4$ powder. This is due to a combination of changes in the cell surface and size with cell cycle phase.

The association of particles with a given cell can be expected to depend, in part, upon the available surface area of that cell. To explore this factor we measured the diameters of many cells in the various morphologies. Many of the cells were ovoid, of major and minor diameters a_i and b_i. Since the area of a prolate spheroid is proportional approximately to the product of a_ib_i, we approximated the total area of a single cell by this product, and that for a budded or doubled cell by the sum of the measured diameter products, $\approx \sum_i a_ib_i$. These values are shown in Table IV, along with the term, $\Delta(n/p)/\sum(a_ib_i)$ which is the difference in the n/p values for the high and low *DK* powders, divided by the area factor, $\sum(a_ib_i)$ to correct roughly for the cell areas at different morphologies. As can be seen from the values in the last column, the "corrected" values still peak at the stage of cell cycle growth of the small-budded cells, i.e. in the region where mitosis is present. We can understand this result in terms of considering the *rf* oscillations of the cells being a maximum during the mitotic stage. A plot of the "corrected" $\Delta(n/p)/\sum(a_ib_i)$ versus the nominal cell cycle fraction is shown in Fig. 7.

Table IV. Micro-Dielectrophoresis Events During the Life Cycle of Yeast (S. cerevisiae)

Nominal life cycle fraction	Morphology	2 μm Test particle type	(n/p)	σ_{n-1}	No. runs	$\Delta(n/p)$ High DK less low	$a_1 \times b_1$	$a_2 \times b_2$	$\Sigma a_i b_i$	$\Delta(n/p)/\Sigma$ ($\times 10^3$)
0.15	single cell	BaSO$_4$	0.025	0.008	8	—	20.4±6	—	20.4	—
0.4	small budded	BaSO$_4$	0.050	0.031	4	—	25.3±9	2.3	27.5	—
0.6	medium budded	BaSO$_4$	0.045	0.021	4	—	29.8±5	9.0	38.8	—
0.8	double cell	BaSO$_4$	0.025	0.014	3	—	29.5±6	15.6	22.6 (av.)	—
0.15	single cell	BaTiO$_3$	0.076	0.013	11	0.051	20.4	—	20.4	2.50
0.4	small budded	BaTiO$_3$	0.185	0.039	4	0.135	25.3	2.3	27.5	4.89
0.6	medium budded	BaTiO$_3$	0.139	0.019	6	0.094	29.8	9.0	38.8	2.42
0.8	double budded	BaTiO$_3$	0.093	0.010	5	0.068	29.5	15.6	22.6 (av.)	3.01
0.15	single cell	NaNbO$_3$	0.095	0.015	8	0.070	20.4	—	20.4	3.43
0.4	small budded	NaNbO$_3$	0.170	0.028	4	0.120	25.3	2.3	27.5	4.36
0.6	medium budded	NaNbO$_3$	—	—	—	—	—	—	—	—
0.7	double budded	NaNbO$_3$	0.110	0.025	4	0.085	29.5	15.6	22.6 (av.)	3.8

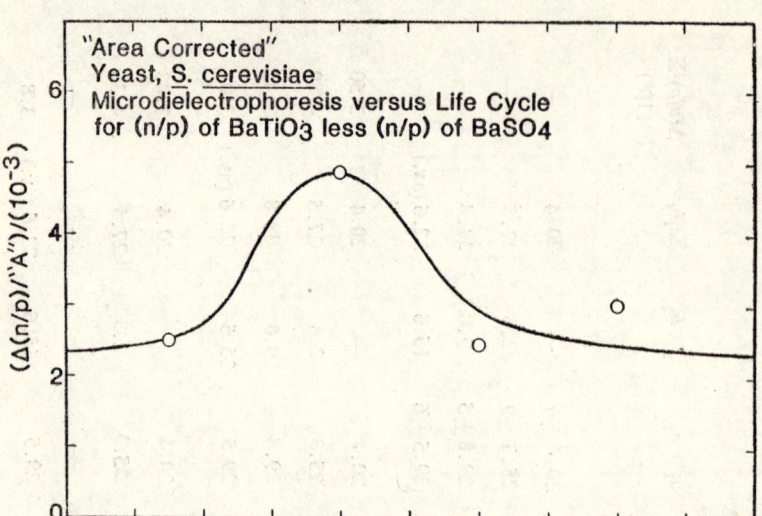

Fig. 7. The life cycle variation of the micro-DEP of yeast cells for the high dielectric constant powder, $BaTiO_3$. In this plot an attempt has been made to correct for the effective area of the several cell morphologies. The n/p values were divided by a factor proportional to the cell surface areas as described in the text.

A Model for the Origins of the Natural rf Oscillations of Cells:

The presence of periodic reactions in living system is well known[18, 20–22] and has been the subject of increasing study. It is important to appreciate that periodic reactions can be periodic in both time and *space*. This is clearly evident in the studies of Winfree,[23] for example, and has been examined from the theoretical viewpoint by Schmidt and Ortoleva[24]. There it is pointed out that such space-varying reactions can produce charge waves. This is a logical outcome if one recalls that in the several oscillating reactions studied, there is one phase of the reaction alternation dominated by ionic processes and another dominated by free radial branching. It is reasonable to assume, therefore, that the observed natural rf oscillations of the electrical fields about cells are to be associated with the presence of periodic charge waves in the cell. To this picture of periodic charge waves arising from a multiplicity of regions within the cell, we add the realization that the cell interior is not simple, but interrupted by various structures such as the endoplasmic reticular sheets, the laminae of mitochondria or chloroplasts, and the mitotic spindle apparatus. A reaction charge wave spreading out from some point in the cell would be restricted in its spread by the presence of such structures and tend to be collimated thereby, and be given a preferred direction of spread. Should these partially collimated charge waves, originally moving in a relatively incoherent fashion, become even partially coherent because of interaction of their electric fields, then an overall collimated and coherent electrical oscillation would arise. This, in our model, is the source of the natural rf oscillations.

In Fig. 8 we show diagrammatically the stages in the development of such a set of collimated and coherent charge waves capable of generating

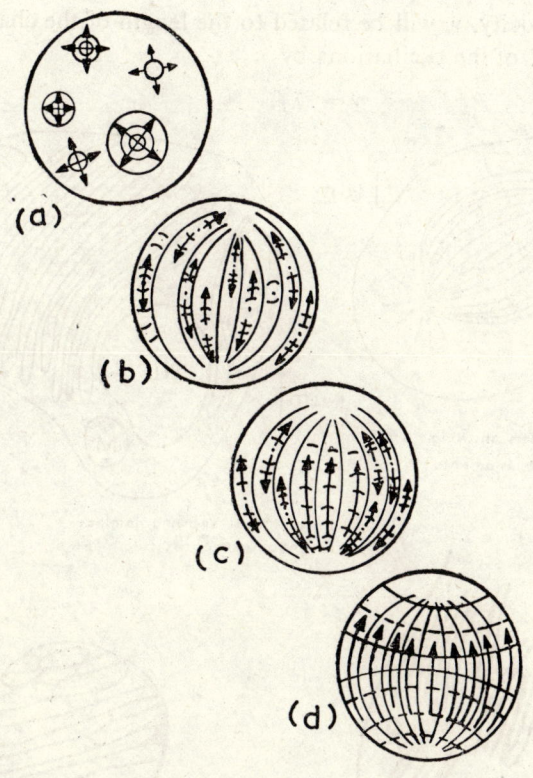

Fig. 8. A model for the origins of the natural rf oscillations of cells. The oscillating reaction, in near steady-state conditions, is presumed to produce cyclic alterations in ion concentration. These alterations produce charge waves if the diffusional speeds of the positive and negative ions are unequal. In (a), a number of charge waves are pictured as arising within a cellular volume at random times and points. In (b) the presence of some cellular structures, e.g. the mitotic spindle apparatus, the endoplasmic reticulum layer structures, or layer structures such as the granae of chloroplasts or the cristae of mitochondria bring about collimation of these charge waves. In (c) some of the charge waves are picture as partly coherent. In (d) the charge waves are both collimated and coherent. In such a circumstance, there would exist an oscillating *ac* field about the cell.

a natural rf oscillation, and an oscillating electrical field which might be sensed for short distances outside the cell. Its dipole nature ensures that it will not be strong far away from the cell. In Fig. 9 we show diagrammatically several layered or parallelized structures known to be present in cells, and which suggest themselves as natural collimators of such reaction charge waves. Indeed, it is interesting to consider the estimate of the charge wave velocity which can be made from the known dimensions of the several types of parallelized structures mentioned above. The charge wave velocity, v, will be related to the length of the channel, λ, and the frequency, f, of the oscillations by

$$v = 2\lambda f.$$

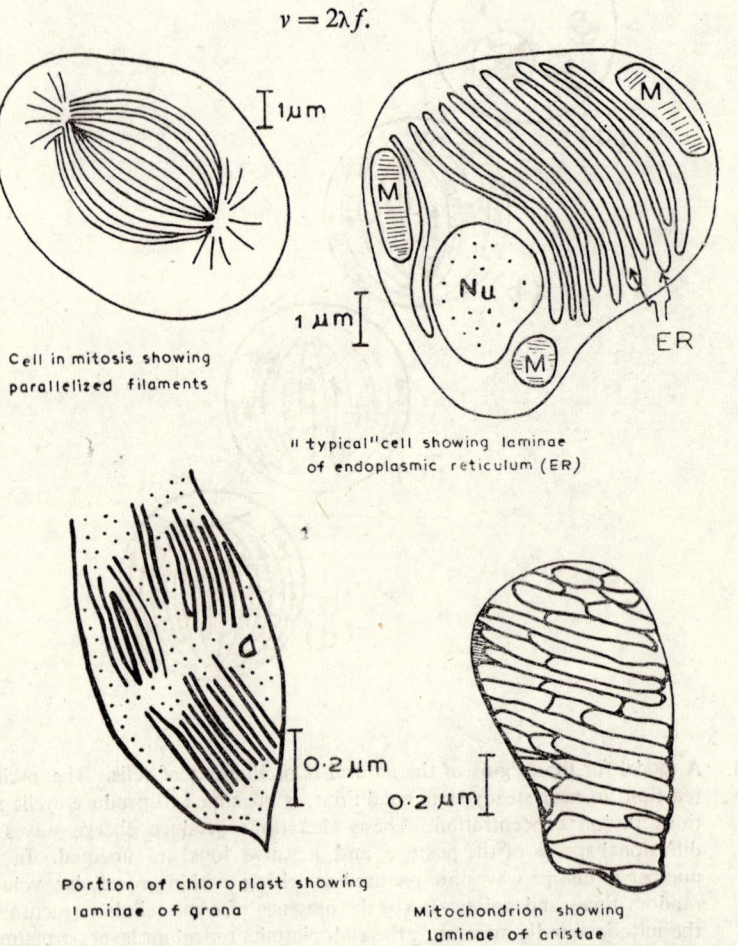

Fig. 9. Drawings of cellular structures which have a lamellar or parallelized morphology. (a) Cell in mitosis showing parallelized filaments. (b) "Typical" cell showing laminae of endoplasmic reticulum (ER). (c) Portion of a chloroplast showing laminae of grana. (d) Mitochondrion showing laminae of cristae.

From cellular spin resonance experiments on yeast cells, one sees narrow peak responses at 80 kHz and at 2 MHz, with a suggestion of a peak at about 200 kHz. Provisionally assigning these frequencies as in the table below, one observes that the velocity of the postulated charge wave would appear to be relatively constant, at about 80 cm s^{-1}.

System	Frequency, Hz	λ, μ m	Charge wave velocity, v cm/sec.
Mitochondrial laminae of cristae	2×10^6	≈ 0.2	80
Endoplasmic reticular laminae	2×10^5	≈ 2	80
Mitotic figure	8×10^4	≈ 5	80

In our model, we view the causative oscillating reaction to be capable of arising at numerous points in the cell. In the absence of other constraints they would produce charge waves emanating from these points in a relatively random pattern both in time and space (Fig. 8a). The presence of parallelized structures in the cell (illustrated, for example, as the mitotic spindle apparatus in Fig. 8b, c, d) would serve to restrict the direction of the charge waves development and tend to collimate them as shown in Fig. 8b. Should several neighboring charge waves "lock in" and become coherent, (Fig. 8c) these separate oscillating system would tend to increase their coherence in a "Bose-like" condensation and end up having a least-energy configuration by maximizing their coherence. This would follow from the ideas of Fröhlich who developed a most stimulating and fruitful theory for loosely coupled oscillators applicable to biological systems.[1-4]

The present model would suggest that the application of an external dc field pulse could help to increase the coherence of those cells which had only partially coherent charge waves. The experiments on dc pulsing described above tend to support this, but do not provide a proof, for it is possible that the momentary dc pulse only increased temporarily the mutual dielectrophoresis of the cells and the high DK test particles. Further study of this point is needed, such as by comparison of the pulsed field effect at various phases of the cell cycle, or upon the degree of the presence of laminar structures in cells.

Fröhlich[1-4] has pointed out that insight into the reasons that biological systems form cooperative phenomena may be had by considering a model for the assembly of weakly coupled oscillators. In this model long range correlation resembles that of a gas of bosons forming an Einstein condensation, in which the boson particles "condense" into a single quantum state at low temperature. He suggested that electrical vibrations or oscillations may be excited coherently (i.e., in phase) in active biological materials that provide energy through metabolic processes. Under appropriate conditions, Fröhlich[1] suggested that a phenomenon quite like a Bose condensation might occur in a system containing numerous longitudinal electrical modes. He then demonstrated that if (chemically derived) energy were fed into those modes and then transferred into other degrees of freedom of the system, a metastable stationary state will be attained in which the energy of the modes is in a steady state, yet with the energy in the modes larger than that at equilibrium. Furthermore, this excess energy would be preferentially channeled into a single lowest mode, reminiscent of the Bose condensation, provided that the specific power level, S, exceeds a critical value, S_0. Under these circumstances the supply of energy of random nature is notal immediately thermalized, but is partly diverted into supporting a coherent electric wave in the system. This is just the view needed to understand the natural rf oscillations when considering our model of collimated and coherent charge waves associated with parallelized regions in the cell. In the Fröhlich model, a band of polarization waves, taken here as corresponding to the collimated charge waves in a cell in our model, has a frequency associated with the individual electric dipole oscillators, ω_0. Long range (weak) Coulomb interaction between these individual oscillators will give rise to a branch of electrical modes with a frequency range

$$\omega_1 \leqslant \omega \leqslant \omega_z$$

which may be shifted from ω_0, but is narrow. The modes of this branch describe electrical waves in the system as a whole. Oscillations of individual units can then be described as appropriate superpositions of these waves. The band of polarization waves having a frequency ω_k ($k=1, 2, 3...Z$) emits quanta $\hbar\omega_k$ to the heat bath. At equilibrium, the number of states is described by the Planck distribution,

$$n_k = \frac{1}{[\exp(\hbar\omega_k - \mu)/kT - 1]}$$

for a power input S_k to the kth mode, the population changes as

$$dn/dt = S_k - \theta(T)(n_k[\exp(\hbar\omega_k)/kT] - n_k - 1) + \chi(...) + \text{etc.},$$

i.e. dn/dt=input power less the one and two phonon, etc losses. Fröhlich finds that if $S > S_0$; $\hbar\omega_1 \to \mu$, the chemical potential. Then,

$$n_1 = \frac{(\text{Const.})}{[\exp(\hbar\omega_1 - \mu)/kT - 1]} \to \infty,$$

in a Bose manner, and one has a "condensation" to the lowest mode, $\hbar\omega_1$. There is a well-known physical analog of this behavior. A certain critical level of power input, S_0, is required for a laser to operate in a steady state. In the present case, that for the natural rf oscillations of cells, the molecular nature of the oscillators is as yet not known. Judging from the high speed implied by the earlier analysis from the observed frequencies and the dimensions of the probable cellular laminar or other reaction collimating components, about 80 cm sec^{-1}, the reaction waves might be due to any of several mechanisms. Since it is doubtful that direct ionic diffusion is capable of such high speeds in water, one must consider other mechanisms such as thermally activated (polaron) tunnelling, or possibly Grotthuss-type[25] conduction, involving the displacement of ions in the ionic double layers of the cellular makeup. The theoretical problem, then, is of two facets. On the one hand, one needs to understand the possible origins of the electrical oscillators, and secondly, one needs to understand how these individual oscillators could be coupled in an organized state. At the moment, it would appear that collimated oscillating charge wave mechanisms could produce sources of dipole oscillators, and that the Fröhlich mechanism or a similar population theory of cooperative phenomena[26,27] could be understood to provide means for the "condensation" of the individual oscillators to a collective mode. Much remains to be done.

Conclusions

The presence of natural rf oscillations which produce oscillating electrical fields about cells is evidenced by the selective attraction shown by lone cells for test powders of high dielectric constant. This phenomenon is readily detected by direct microscopic examination under conditions permitting the action. We can understand the phenomenon in terms of a process of micro-dielectrophoresis, wherein the nonuniform electric field about a cell can cause a small but appreciable force upon nearby polarizable but neutral matter. The μ-DEP, it is to be emphasized, is a differential process, one acting upon both the suspended particles and upon the suspending medium, and, therefore, depends upon differences in the volume polarization of the particles and the medium. Said another way, the μ-DEP force on a particle depends upon the difference in the *effective dielectric constants* of the particle and medium. The effect can be suppressed by the simple addition of conductive salts to the medium, as we have shown.

A variety of test powders of high *DK* will serve to demonstrate μ-DEP of living cells, including $BaTiO_3$, $NaNbNO_3$, and the organic dielectric polymer, DPIA. The μ-DEP of cells is likewise absent (or even slightly negative) for test powders of low *DK* such as $BaSO_4$ or SiO_2. It is quite unlikely, therefore, that the observed effects are due to, say simple surface chemistry interactions, but rather are due to the postulated electrical ones, especially in view of the demonstrated ability of simple salts such as NaCl or KCl to suppress the observed μ-DEP in the manner and degree quantitatively predicted by dielectric theory.

The rate of μ-DEP of the yeast and mouse cells for test powders having nominal diameters of 2 μm is observed to be relatively rapid, allowing sampling of the cell-test powder mix after only 2 or 3 minutes. The μ-DEP of cells is clearly dependent upon their physiological state. Cells killed by heat or by poisons lose their ability to selectively attract high *DK* materials, and revert to behavior of cells towards low *DK* powder materials. Inhibitors such as NaN_3 or NaCN that act on the cytochrome system in oxidative phosphorylation and respiration may not suppress μ-DEP except in proportion to the percent of cells killed.

Perhaps the most interesting aspect of the μ-DEP of cells is that it is observed to be associated with the reproductive phase. We observe μ-DEP for test powders of high *DK* upon examining bacterial, yeast, and mammalian cells, hence it is probably universal among living cells. It is most particularly evident in cells that are in the reproductive cycle. We note that there is a preference of cells for high *DK* particles in the following cases: (a) mitotically active *Bacillus ccreus*, (b) mitotically active yeast cells, (c) rapidly dividing fetal mouse fibroblasts, (d) murine ascites tumor cells, and (e) rapidly dividing murine "L" fibroblasts. On the other hand, normal non-preferential pick-up by cells, e.g. confluent murine "L" fibroblasts, or for the (enucleate) mouse erythrocytes. A strong maximum in μ-DEP is observed during the life cycle of yeast cells. It occurs when mitotis is presumably present, at or near the production of the small bud. This can be understood in terms of our model which postulates the presence of oscillating chemical reactions that form charge waves and are first collimated and later become coherent to become gross electrical oscillations in parallelized domains within the cell.

At the present moment we do not feel that these natural rf oscillations are of high enough intensity to provide intercellular interactions of much size. On the other hand, it is relevant to ask if these electrical oscillations are important intracellularly. Do they represent cause or effect? Are they necessity or frill? If one looks at the fact that they are seen in such a wide variety of organisms as bacteria, yeast, and mammalian cells, then the implication must be that they are universal and have been carried on for billions of years. Since it is unlikely that life with its continually evolving changes would continue to carry along an unimportant or unnecessary process throughout such a long period, we are left to infer that the natural rf oscillations are in some sense necessary and important for cell replication. It remains to be found why and how these natural rf oscillations play an essential role in cellular replication. As we learn the answers to these questions we shall learn more about factors governing the replication of cells during tumor growth, somatic repair, and embryonic growth.

Acknowledgements

Support of this research by the National Science Foundation (Grant No. NSF PCM76-21467) is sincerely appreciated. Stimulus was given this research by discussions with Professor Albert Szent-Gyorgyi and with Professor Herbert Fröhlich.

The National Foundation for Cancer Research has also provided support for this research and this support is sincerely appreciated.

References

1. H. Fröhlich, Int. J. Quant. Chem. **2** (1968) 641.
2. H. Fröhlich, Phys. Lett. **26A** (1968) 402.
3. H. Fröhlich, Nature **228** (1970) 1093.
4. H. Fröhlich, Coop. Phenomena **1** (1973) 101.
5. H. A. Pohl, in BIOELECTROCHEMISTRY, pp. 273-95, edited by H. Keyzer and F. Gutmann, Plenum Press, NY (1979).
6. H. A. Pohl, J. Biological Physics 7 (1979) 1-16.
7. H. A. Pohl, Int. J. Quantum Chem. 7 (1980) 411-31.
8. H. A. Pohl, DIELECTROPHORESIS, Cambridge University Press, London, NY (1978).
9. H. A. Pohl, "Dielectrophoresis: Applications to the Characterization and Separation of Cells", in METHODS OF CELL SEPARATION, Vol. I, pp. 67-169, Plenum Press, NY (1978), edited by N. Catsimpoolas.
10. H. A. Pohl and J. R. Wyhof, J. Non-Crystalline Solids **1** (1972) 137.
11. A Coehn, Ann. Phys. **64** (1898) 217.
12. C. F. Gallo and W. L. Lama, J. Electrostatics **2** (1976) 145.
13. A. E. Galigher and E. N. Kozloff, ESSENTIALS OF PRACTICAL MICROSCOPE TECHNIQUE, Lea and Febiger, Phila. PA (1964) p. 62.
14. H. A. Pohl, "Natural rf oscillations from cells", J. Bioenergetics and Biomembranes (in press).
15. M. E. Hunter-Szybalska, W. Szybalski, and E. D. de Lemater, J. Bacteriol. **71** (1956) 17.
16. W. Szybalski and M. E. Szybalska, Bacteriol. Proc. (1955) 56.
17. L. H. Hartwell, J. Bacteriol. **104** (1970) 1280.
18. L. N. Edmunds, Jr. and K. J. Adams, Science **211** (1981) 1002.
19. R. W. Shulman, Chap. 3 in METHODS IN CELL BIOLOGY, Vol. 20 (1978), edited by D. M. Prescott, Academic Press, NY.

20. J.E. Treherne, W.A. Foster, and P.K. Schofield, "Cellular Oscillators" in J. Exper. Biol. **81** (1979) (review volume).
21. M. J. Berridge and P. E. Rapp, ibid. **81** (1979) 217.
22. P. E. Rapp, ibid. **81** (1979) 281.
23. A. T. Winfree, Sci. Amer. **230** (1974) 82.
24. S. Schmidt and P. Ortoleva, J. Chem. Phys. **71** (1979) 1010.
25. T. von Grotthuss (1805), cf. S. Glasstone, PHYSICAL CHEMISTRY, D. van Nostrand, NY (1946) pp. 887 and 898.
26. H. Haken, SYNERGETICS, Springer-Verlag, NY (1978).
27. H. Haken, COOPERATIVE PHENOMENA, Edited by H. Haken and M. Wagner, Springer-Verlag, NY (1973) pp. 363–72.

Solitons in Biology and Muscle Contraction

A. S. Davydov

Member of Ukrainian Academy of Science
Kiev, USSR

I should like to discuss some theoretical investigations of an energy transfer along alpha-helical protein. These investigations have been performed at ITP in Kiev by my coworkers and by myself.

The background problem was essentialy to understand the nonlinear dynamics of energy propagation along the alpha-helical protein. So, let me begin to orient a little in structure of alpha-helical protein molecules. All the protein molecules are formed by linear polymerization of 20 different aminoacids and represent the polypeptide chains with periodic repetition of peptide groups which contain the HNCO atoms.

The polypeptide chains can be in different spatial configurations among which the structure called alpha-helix is of special interest. In these molecules the polypeptide chains are coiled into long helices. Hydrogen bonds between the peptide groups (PG's) hold the helical structure. The PG's are situated along the three chains of hydrogen bonds at equal distances forming periodic structure.

It is known that the energy released in the hydrolysis of ATP molecules is universal unit used in a great number of energy transformations in biological systems.

In 1973 a conference devoted to an investigation of the mechanism of energy transport in biological systems was held in New York. The following three problems were discussed: (i) Is there a crisis in bioenergetics? (ii) What is the nature of the crisis? (iii) How can the crisis be resolved?

Taking into account the small energy (about half eV) released under the ATP hydrolysis some scientists (Mc Clare, Green) held that the energy of vibrations of carbon monoxide of PG is transferred along protein molecules. The quanta of energy of vibration $C = 0$ is equal to about two tenth eV.

Weber and others disputed this opinion asserting that due to very short life time (of the order of 10^{-12} s) of vibrational excitations of individual PG they can not take part in the excitation transfer. So the problem concerning the crisis in bioenergetics remained open.

Theoretical investigations conducted at the ITP were aimed to solve this problem. It was shown that the energy of the hydrolysis of ATP can be transfered without losses along alpha-helical molecules as special collective excited states—SOLITONS.

The quasi-periodic structure of an alpha-helical protein provides the collectivization of vibrational excitations of individual PG. It was shown that it is very important to take into account the displacement of equilibrium position of PG's along hydrogen bonds.

Nonlinear interaction between longitudinal displacements and vibrations of carbon monoxide of PG could give an excitation which are described by a set of differential equations.

For the simplicity now I shall consider only excitations of one chain. In this case the set of equations is reduced to an nonlinear Shrödinger equation which describes a motion of excitation along chain with a constant velocity (v).

$$\left\{i\hbar \frac{\partial}{\partial t} + \frac{\hbar^2}{2m} \frac{\partial^2}{\partial x^2} + G|\psi(x,t)|^2\right\}\psi(x,t) = 0 \qquad (1)$$

Here $G = 4\chi^2/H(1-s^2)$ is a nonlinearity coefficient; H is longitudinal elasticity of molecule. χ is coupling of $C=0$ vibration with displacement of the PG. $S = V/V_0$ is ratio of velocity excitation V to velocity V_0 of the longitudinal sound.

This equation describes two types of collective excitations:

Excitons and Solitons

(1) When $V > V_0$ we have only *excitons*. They are described by wave packets smearing with the time. Owing to the rapid motion of an exciton, the local deformation cannot follow it. Excitons carry along the molecule only energy of intrapeptide vibrations. Its energy may be written as

$$E_{ex}(V) = E_{ex}(0) + \tfrac{1}{2}m_{ex}V^2$$

where
$$m_{ex} = \hbar^2/2Ja^2$$

is the effective exciton mass.

(2) When excitation travels along the chain with velocity lower than that of longitudinal sound waves, the coupling with a deformation can take place. The deformation then follows the motion intrapeptide vibrations forming a solitary wave—*soliton*. The time evolution of the probability of intrapeptide vibrations along chain is given by bell-like function

$$|\psi(x,t)|^2 = \frac{a\mu}{2\cosh^2[\mu(x-x_0-Vt)]}, \quad \mu = \frac{\chi^2}{2HaJ(1-s^2)}$$

Solitons transfer the following energy along the chain

$$E_{sol}(V) = E_{sol}(0) + \tfrac{1}{2}m_{sol}V^2$$

where
$$E_{sol}(0) = E_{ex}(0) - \chi^4/3H^2 J$$
is the internal soliton energy.
$$m_{sol} = m_{ex} + \frac{4\chi^4(1 + \tfrac{3}{2}s^2 - \tfrac{1}{2}s^4)}{3H^2(1-s^2)JV_0^2},$$
is the effective soliton mass.
$$V_0 = a\sqrt{H/M}$$
is the velocity of longitudinal sound waves in the chain.

The probability of a soliton to be excited directly by light is very low since the excitation ought to produce a local deformation of the chain.

In real alpha-helical molecules there are three nearly parallel chains of PG's. As shown at ITP in this case three types of solitons can travel along the protein molecule.

The first type which we called symmetric solitons correspond to a synchronous transfer of local excitation along three chains of PG's. In this case the helix diameter in the excitation region increases and the distances between neighbouring PG decrease.

The second and third types of solitons correspond to the transfer of local excitation along three chains of PG with phase shifts. In this case in the region of excitation molecule is bend and its diameter increases.

I have discussed the analytical expressions which were obtained in the continuum approximation when discrete chains were replaced by continuous chain of an infinite length.

In 1979 Hyman, McLaughlin and Scott obtained the numerical solutions for our equations for discrete chain containing 200 PG's. They use the computer of Los Alamos scientific lab. The numerical calculations support the results have obtained analytically.

The numerical calculations were repeated by Chris Eilbeck and A. Scott using another program at Atas Computing Division Rutherford Laboratory of Heriot-Watt University in Edinburg and a film was produced which clearly *demonstrated* the high stability of solitons relative to acoustic noises appearing in the system.

So, *as was shown the energy of hydrolysis can transfer without the loss along alpha-helical protein in the form of solitons.*

In this part of my discussion I should like to consider a new hypothesis of the mechanism of muscle contraction.

At present the transversely striated muscles of vertebrates have received most attention. The contractive elements of muscle fibers are the bundles of parallel packed myofibrils as is well known Myofibrils are surrounded by the Sarcoplasm and the system of transverse and longitudinal tubules acting through which a nerve impulse, initiating a muscle contraction is transfered inside the fiber.

Each myofibril is divided into repeating segments, or sarcomeres, which are separated, or Z plates. Each sarcomere is *made up* of parallel packed filaments of two types: *thick* and *thin one*.

The thick filaments occupy the central part of a sarcomere. The thin filaments are attached to Z plates.

The electron-microscopic research showed that the contraction of the muscle fiber was due not to the change in length of thick and thin filaments but to the sliding of thin filaments relative to the thick filaments.

As can be visualised the thin filaments move in the space between thick filaments to the sarcomere center. So the length of each sarcomere, is shortened.

How to explain the mechanism of sliding at the molecular level? What forces cause such a sliding? And how is the energy, released in ATP hydrolysis, used?

At present there is a widespread opinion that the sliding of thin filaments relative to thick ones in contracting muscle is due to an active motion of the heads of myosin molecules from which the thick filaments are formed.

Myosin molecule is very big with molecular weight about 500 000. In the thick filament they are placed in such a way that their head directed towards both ends of the filaments and their tails to its middle part.

By the ATP hydrolysis the head of the myosin molecule lengthens and forms a link (bridge) with the thin fiber, and then turns, displacing the thin fiber relative to the thick one. Next, it detaches from the thin fiber, regaining the previous size and position in a thick filament.

Having joined the new ATP molecule, the head of myosin molecule can repeat the whole cycle again.

However, such concept does not elucidate the nature of the contraction. The questions remain:

How is energy, released in ATP hydrolysis, used in the lengthening of the head of myosin molecule, the cross-bridge formation, the pull force, and the cross-bridge breaking?

What is the molecular mechanism of changes at the myosin molecule head that leads to these phenomena?

And, finally, why does only the head of the huge myosin molecule take an active part in the sliding mechanism?

Using the theory of solitons, which correspond to intrapeptide vibration, I proposed a *new molecular explanation* of the mechanism of sliding of thin filaments relative to thick.

When a nerve impulse reaches the muscle fiber electric depolarization of its membrane occurs. As a result the Ca^{2+} ions are thrown out from the sarcoplasmic tubes into sarcoplasm which surrounds thin and thick filaments. The calcium ions, reaching by diffusion the heads of myosin molecules at the ends of thick filaments, initiate the process of hydrolysis of ATP molecules, which are at the heads of myosin molecules. In myosin molecules arise solitons which move from the heads of molecules, where they were formed, to their tails.

The motion of solitons is accompanied by a local swelling and bending of a molecule. Therefore in region of moving solitons the thick filament gets swollen.

Due to the swelling of thick filament the myosin molecule heads become closely attached to thin filaments. When swollen parts of thick filaments move towards the sarcomere centre they displace thin filaments.

According to this model the myosin molecule heads attach themselves to thin filaments, pull them at small distance and detach them, as in the model of cross-bridge formation. However, this movement is due not to the lengthening, turning and contraction of the heads themselves, but to the movement of solitons inside the thick fibre. In this model all parts of the myosin molecule, but not only its heads are active contractive elements.

Now I am going to a third part of my discusion. As you know an important role in bioenergetics is played by the processes connected with the electron transfer along protein molecules. The problem concerning the possibility of electron transfer from a donor to an acceptor molecule through the protein molecule has often been discussed in the literature,

Proteins are dielectric rather than semi-conductor. The forbidden band width of the *intrinsic electrons* of the protein molecule is of the order of 4-6 eV. But erlier one did not take into account that in α-helix protein molecules for an extra electron transferred to a protein molecule from a donor, the conduction band is generated by constant electrical dipoles of PG's.

Each PG in the protein molecule has a large constant dipole moment (3-4 debye). Turner, Anderson and Fox in 1968 showed that in the field of such a dipole the electron may be in a bound state with a binding energy about one eV.

The overlapping of the wave functions of the electron ground state in the neighbouring potential wells causes the collectivization of these states. The conduction band arises. The extra electron causes local displacements of equilibrium positions of PG's.

As was shown by myself in 1979 taking into account these displacements the motion of an external electron along the chain of PG's described by a nonlinear Schrödinger equation. Its exact solution shows that the *total energy of the electron and the deformation energy of the displacements* decrease relative to the *bottom energy of conduction band* by the value

$$\Delta E = m a^2 \sigma^4 / 12 H^2 \hbar^2$$

σ is the parameter of the deformation potential causing by the local displacement. m is the effective electron mass in the conduction band. H is the coefficient of the elasticity of the chain.

The probability of electron distribution in the chain is determined by a bell-like function

$$W(\xi) = aq/2 \cos \hbar^2(q\xi), \; \xi = x - x_0 - Vt.$$

The electron travelling along the chain together with a local deformation may be called an electrosoliton. Its effective mass

$$M_{sol} = m(1 + M\sigma^4/6H^2\hbar^2).$$

considerably exceed the effective electron mass in the conduction band.

The energy of electrosoliton

$$E_{sol}(V) = \xi_0 - \frac{ma^2\sigma^4}{12H^2\hbar^2} + \frac{1}{2} M_{sol} V^2, \; s^2 \ll 1$$

The electrosoliton can travel through a helical protein molecule only with a velocity V less than the velocity of a longitudinal sound V_0. Therefore, it does not lose its energy for emitting phonons.

Thus, owing to the autolocalization of an electron its motion along the alpha-helical protein molecule is stabilized and assumes a superconducting nature. This is a special type of superconductivity without pairing which is characteristic only of one-dimensional systems. Thus, alpha-helical protein molecules may be *ideal guides in the electron transfer from a donor to an acceptor molecule.*

In conclusion I should like to make some historical remarks.

The qualitative scientific description of solitary waves on the water surface in a narrow channel was first made by John Scott-Russell nearby at Edinburg more than 140 years ago. He observed the boat which was drawn along a channel by a pair of horses.

When the boat suddenly stoped a large solitary wave was formed, assuming the form of a large solitary elevation, a rounded, smooth and well-defined heap of water, a solitary wave rolled forward along the channel. "I followed it on horseback", Scott-Russell wrote, "and overtook it still rolling on apparently without change of form or dimination of speed". "Such, in the month of August 1834, was my first chance interview with that singular and beautiful phenomenon".

Only in 1895 Korteweg and de Vris obtained equations which describe solitary waves on the water surface in narrow chanel. In 1965 Kruskal and Misra in U.S.A. obtained numerical solutions for these equations and showed that solitary waves have properties like a quasiparticle. From this time the word "Soliton" is used in scientific literature.

Molecular and Global Perspectives on Life

Harold J. Morowitz

Yale University, New Haven, Conn. USA

The living state may be viewed from both a global and a local perspective. The latter approach is that of molecular biology and biochemistry while the former emerges from a consideration of ecology in its broadest aspects. This paper will first present the global approach in a somewhat more directed way than is conventional. Using this view as a framework the origin of the living state will be viewed as a problem in nonequilibrium physical chemistry.

We begin the global analysis with the geochemist's division of the earth's surface into lithosphere, hydrosphere, atmosphere and biosphere. These compartments are connected by fluxes, many of which are parts of major global cycles. The flows so interconnect the geospheres that the activity of any one can only be understood in terms of its relation to the others. It is not possible to study the processes in a single geosphere independently of the activities of the others.

By way of example consider the biosphere and the atmosphere. Carbon dioxide is taken up by plants and oxygen is released. Animals consume oxygen and release carbon dioxide. Bacteria both produce and consume methane. Nitrogen (N_2) is fixed to ammonia and the complex nitrogen cycle involves many forms of that atom in different oxidation states. The chemical composition of the atmosphere is regulated by biological processes, high altitude photochemistry, and meteorological mixing processes. In every way the biosphere and atmosphere are completely chemically interrelated. This relationship has been explored by Lovelock as the Gaia hypothesis (1).

A somewhat less well known example is the relation of the biosphere to the lithosphere. The leaching of minerals from the land areas by the hydrological cycle is strongly influenced by soil biota. The precipitation of minerals in the ocean sediment is controlled in part by marine organisms. The ocean floor is subducted in tectonic plate movement. Lithospheric cycles involving vulcanism and uplift are required to return the vital minerals to the biosphere. Thus there is flow and interchange between biosphere and lithosphere. A planet without tectonic activity could not maintain life as we know it since, without subduction and return, vital elements would be permanently lost to the biosphere (2).

What these cases and many others demonstrate is the concept that continuing life is a property of the entire planet rather than an isolated molecular feature of small subsystems. Although much of this discussion will deal with prebiotic chemistry it should be thought of within the framework of planetary interrelationships shown below.

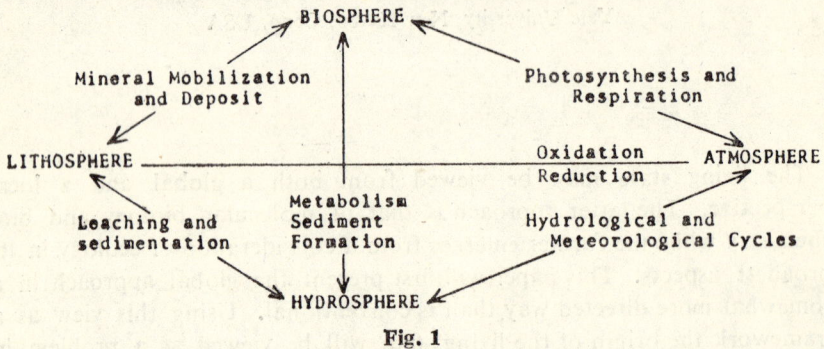

Fig. 1

A more specific relation among the geospheres is seen in the nitrogen cycle. The obvious role of the biosphere emerges in the fixation of N_2 to NH_3 and the many oxidations and reductions involving nitrite, nitrate, amines, uric acid, urea, nitrous oxide and others. The biosphere thus interacts with the atmosphere, dissolved nitrogen compounds in the waters, and nitrogen compounds in the soil and rocks. A more subtle kind of interaction shows up when we realize that nitrogenase and nitrogen oxidate, the two primary redox enzymes for nitrogen, are both molybdenum containing enzymes (3). There is thus a necessary association between N and Mo so that the geochemistry of molybdenum becomes an integral part of planetary ecology. Global geochemistry is related to biochemical process through the obligatory role of molybdenum atoms in the oxidation and reduction of nitrogen. The cause of the nitrogen molybdenum relationship is unknown, but it stands as an empirical generalization.

The nitrogen reactions are part of the great global cycles, which allow us to focus on the molecular macroscopic interactions. The cycling theorem (4) establishes, from very general considerations of nonequilibrium theory, that a system undergoing an energy flux will be subject to material cycles. If the inputs are of sufficiently high frequency for electronic transitions to accompany absorption, then the steady states of the systems will be characterized by a flow of matter from low energy states to higher chemical potentials and back to the ground states through available intermediates. The cycles must begin with molecular species present in high concentration when the system is allowed to decay to equilibrium in the absence of an energy flow. These cycling theorems come strictly from physics with no assumptions about living organisms. Therefore, the major global ecological cycles are not strictly biological phenomenon; we would expect similar kinds of cycling in the absence of biota. Any planet receiving radiation in the visible and ultraviolet would be expected to have chemical cycles consistent with its chemical composition, the gravitational force, and the temperature. The last two determine the atmospheric partition and the leak.

Given the size and density of the earth we are guaranteed an atmosphere. The chemical composition and temperature necessitated a hydrogen leak and required that the atmosphere became more oxidizing as the planet aged. Such a planet would have had major chemical cycles before the advent of life. The cycles are a dynamic set of boundary conditions within which life originates and evolves. Life is not a sudden isolated event but arose within a framework of planetary events. The cycles are the best understood of these processes because their existence and certain of their features can be predicted from quite standard physical considerations.

The ecological cycles are interactive with the hydrological, meteorological, and lithospheric cycles. From the very beginning all the geospheric fluxes were highly linked giving rise to a kind of total planetary activity.

In assessing the development of the present-day biosphere we must consider three phases: physics, biology and history. The physics stage is the domain that Schrödinger has described as "order from disorder" (5). It is the domain of thermodynamics, chemical kinetics, and ordinary macroscopic description. As such it is deterministic and subject to attempts at experimental verification. It covers the stage up to the encoding of genetic information in macromolecules. At that point we move into the domain which is dominated by the fact that microscopic thermal fluctuations give rise to macroscopic consequences; a single mutational event can transform an entire ecosystem.

The domain of thermal noise is clearly history and must be treated by strategy and game theory considerations (6). Between physics and history comes biology proper, the origin of the code. Here we simply do not know enough at the moment to say if the code is deterministic or arbitrary. The approach of choice is to study the problem of biogenesis in the deterministic domain of statistical physics. Only by pushing the deterministic aspect of the problem to its full limits can we assess where either chance takes over or new physical principles are called for.

A radically new approach to biogenesis has been emerging in recent years and its present status can be seen in *A Symposium in Print on Molecular Origins and Evolution of Photosynthesis* edited by H. Baltscheffsky (7), and *Light Transducing Membranes* edited by D. Deamer (8). What distinguishes the new viewpoint is the emphasis on bioenergetics as the unifying principle in biogenesis. Previous theories have stressed organic reaction schemes and the development of genetics and have assumed that there was plenty of energy available. Energy as an effective organizing factor was virtually ignored except to generally recognize the need of energy flux for all organization. The newly emerging view is more ecological at the outset and recognizes that life, whatever else it may be, is the interface between solar electromagnetic energy and terrestrial chemistry. We must either assume that the relationship is primary or preexisting life forms at some point shifted to using the sun's energy. The principle of continuity dictates that we give serious consideration to the primary nature of solar energy coupling.

In formulating a theory of biogenesis we utilize two main classes of information, the generalizations of molecular biology and the self organizing principles of physics. The rather satisfying theories of synergetics (9), dissipative structures (10) and energy flow (11) assure us that the flux of energy can and necessarily does organize systems that are intermediate between sources and sinks. For the specific kinds of organization that are called for we must look to the general features of living cells. In the past, people interested in biogenesis have focused on genetic mechanisms. Recent developments of the extended Mitchell hypothesis (12, 13, 14) reveal other ubiquitous features. The assumption we shall explore is that metabolic networks and energy processing devices occurred before genetic information storage; hence the character of the code and macromolecular system emerge from cellular energetics rather than vice versa.

In one other major respect this discussion will deviate from the more widely discussed views of biogenesis, which assume that solution phase chemistry preceded cellularity. If one, however, starts from bioenergetic considerations the transmembrane events become so crucial that it is necessary to postulate the existence of closed vesicles as a very early event in biogenesis. We argue for the membrane as the primary and most primitive biological structure. Although this seems obvious in many ways, we shall formalize the argument. There are two general types of structures in biology, macromolecular and membranous. The former are highly ordered and require an exact specification of monomeric sequence as well as constraints on secondary and tertiary structure. Membranes are by comparison rather unprecise structures. They vary in fatty acid composition, amounts of various polar lipids, and contents of dissolved non-polar components. Membranes are a phase system with orders of magnitude less requirement for specificity than macromolecules or aggregates of macromolecules. Membranes can form spontaneously; they are macroscopic, statistical structures reflecting order from disorder. If biogenesis moved from the statistical to the precise macromolecular domain, it is reasonable to assume that membranes occurred early and before macromolecules.

There are two main types of theories of biogenesis. One supposes that liquid phase chemistry led to metabolism, macromolecular synthesis, and coding followed by capture of these functions in a cell. The second viewpoint is that cellularity in terms of phase separation from the environment is the vital first event. This distinction is sharp and we are arguing that the second is more consistent with bioenergetics, theory of catalysis, and continuity of development. We turn to some general features of phase separation.

While the equilibrium theory of phase separation follows from the theory of Gibbs (15), a more molecular approach is possible in dealing with aqueous environments. The critical feature is polarizability, generally reflected in the dielectric constants of materials. The popular statement of the relevant principle is that oil and water don't mix. High dielectric constant is associated with large permanent dipole moment and high electronic polarizability along bonds or sequences of bonds. Mixtures of electrically asymmetric molecules will be energetically stabilized by electrostatic interactions while electrically symmetric molecules will destabilize the hydrogen-bonded structure of water. As a result, non-polar organic molecules tend to have low solubility in water and form a non-aqueous or oily phase.

The carbon-hydrogen and carbon-carbon bonds tend to be dominant features of low dielectric constant organic materials. In a very highly reducing environment methane, CH_4, will predominate as a gas phase molecule. As one moves toward less reducing environments, with the loss of hydrogen, there is a general tendency toward hydrocarbon formation.

$$CH_4 + CH_4 \rightarrow H_3C-CH_3 + H_2 \uparrow$$
$$H_3C-CH_3 + CH_4 \rightarrow H_3C-CH_2-CH_3 + H_2 \uparrow$$
$$\text{etc.}$$

Hydrocarbons lead to a two phase oil-water system.

Mixed dielectric constant system also admit of a third class of molecules with very rich structure forming potentials. These are the amphiphiles which possess highly polar and highly non-polar portions. Examples are fatty acids and non-polar amino acids.

$$CH_3(CH_2)_n - \overset{O}{\underset{\|}{C}} - O^-$$

$$NH_3^+ - \underset{\underset{H}{|}}{C} - \overset{O}{\underset{\|}{C}} - O^- \quad \text{with} \quad \underset{CH}{\overset{CH_3 \quad CH_3}{\diagdown \diagup}}$$

In aqueous environments such molecules may form colloids, coacervates, membranes multilayers, and a wide array of possible equilibrium structures. Amphiphiles may be formed in energy flow experiments (16). In short, given enough $-CH_2-$ groups a very rich variety of structures is possible in the domain from nanometers to microns. Among these structures are closed membranous vesicles made of bimolecular leaflets of polar lipids. Such vesicles are candidates for protocells. They have a number of properties which are of special interest for biogenesis.

1. They are statistical objects, belonging to the order from disorder domain.
2. The membranes themselves have a low dielectric constant non-polar interior with polar groups on both sides.
3. Non-polar molecules may partition into the membrane interior.
4. The membranes will be relatively impermeable to water soluble polar molecules.
5. The membranes can grow in size by the insertion of amphiphilic molecules.
6. The membranes partition the system into an interior aqueous phase, a non-polar phase, and an exterior aqueous phase.
7. The membranes have a high electrical capacitance of about one μ farad/cm^2, therefore a small amount of charge separation leads to substantial electrical potentials.

Developments in the theory of bioenergetics in recent years (12, 13, 14) have made it clear that the major components are:

1. Electrochemical processes.
2. Protochemical processes.
3. Transmembrane charge separation.
4. Chemical reactions linked to electron or proton flow.

All of the components require a membrane and the resulting phase separation.

In mitochondria, oxidation reactions lead to proton transport across the membrane against a gradient of electrochemical potential. The back flow of protons is coupled to the synthesis of ATP. In chloroplasts the original proton transport is driven by photochemical reactions. Other than the primary photochemical reactions the system acts in a near-to-equilibrium domain so that both oxidative phosphosylation and reductive diphosphorylation may occur. The later leads to reduced NAD or NADP.

There is a reciprocal character between electrochemistry and protochemistry that is characteristic of bioenergetics. Thus for redox reactions

I.   A reduced$\rightarrow$A oxidized+electron

    B oxidized+electron$\rightarrow$B reduced

For acid base dissociations
II. A acid→A base+proton
B base+proton→B acid

Because of the existence of electron conducting wires, electrochemistry has utilized half cells to use reactions I for investigating chemical potentials. Shedlovsky (17) has demonstrated that with proton conductors an exact parallelism exists, with acid-base dissociations in protochemistry being analogous to oxidation-reduction reactions in electrochemistry.

Electron transport in oxidative phosphorylation presumably involves electron tunneling from a heme group to a neighboring heme group. The flux of protons can take place in ice-like channels of contiguous hydrogen bonds (18). The two processes are linked by a common transmembrane electrical potential and can be more closely connected if there is direct coupling between the electron and proton charge distributions.

Within the membrane, oxidation-reduction reactions or photochemical processes lead to asymmetries in electron distribution giving rise to dipoles and higher order fields. The linkage of proton conducting channels to these electrical asymmetries allows the pumping of protons across the membrane (19). This transport of protons gives rise to a difference of electrochemical potential coupled to the synthesis of energy-rich phosphate bonds (14).

The early prebiotic synthesis could certainly have given rise to lipophilic chromophores as can be seen from the water insoluble light absorbing material formed in a wide variety of energy driven synthesis under reducing conditions. Within a bilayer vesicle, this material will orient asymmetrically with respect to the radial vector. The membrane asymmetry arises in the first instance from the fact that the radius of curvature is different on the two sides of the membrane. Under a light flux such a membrane will become electrically polarized. If this polarization can be coupled to proton pumping then a difference of electrochemical potential will be available for chemosynthesis. A more highly developed system of the type we are postulating now exists in *Halobacterium halobium* purple membrane (20). Under the appropriate conditions these bacteria experience a light driven proton transport against a gradient of electrochemical potential. The back flow of protons leads to the synthesis of ATP.

Present-day biological processes are driven almost entirely by the hydrolysis of energy-rich phosphate bonds. This is such a pervasive characteristic of life that it cannot be ignored in searching backward from the ubiquitous features of modern organisms to the origin of life. In a sense biochemistry can be described as follows.

1. It is primarily the chemistry of carbon, hydrogen, nitrogen, oxygen, phosphorus, and sulfur.

2. Reactions are driven by phosphate transfers requiring a continuou input of energy-rich phosphate bonds.

One other feature of present-day biochemistry deserves consideration. The major features of the metabolic chart and macromolecular synthesis must date back to the origin of genetic coding, or somewhat over 3×10^9 years. Ubiquitous features must reflect: common ancestry, parallel mutation in lines leading to all species, or promiscuous genes. The second of these seems to have a vanishingly small probability. In the absence of evidence for the third possibility it seems reasonable to assume that ubiquitous features of biochemistry are ancient. Phosphate driven bioenergetics must predate the origin of life.

Given the preceding we can formulate a scenario for the origin of life which has the following features:

1. It is consistent with known geophysical data.
2. It makes contact with physical theories of self organization and generalizations of molecular biology.
3. It is consistent with the postulate of continuity.
4. It is subject to experimental approaches at several points.

The sequence of events we postulate is as follows:

1. A primordial earth with appreciable amounts of carbon in the form of intermediate and long chain hydrocarbons.
2. The formation of amphiphilic molecules under the flux of solar energy and terrestrial sources.
3. The synthesis of molecules absorbing visible light and ultraviolet radiation. This synthesis was also driven by various energy fluxes.
4. The condensation of biomolecular polar lipid membranes containing chromophores dissolved in the lipid domain.
5. The closure of membranes into vesicles.
6. The development of transmembrane potentials due to the absorption of photons by the molecules dissolved in the membrane.
7. Photochemical and redox reactions coupled to proton channels lead to transmembrane proton pumping.
8. The back flux of protons leads to the synthesis of pyrophosphate and higher polyphosphates; the process may be described as anaerobic photophosphorylation.
9. The continuous source of chemically specific energy in energy-rich phosphate bonds determines the emerging chemical networks inside the protocells. This is the most novel feature of the theory. Random synthesis is replaced by directed synthesis because of the chemical specificity of the inputs.
10. Some of the molecules synthesized by the emerging network absorb onto the inside of the membrane and act as catalysts.

11. Certain pathways are thus enhanced and the chemical specificity increases even further. The emergence of this order following the closure of the vesicles is very, very fast on a geological time scale.
12. A coded information system develops within the cells. This last feature makes much less contact with the deterministic underlying physics, but is introduced to look forward to modern cellular processes.

Steps 2–11 listed above are subject to experimental approach. Indeed, steps 2–5 and 9 have considerable experimental support (8). All of the major features of this model are open to laboratory study with the exception of knowledge about the early state of the planet.

What is clear is that the origin of life may be studied by the ordinary methods of scientific inquiry. We must coutinue to explore the generalizations of molecular biology to extract the necessary features of life. From the point of view of physical chemistry we must refine our theories of non-equilibrium self organization to be able to deal in a more organized way with the details of chemical networks. A challenging field of inquiry lies ahead, but the outlines suggest that the problems are soluble.

References

1. J. E. Lovelock, "Gaia, A New Look at Natural History", Oxford University Press, Oxford, 1979.
2. J. Tuzo Wilson, Editor, "Continents Adrift and Continents Aground", W. H. Freeman and Co., San Francisco, 1976.
3. A. Lehninger, "Biochemistry", Worth Publishers Inc., New York, 1975.
4. H. J. Morowitz, "Foundations of Bioenergetics", Academic Press, New York, 1978.
5. E. Schrödinger, "What is Life?", Cambridge University Press, Cambridge, 1967.
6. T. Smith and H. J. Morowitz, "Between Physics and History", In Press, Journal of Chemical Evolution, 1982.
7. H. Baltscheffsky, Editor, "Symposium in Print on Molecular Origins and Evolution of Photosynthesis", Biosystems **14**, 1–148, 1981.
8. D. Deamer, Editor, "Light Transducing Membranes", Academic Press, New York, 1978.
9. H. Haken, "Synergetics", Springer-Verlag, Berlin, 1978.
10. P. Glansdorff and I. Prigogine, "Structure, Stability and Fluctuations", Wiley-Interscience, London, 1971.

11. H. J. Morowitz, "Energy Flow in Biology", Ox Bow Press, Woodbridge, Conn., 1980.
12. P. D. Boyer, B. Chance, L. Ernster, P. Mitchell, E. Racker and E. C. Slater, Annu. Rev. Biochem., **46**, 955–1015, 1977.
13. H. T. Witt, Biochem. Biophys. Acta., **505**, 355–427, 1979.
14. H. J. Morowitz, Am. J. Physiol., **235**, R99–R114, 1978.
15. J. W. Gibbs, Trans. Conn. Acad. III, 108–248, 343–524, 1978.
16. C. Folsome and H. J. Morowitz, Space Life Science **1**, 538–544, 1969.
17. T. Shedlovsky, Science **113**, 561–562, 1951.
18. J. F. Nagle and H. J. Morowitz, PNAS, **75**, 298–302, 1975.
19. J. F. Nagle, M. Mille, and H. J. Morowitz, J. Chem. Phys., **72**, 3959–3971, 1980.
20. W. Stoeckenius, R. H. Lozier and R. A. Bogomolni, Biochem. Biophys. Acta, **505**, 215–230, 1979.

A Model of the Origin of Life

Hans Kuhn

Max-Planck-Institut für biophysikalische Chemie (Karl-Friedrich-Bonhoeffer-Institut) Molekularer Systemaufbau, D 3400 Göttingen-Nikolausberg

How did life on earth arise? What conditions induced the formation of the living machinery? Can the crucial event be described as a spontaneous selforganization of a homogeneous or quasi homogeneous medium and formation of dissipative structures (Prigogine) or hypercycles (Eigen and Schuster)? Is it a cooperative phenomenon in a multi-component system (Haken)? Is the essential process the formation of a functionalized lipid membrane structure serving as photocatalyst to produce energy rich compounds (Morowitz)? Is the physical paradigm insufficient to grasp main aspects (Kothari)?

In the present view the crucial event is a very particular cooperative phenomenon: the formation of an adaptable device, possessing a self-replication and translation machinery. This requires a particular spatial and temporal structure assumed to be present at particular points on the prebiotic earth, where the availability of appropriate energy rich compounds should be no basic problem.

In this view then the question of the origin of life can not be answered by a search for general conditions that produce structure in an unstructured medium. On the contrary the genesis of self reproducing machines requires confinement and periodic changes necessary to stimulate the formation of selfproducing entities. A model path consisting of many small physically and chemically plausible steps is considered. The specific model reveals the logical framework and the organizational structure of evolutionary processes, the nature and location of fundamental difficulties as well as the means by which they might be overcome. In more general considerations it is easy to overlook the crucial difficulties, and the detailed consideration of an imaginable path constitutes a method of avoiding this problem. However, it cannot be expected that the model steps considered furnish an accurate description of the events that actually occurred, but indicate a way to approach the fundamental enigma by discussing possibilities that are experimentally testable.

Basic Concepts of the Approach

The origin of life is frequently associated with the spontaneous appearance of dissipative structures in a homogeneous system far from an equilibrium that is the result of an inner instability. Many natural phenomena in physics, biology and sociology were shown to be examples of spontaneous selforganization in homogeneous systems (1). The origin of life is frequently considered as being one among these processes. In contrast, in our view (2), (3) the crucial event in the origin of life is based on a very particular process which differs in its quality from the others; the appearance of systems that reproduce themselves and that eventually change by errors during the copying process. In general, the changed copies are handicapped and cannot survive, and correct copies remain. In rare cases errors will lead to improved survival chances of the changed individuals and this is repeated again and again. Individuals evolve that are best adapted to the environment. Such progressive adaption to the environment represents a process of learning.

The appearance of the first systems capable of learning represents a jump in quality, in which a fundamental property of matter suddenly manifests itself. Systems begin to be carriers of information, of a meaningful message, with a content capable of growing as the learning process advances. Prior to this not even the faintest trace of this property existed, but once the breakthrough has occurred, the process of learning goes on inexorably via the continued confrontation of evolving systems with their surroundings and their adjustment to environmental changes by multiplication, mutation and selection.

It seems difficult to find possibilities of how a machinery with the capability of learning could have originated in a spontaneous process without confrontation with a complex spatial and temporal environment. A cardinal point in the present approach is that a particular pre-existing external spatial and temporal structure is essential for the formation of the living machinery. A machine is constructed by fitting its interacting parts together via external directed action. On the primordial planet the role of the external designer is replaced, in a way by the enormous variety of environmental influences. Given such a starting position questions about the thermodynamic conditions for the spontaneous appearance of dissipative structures in a homogeneous system caused by inner instability do not arise. One does not ask for the conditions of the existence for a cyclic reaction sequence that must be satisfied everywhere in a solution. However, one asks for conditions that must prevail in a specific location so that appropriate aggregates consisting of a small number of macromolecules can form and multiply. Questions arise that are primarily concerned with the problem of finding the logical framework and organizational structure of evolutionary processes: By what principles can systems emerge that are capable of learning? What fundamental barriers confront such systems that can learn once they exist? What fundamental possibilities exist to overcome these barriers? Secondly these questions are concerned with the feasibility: What fundamental possibilities satisfying the organizational requirements exist in physical chemical models, that is, by remaining within the framework of physical and chemical laws?

The methodological program followed here for recognizing and understanding the grand connections in the process of selforganization is to consider a specific path consisting of many small physically and chemically plausible model steps that lead to a selforganization of matter. It is driven by periodic temperature changes and a multifaceted spatial environment, both of which were available in suitable locations on the primordial planet.

It was pointed out (4) that a great variety of conditions can have initiated the evolution of life that it is therefore improbable to find exactly the historic path by inventing a sequence of logically consistent model steps. However, these steps have not been devised with the hope of giving an accurate description of the actual path followed by nature, but rather with the intention of understanding the fundamental aspects and difficulties as tangible as possible and to find ways by which they might be overcome. In more general considerations on selforganization it is easy to overlook the crucial difficulties, and the detailed consideration of an imaginable path constitutes a method of avoiding this problem.

In this detailed model path each step only leads to the next and so the overall logical structure of the model is not apparent until the very end. It is this overall structure that represents the essence of the model. While individual steps can always be replaced by others that are somewhat different, without thereby destroying the logical connections between the steps, the fundamental changes in the system structure of the explored model seem to be fixed necessities.

This detailed model path was described elsewhere (2, 3, 5) and shall not be considered here. We will illustrate the procedure by indicating some model steps and describe the overall organizational structure.

For studying the important question of how a learning device can originate, it is useful to discuss first some general features of the learning process. The zig-zag-lines in Fig. 1a were obtained by proceeding in equal steps from left to right, each time deciding by tossing a coin whether the step should go up or down. Afterwards zig-zag-lines are cancelled according to a rule, until half of the original number is present (selection phase). According to this rule the zig-zag-lines to be cancelled are determined by the dice in such a way, that each line that fits better into a given form than another has the better chance to survive. (In the case of Fig. 1a, b) zig-zag-line 4 has be best survival chance.) The surviving zig-zag-lines are copied. Eventually copying errors are introduced by drawing lots. A step may go up instead of down (Fig. 1c) or the chain may add a step or loose a step. This multiplication phase is followed by a selection phase defined by the rule described above, and the process is repeated again and again. Zig-zag lines are obtained that are adjusted to the given form, since such lines have the best chance to survive. If the given form is slowly changed in shape, the zig-zag-lines will adjust to the new shape in a sufficient number of generations (Fig. 1d).

This example demonstrates that the adaption process is only possible if the frequency of copying errors is neither too large nor too small. In the case of too large a copying error rate zig-zig-lines cannot adjust to the form, since a zig-zag line of appropriate shape is likely to suffer a change during the time needed for its selection. In the case of too small error rate adjustment to a form that is slowly changing in shape is not possible. The system of evolving zig-zag lines is not sufficiently adaptable. For not loosing the message given by a zig-zag line this line must be copied error free with sufficient probability, and therefore a longer line (an increased amount of genetic information) requires a smaller copying error rate per step. Most barriers that must be surmounted in the evolution of life are related to the difficulty of suppressing copying errors.

First Steps in the Origin of Life

A simple learning system with nucleotides can well be imagined to have evolved on a planet such as the primeval earth. It has been demonstrated that energy rich nucleotides and oligo-nucleotides can be obtained by simulating conditions thought to have existed on prebiotic earth. The prebiotic synthesis required solid state reactions as well as reactions in aqueous solution and in the gas phase. It is plausible that such substances would have accumulated on primeval earth in particular locations where a multitude of special conditions were fulfilled that allow a succession of very different reactions to occur, requiring highly diversified and structural regions. Orgen and Lormann (6) have succeeded in the enzyme free polymerization of nucleotides on nucleic acid templates, more than 90% of the nucleotides of the replicate strand being complementary to those on the template strand.

We may thus imagine a short nucleic acid strand on primordial earth containing the two complementary bases guanine and cytosine that are anchored together in the strand by ribose and phosphate groups and are capable of interlinking by three hydrogen bonds. It is further supposed that this short strand has diffused into a special region, after having been formed elsewhere by accidental condensation during the drying of a solution of monomers. In this strand, all of the monomers are presumed to be linked in such a way that the strand can serve as a precise template for replication. By considering a solution of the different monomers that may have existed on earth at the location considered, it

A Model of the Origin of Life 239

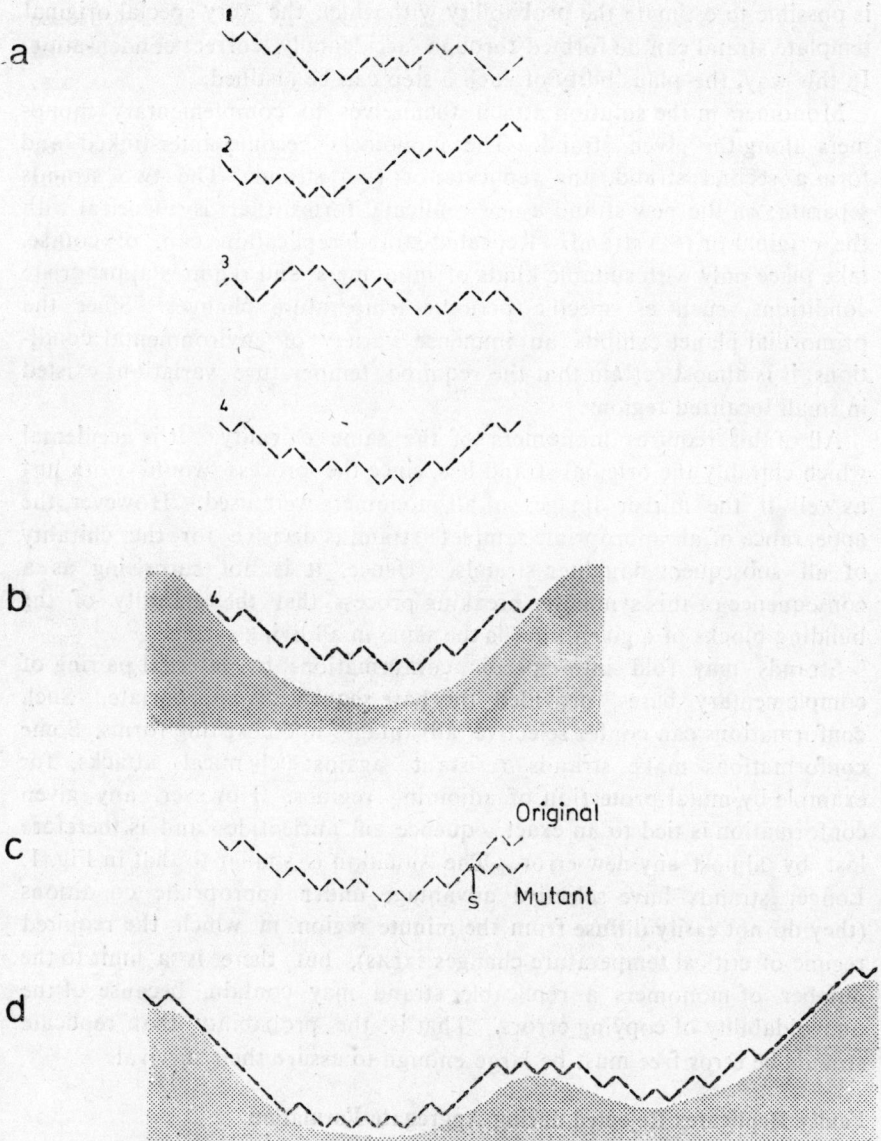

Fig. 1. Arrangement to demonstrate process of learning
(a) Arbitrary zig-zag lines 1–4
(b) selection of zig-zag line 4
(c) copying error in step *s*
(d) zig-zag line adapted after many generations

is possible to estimate the probability with which the very special original template strand can be formed through accidentally correct condensation. In this way, the plausibility of such a step can be justified.

Monomers in the solution attach themselves to complementary monomers along the given strand. The monomers become inter-linked and form a second strand, the replicate or (—) strand. The two strands separate; on the new strand a new replicate forms that is identical with the original or (+) strand. Repeated strand replication can, of course, take place only with suitable kinds of monomers and requires appropriate conditions, such as specific periodic temperature changes. Since the primordial planet exhibits an immense variety of environmental conditions, it is almost certain that the required temperature variations existed in small localized regions.

All of this requires monomers of the same chirality. It is accidental which chirality the original strand has, since the process would work just as well if the mirror images of all monomers were used. However, the appearance of an appropriate template strand is decisive for the chirality of all subsequent daughter strands. Hence, it is not surprising as a consequence of this symmetry breaking process, that the chirality of the building blocks of a given kind is the same in all living systems.

Strands may fold into specific conformations by internal pairing of complementary bases, provided the base sequence is appropriate. Such conformations can confer selective advantages upon existing forms. Some conformations make strands resistant against chemical attacks, for example by mutal protection of adjoining regions. However, any given conformation is tied to an exact sequence of nucleotides and is therefore lost by almost any new error. The situation is similar to that in Fig. 1. Longer strands have selective advantage under appropriate conditions (they do not easily diffuse from the minute region in which the required regime of critical temperature changes exists), but there is a limit to the number of monomers a replicable strand may contain, because of the unavoidability of copying errors. That is, the probability that replicate strands be error free must be large enough to assure their survival.

Faulty Replicates Rejected during Aggregate Formation

In the model the resulting impasse is broken by a mechanism in which the strands convolute into a conformation that allows their interlocking into aggregates. Not only is diffusion of such aggregates slowed by their size, but it is of the greatest importance that practically all erroneous copies are rejected during aggregation, because of the precision required for interlocking. Aggregation therefore serves the function of an error filter.

The survival chances of strands in an aggregate are increased by their mutual protection against chemical attack, or because it may be less easy for aggregated strands to leave the favourable region. Aggregates may undergo self-reproduction. Suitable temperature and other environmental changes may cause their disassembly into component strands. The individual strands may replicate and again assume folded conformations. They move about at random and again aggregate through favourable accidental collisions, increasing the number of aggregates. All of this requires a very specific and highly detailed program of changes of temperature and other parameters (e.g. ionic composition of solution) that must be periodically repeated for repetition of the process described. Convolution of strands and aggregation are favoured by cooling, disassemblage of aggregates and deconvolution of strands by warming.

It is of importance that the component strands do not diffuse too far from each other during the multiplication phase, because they would be lost and reaggregation could no longer occur. The phenomena discussed must therefore take place in a confined space, such as in small pores of a rock formation. The pore walls keep the strands together during their diffusion, so they are able to locate each other again and can form new aggregates. The rock formation is presumed to be inundated by a solution of suitable energy-rich monomers that can easily diffuse through the pore channels, while the strands formed are largely retained by them.

A great selective premium is associated with the evolution of machinery that facilitates the aggregation of interlocking *subunits* $B_1, B_2, B_3, \ldots$. A model of such machinery in its simplest version can be unfolded strand C. The function of C is that of a *collector strand* for the subunits $B_1, B_2, B_3, \ldots$ that are guided by this strand to the growth region of the aggregate.

Consider strands B_1, B_2 that are capable of assuming the conformation of hairpins (Fig. 2a), with legs twisted into a double helix. Such hairpin strands would be capable of interlocking into picket-fence like aggregate (Fig. 2b), stabilized perhaps by bivalent cations, such as Mg^{+2} or Ca^{+2} ions, that form links between negatively charged phosphatidyl groups on the outside of neighbouring hairpin strands.

Model of Emergence of Premordial Translation Machinery

A detailed model with GC RNA strands in the conformation of a left-handed double helix analogous to the Rich-Arnott-Dickerson Z conformation (7) exhibits an astonishingly good fit between neighboring hairpin subunits (Fig. 2c). Equally excellent is the steric fit between "anticodon" loops that form the hairpin bends and "codons" along the collector strand. The bases in these complementary triplets are stacked, providing stabilization, and the 3'5' directions of the strands connected by complementary base triplets are opposite.

Comparative studies by Eigen (8) of t-RNA sequences indicate that all t-RNA's developed from a common ancestor which may have consisted mainly of G and C nucleotides. It is therefore not implausible that at

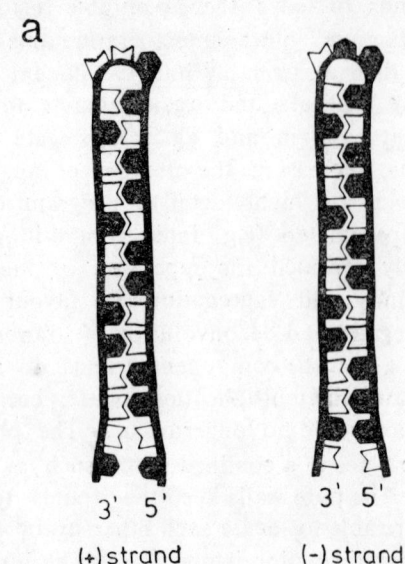

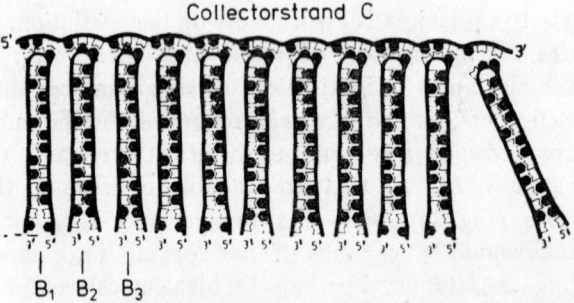

Fig. 2 (a) Possible arrangements of bases G (filled symbols) and C (open symbols) that permits a hairpin conformation. $(+)$ and $(-)$ strand.

(b) Schematic view of collector strand C, and subunits B_1, B_2, B_3...forming picket-fence like aggregate. Attachment of subunits to collector strand by base triplets. In the present case B_1 is a $(+)$ strand, B_2 and B_3 are $(-)$ strands.

first there were G and C nucleotides only. Moreover, the ancestral t-RNA sequence derived by Eigen (8) from comparisons among the known nucleotide sequences of many present-day t-RNA's is such that only a few changes are needed to allow a perfect hairpin conformation.

An attractive feature of the hairpin conformation is that $(+)$ and $(-)$ strands are identical except for the nucleotide in the middle and the nucleotides at the open ends (Fig. 2a), a feature that permits great economy in the usage of strand material and the assemblage of aggregates. The

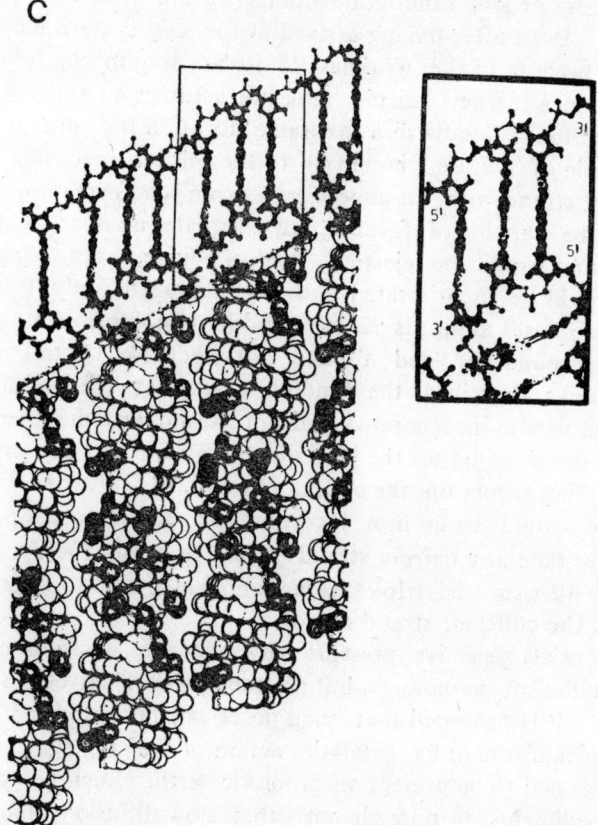

Fig. 2 (c) Section of collector strand and attached subunits. For clarity nucleotides located along the collector strand and in the "anticodon" loops are shown as ball-and-stick models, the remaining portions of the hairpin strands are shown as space-filling models. The ends of the hairpins are shown bonded to each other in this figure. The inset depicts details of the bonding between "codon" and "anticodon".

nucleotide in the first position of the triplet at the hairpin bend can be C in (+) as well as (−) strands; in the third position it is G in both strand types. At corresponding first positions of the collector strand there must therefore be G and at third positions C. The midpositions can randomly be occupied by G or C.

It is known that the bonding energies of base pairing by hydrogen bonds are insufficient at room temperature to establish stable bonding between two triplets of complementary bases. The model would therefore permit the required one-dimensional diffusion of subunit pickets along the collector strand. There is, of course, little constraint on orientation, and any new picket that may be on its way towards being incorporated will point in more or less random directions during its travels to the growing aggregate. Even after having arrived at the aggregate it may still become detached, because of the weakness of its bonding to the collector strand. However, once the new hairpin molecule assumes an appropriate orientation by chance, it is now in a favorable location for joining up, provided it is capable of a close enough fit to the picket preceding it along the strand. A strand with an imperfect hairpin conformation would, however, not be capable of forming an adequate number of $O-Me^{2+}-O$ linkages and would be rejected. And of course, even a perfect hairpin strand would be incorporated into the aggregate only if it carried the appropriate bases along its "anticodon" loop, complementary to the bases of the corresponding "codon", and not otherwise. If this mechanism is to function as described, the binding energy for the lateral interactions between a newly incorporated hairpin strand and the one preceding it must be about equal to the binding energy for its base triplet and the corresponding triplet on the collector strand. If this energy were smaller then there would be no firm incorporation at room temperature. If it were larger then any hairpin strand would be firmly incorporated, regardless of whether its base triplet were complementary to the corresponding triplet on the collector strand or not.

There exists selective pressure towards the invasion of pores that contain sufficient monomers, but are too large effectively to curb strand diffusion. It is proposed that such pores can be colonized by the aid of polypeptides formed, by catalytic action of the aggregates, from amino acids presumed to be present on prebiotic earth. Such polypeptides could form impediments in pore channels that slow diffusion; later they could coalesce into confining "cellular" envelopes that keep strands together as aggregates disassemble into strands which replicate and again aggregate. A strong evolutionary gradient would exist in this manner.

In our model the aggregates described develop catalytic properties in the following way. The open ends of the hairpin strands possess or develop affinity towards activated amino acids that are thereby brought into favorable proximity of each other and enabled to form polypeptide bonds; that is, the hairpin strands become adapters of amino acids. Open ends of the hairpin strands would be of help in strand replication as well, because replication could automatically begin at an end of a strand only and offer selective advantages in this way.

Both $(+)$ and $(-)$ strands would be capable of serving as adapters. To fix ideas, assume that the midpositions of a $(+)$ strand are occupied by bases C, the ends by bases G; then there must be bases G at the midpositions and bases C at the ends of a $(-)$ strand (Fig. 2a). The two kinds of strand ends could then have different affinities for two kinds of activated amino acids. Amino acids bonded to nucleotides could have existed prebiotically, for example, a_1 combined with C and a_2 with

G. In this way (+) strands would serve automatically as adapters Ca_1 for a_1 and (−) strands as adapters Ga_2 for a_2.

A possible explanation of how a specific linkage of an amino acid to a nucleotide could have arisen is illustrated for (+) strands in Fig. 3a. The amino acid a_1 is activated by a purine nucleotide (G) (cf. Fig. 3a, step 1). Intercalation and complementary base pairing would then mediate specific

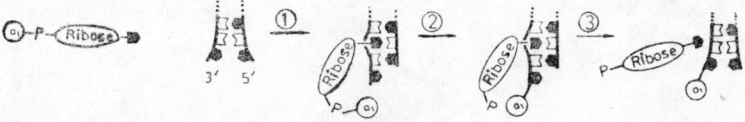

Fig. 3a. Possible mode of attachment of amino acid a_1 that has been activated by a purine nucleotide to the end of a (+) strand.

linkage and place the amino acid in such a position that reaction with the 2′–OH group of the ribose of the terminal nucleotide of the strand could occur. Our molecular model shows that the steric condition for reaction is only given for intercalation of the strand with the 3′ end, and that the whole sequence of steps described would be sterically quite plausible, permitting reaction with the 2′–OH group of the ribose in question. This would correspond to the same linkage that is found in today's charged tRNA. The steric fit and the possibility of hydrogen bond formation between the amino group of the amino acid and the 3′ OH-group of the ribose can only be realized with L-amino acids (Fig. 3b). This may indicate the origin of the stereospecificity of amino acids in biological systems and explain the astonishing fact that each amino acid selected in nature for the genetic machinery is L (not D). Amino acid a_2 could be activated in a corresponding way by a pyrimidine nucleotide (C) and then linked to the terminus of a (−) strand. The open ends with the amino acids have sufficient freedom of motion so peptide bonds can be formed in the aggregate since the two bases at the 3′ end are not paired with the corresponding bases at the 5′ end. Our molecular model shows a steric fit allowing peptide bond formation (Fig. 3c). One can imagine possibilities for a specific activation of amino acids under prebiotic conditions. Dr. Lehmann in our laboratory has shown by chromatographic separation of amino acids and nucleotides using salt solution as eluent, that glycine and the guanine nucleotide are eluted at the same speed, while alanine and the cytosine nucleotide eluted at another significantly different speed. This offers the possibility of in situ formation of the proper activated forms.

It should be emphasized that the study of all possible conformations of nucleic acid double helices is still a very new field, and further stable conformations may well be formed (see e.g. the DNA conformations recently proposed by Hopkins (9) and may be important in the search for better models. Not more is intended with the present model than to discuss

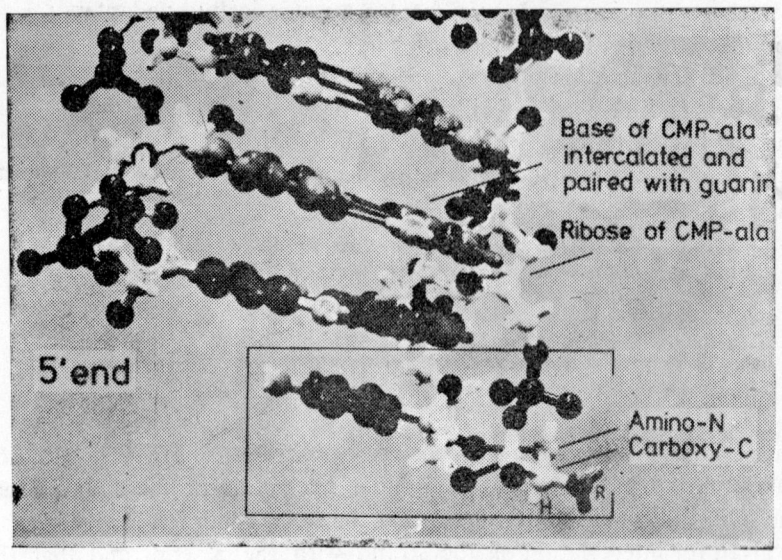

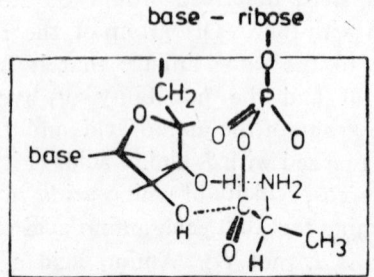

Fig. 3b. Model showing situations (2) in more detail.

possibilities. The essential point is the cooperative stabilization of a complex of collector strand and of at least two adjacent adaptors, and the close fit of codon and anticodon, regardless of whether this can be achieved by left-handed or right-handed double helices or by other conformational details. All that is needed is a primitive translation device based on a mechanism for rejection faulty replicates.

Eventually a sequence along the collector strand will arise that corresponds to a polypeptide endowed with enzymatic properties, acting as a primitive "replicase". Such a "replicase" would have to decrease the frequency of replication errors sufficiently so the information coding for it would not be lost during the number of generations required to fix it by selection. It can be shown, for example, that if the survival rate is increased by just 10% by the advent of a "replicase" consisting of ten amino acids, a reduction of the replication error rate per base from 1/100 to 1/300 is sufficient for fixation by selection (5). The attainment of such a "replicase',

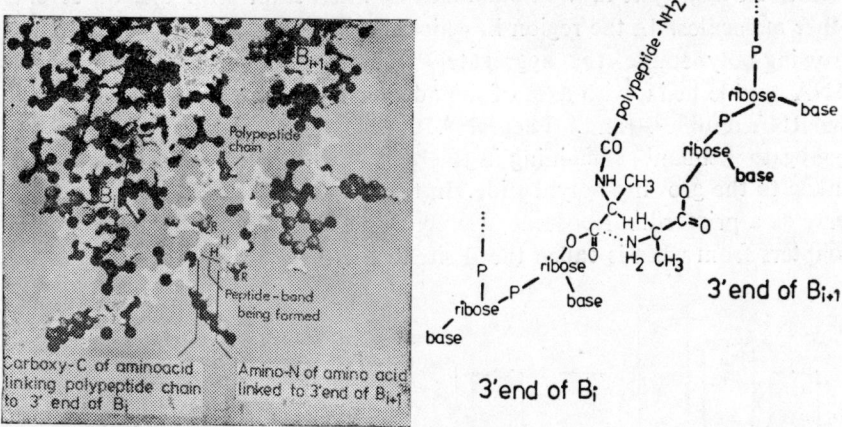

Fig. 3c. Model showing peptide bond formation in more detail. Polypeptide linked to subunit B_i and amino acid linked to subunit B_{i+1} at 3' end by ester formation with the ribose 2' OH group. The C—O bond linking the polypeptide to B_i is opened while a peptide bond is formed between carboxy C of polypeptide and N of amino acid. By this process the polypeptide chain moving from subunit B_i to subunit B_{i+1} adds one chain member. The steric fit in Figs. 3b and 3c is no more present in the case of D amino acid (exchange of R and H).

represents a major breakthrough, the stabilization of machinery for reading and translating a code, making possible the acquisition of further enzymes, and conferring stability upon the code.

One can imagine, for example, that such a "replicase" is a suitable short polypeptide of even just ten amino acids that fits into a notch of the double helix which exists during strand replication, in the region where the new strand is formed, slipping along in the notch to remain in the region of replication as the new strand is lengthened*. Its presence would stabilize the double-helix conformation during replication and improve the contact between the two strands in that region. Replication would be speeded up and there would be fewer errors in base pairing during replication.

There is a striking similarity between the functioning of the aggregates here described and the operation of present-day genetic machinery, a similarity that is most encouraging if not altogether surprising. The collector strand C would be the precursor of messenger RNA, the subunits B_i those of tRNA's and the "cellular" envelope that of the cell membrane. As is the case today, codons along C are read in the 5'3' direction while the strand direction is opposite for attached subunits B_i.

*Investigations of specific binding of proteins on nucleic acids have shown that a section in a polypeptide of, say, ten amino acids can form a two-stranded antiparallel β sheet that fits into the groove of a double helix (9).

Such an aggregate may be stabilized by interaction with one or several other molecules. In the region in which new amino acids are linked to the growing polypeptide, the aggregate "ribbon" might partially enfold an RNA double helix, with axes of the adapter hairpins parallel to the axis of the RNA double strand. The RNA double helix might glide along the aggregate "ribbon", remaining in the region in which the amino acids are linked to the growing polypeptide. In this way the RNA double helix would serve as a primordial ribosome, a precursor of the device that translocates adapters from what is called the A site to the P site (Fig. 3d).

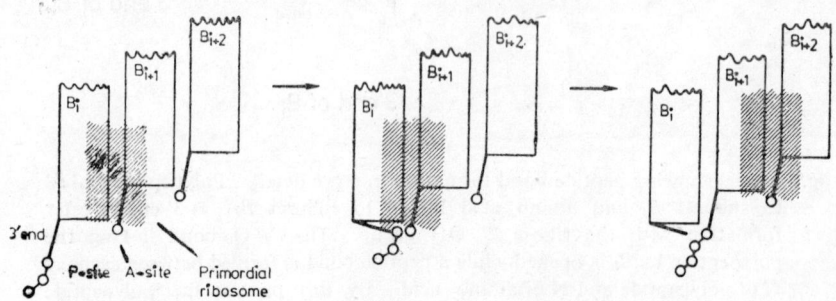

Fig. 3d. Possible role of primeval ribosome. "Ribosome" at line-up elements B_i and B_{i+1} charged with peptide and amino acid respectively (left). P site and A site occupied, ribosome fixed, allowing formation of additional peptide bond (middle). "Ribosome" (free to diffuse on aggregate-ribbon after peptide bond has been formed) reaches line up elements B_{i+1} and B_{i+2} (right). "Ribosome" fixed to line-up elements catalyses again peptide bond formation, and in this manner the polypeptide is formed.

In our molecular model of the stabilized situation in the middle of Fig. 3d the base at the 3' end of the "ribosome" is paired with the base at the 3' end of B_{i+1}, the base at the 5' end of the "ribosome" is paired with the base at the 3' end of B_i and the arrangement is stabilized by stacking interaction with the "ribosome".

The view that the collector strand is the primordial form of the carrier of genetic information is supported by a sequence analysis of the DNA of virus, procaryotes, and eucaryotes by Shepherd (11). The clearly found periodic correlation indicates that originally there existed the reading frame PuNPy (Pu purine, as G; Py pyrimidine, as C; N purine or pyrimidine), vestiges of which are still in evidence.

Experiments that would test the proposed molecular model should be very fruitful and are highly desirable.

Organizational Structures of Evolving Processes

The example of the emergence of a translation apparatus illustrates the method of procedure in our model. Further breakthrough phases appear which are again concerned with the difficulties in increasing the amount of genetic information. These difficulties are due to the logic requirements in the different stages of the evolutionary process. In the following we consider these requirements and indicate their realizations.

Under appropriate circumstances a system appears that is capable of learning (Fig. 4, step 1). The complexity increases but stagnation sets in, because of the accumulation of replication errors. To overcome it, machinery for the discarding of erroneous copies is needed and feasible (step 2). The complexity of the systems can again increase, until stagnation occurs, because the systems are tied to pre-existing compartmentation. To overcome this barrier machinery is needed and achievable that permits a containment of building blocks that is independent of external structure (step 3). Given appropriate conditions this machinery develops by necessity into machinery for code translation and machinery that permits the preservation of the information needed for the forming of translation products (step 4). This development in turn allows an increase in complexity until stagnation sets in at a certain level, because of the accumulation of nonsense products. The surmounting of this hurdle (step 5) requires machinery that permits the bypassing of meaningless production and can be achieved by a reorganization of the systems. Further evolution leads to increasingly complex and thus ever more delicate systems, ending again in stagnation because of inadequate adaptability of the systems. The barrier can again be hurdled by a fundamental reorganization of the systems (step 6). This change permits the achievement of a level of complexity that cannot be raised further because of thermal noise, until another fundamental restructuring of the organization permits a conquest of this barrier (step 7).

There exist analogies between the present model of the evolutionary process and basic computer steps. In both areas information is processed in cycles that are driven from the outside. In a computer the information is contained in the states or positions of many switches, in the evolving system in the sequence of monomers in a strand that carries the blueprint for the evolving form. The operations of the computer are deterministic. In contrast the evolution process must contain an indeterministic ingredient. The probability of copying errors is not arbitrarily small but has an optimal value. On the one hand the probability that no error is made during strand replication must be large enough so a sufficient number of error free descendants is available. On the other hand as many mutants as possible must be generated to make the probability of finding an advantageous form among them as large as possible.

The process of evolution can be described as a phenomenon that starts with a bang, somewhere and sometimes in the development of any suitable planet (Fig. 5). All of a sudden a strand arises that can replicate, thereby creating a system that adapts itself again and again to a changing environment, in many steps of replication and selection. This evolution also proceeds in jumps. Nothing of importance happens for long period

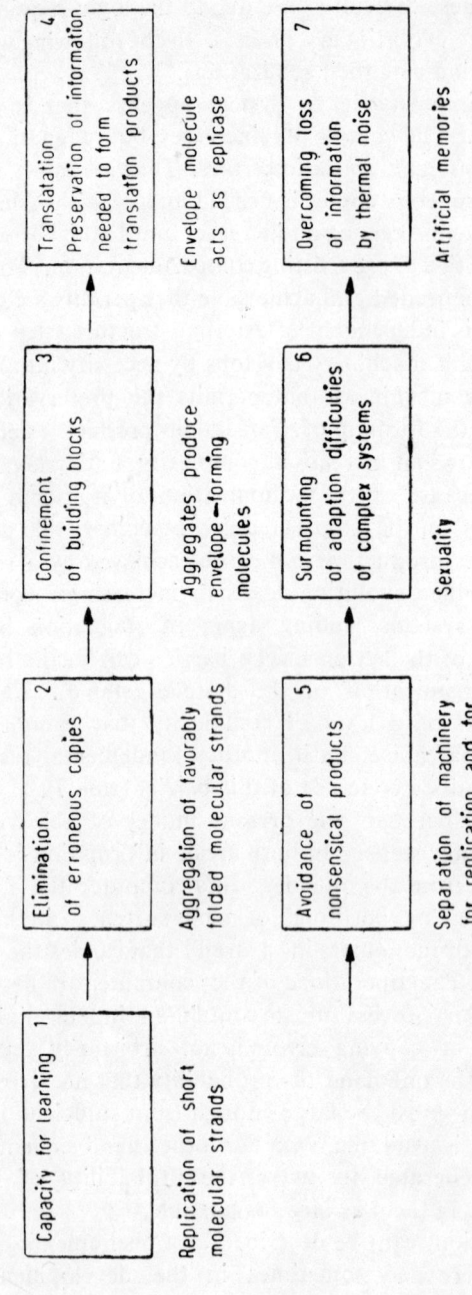

Fig. 4 Organizational framework of the evolution process. Logical requirements (boxes) and their realizations.

A Model of the Origin of Life   251

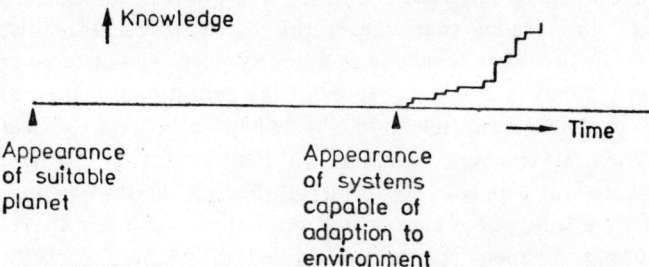

Fig. 5   Emergence of life. Learning systems suddenly appear, without prior traces of this new quality. Knowledge of evolving systems increases discontinuously.

until a reorganization of the system ushers in a rapid new development. In a corresponding way there are discontinuous increases in the contents of the message of which the systems are carriers. An important function increases that is a measure of the evolution of the systems, the knowledge of the systems. We roughly define this function as the information (measured in bits or yes-no decisions) contained in the totality of the blueprints that has to be rejected by necessity until the stage, of evolution considered was reached. Knowledge is a measure of the usefulness of the information accumulated during the course of evolution.

Concluding Remarks

The cardinal problem concerning the origin of life is the question of how one can imagine the development of simplest systems that could adapt to their environment—systems consisting of a few macromolecules capable of mutual cooperation. A necessary requirement for this process is the existence of a spatially and temporally structured environment, needed to keep the potential building blocks of such systems from diffusing away from the region of importance and to drive the replication, assemblage, and disassemblage of the aggregates. A purpose of this account is the stimulation of experiments and of theoretical efforts towards the improvement and the refinement of the model described. The present model should be a possibility demonstrating principle ways to overcome the fundamental barrier. Basic to the approach is its paradigm: The search for detailed models, not for general principles of selforganization. This paradigm is based on the notion that the crucial event is a very particular cooperative phenomenon, the formation of an adaptable device possessing a selfreplication and translation machinery and this phenomenon cannot be grasped by general considerations.

The physico-chemical systems here considered show a holistic, purpose directed behaviour that at the same time appears to be wilful to a certain extent.

This apparent wilfulness is based on the large number of undeterminable parameters of the model that causes the events to be unpredictable in their details. On the other hand the apparently purpose—directed behavior of the model systems is a consequence of the evolutionary mechanism of replication, mutation and selection, by which a process of learning is induced. The first learning systems are restricted to a highly specific environment such as a porous rock, surrounded by shadow-casting objects and soaked by a solution of energy rich monomers and a few short strands that accidentally invaded the region. Eventually there occurs a breakthrough in a new direction that permits spreading into a new ecological niche. Examples are the lengthening of strands by the fusion of shorter pieces, the aggregation of folded strands, the development of machinery to fabricate obstructions and later envelopes, the emergence of a primitive code. The increasing liberation from environment by the emergence of more and more sophisticated machines leads to a continuous evolution of systems which appear holistic, purpose oriented, meaningful, clever more and more ingenious, even the processes can in principle be resolved into a complex sequence of causal steps each describable by the laws of physics.

In this view the physical picture is the framework to describe matter including the material aspects of living systems. A decisive component is lacking in this physical paradigm. The physical picture—1 in the world of objects in space and time and of relations between them—is the framework for the impressions gathered by our sense organs. Our emotions are not accounted for. The framework in which emotional impressions are ordered is less general, but we all see emotions as being induced by entities which we consider as souled, as having an immanence beyond space and time. In the further reflection we see emotions and sensations correlated with processes in the brain. A search for more detail of this correlation is immensly challenging.

But assume for a moment that we would know the connections in the cortical network corresponding to each emotional and sensual impression and that we would even understand the evolution of the scheme of these connections as a consequence of the laws of physics and chemistry. We would not be in the position of explaining the mystery of this correlation. The situation is not different in physics where we know the general laws, but do not understand why they are as they are. Physicists created a pattern of how our sensual impressions can be ordered but there is no answer to the question of why they possess that order. The mystery of consciousness, of immediate awareness of sensation and emotion, which has been so much discussed in this conference, can not be grasped in communicable scientific terms and is therefore beyond scientific scope.

Literature

1. O. Nicolis and I. Prigogine: Self-Organization in Non-equilibrium Systems, Wiley-Interscience, New York 1977; H. Haken (Ed.): Synergetics, Springer, Berlin, Heidelberg, New York 1977, P.M. Allen, M. Sanglier, J. of Social and Biological Structures **1**, 265 (1978).
2. H. Kuhn, Angew. Chem. Intern. Ed. **11**, 798 (1972).
3. H. Kuhn and J. Waser, Angew. Chem. Intern. Ed. **20**, 500 (1981).
4. M. Eigen and P. Schuster, Naturw. **65**, 341 (1978); P. Schuster, Prebiotic Evolution in: H. Gutfreund, ed. Biochemical Evolution, Cambridge University Press, Cambridge (U.K.) (1981).
5. H. Kuhn and J. Waser in: Biophysics—A textbook, eds. W. Hoppe, W. Lohmann, H. Markl, H. Ziegler, Springer Berlin, Heidelberg 1982; H. Kuhn in: Biophysik—Ein Lehrbuch, Springer Verlag Berlin, Heidelberg 1977, p. 187–197.
6. R. Lohrmann, P.K. Bridson and L.E. Orgel, Science **208**, 1464 (1980).
7. A.H.—J. Wang, G.J. Quigley, F.J. Kolpak, J.L. Crawford, J.H. van Boom, G. van der Marel, and A. Rich, Nature **282**, 680–686 (1979);

    H. Drew, T. Takano, S. Takano, K. Itakura and R.E. Dickerson, Nature **286**, 567 (1980);

    S. Arnott, R. Chadrasekaran, D.L. Birdsall, A.G.W. Leslie, R.L. Ratliff, Nature **283**, 743 (1980).
8. M. Eigen and R. Winkler, Naturwiss. 68, 217 (1981).
9. W.F. Anderson, D.H. Ohlendorf, Y. Takeda, B.W. Matthews, Nature.
10. J.C.W. Shepherd, J. Molec. Evolution, in print;

    Proc. Natl. Acad. Sci. USA, 78, 1596 (1981).

DNA Synthesis by Reverse Transcriptase: Role in Evolution

M.R. Das

Centre for Cellular and Molecular Biology
Hyderabad 500 009, India

Introduction

Soon after the discovery 1-5 of reverse transcriptase (RT) in 1970, the functional role of the enzyme with regard to malignant transformation by RNA tumour viruses was clearly established. The mechanism of RNA viral replication through a DNA intermediate explained several unique features of infection by retroviruses: (a) the heritable stability of transformation of normal cells induced by these viruses[3]; (b) the apparent vertical transmission of high leukemia frequency in reciprocal crosses between high and low frequency strains of mice[6]; and (c) the requirement for DNA synthesis in early stages of infection.[7]

In addition to understanding its role in tumourigenesis, the discovery of RT triggered both theoretical speculations and experimental designs with regard to searching possible roles for reverse transcription in other areas. These include (i) gene amplification and other possible roles of reverse transcription in normal cells including that of providing a molecular mechanism for the evolution of RNA tumor viruses from normal cells and (ii) evolution in general. These concepts will be critically examined in this paper.

Structure of the Retroviral Genome and Its Replication

Structure of the viral genome

The genomes of all retroviruses, so far examined, are diploid. The viral genome is made up of two identical subunits of single stranded RNA, each subunit having a molecular weight of 2-3 million. The two subunits are joined at the 5' terminii[8,9] presumably linked by hydrogen bonds between complementary nucleotide sequences.[10]

The essential general features of a haploid subunit are shown in Figure 1.

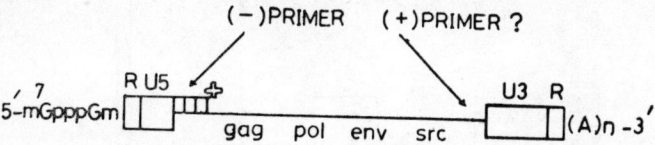

Fig. 1 Essential features of a haploid sub-unit genome of a typical retrovirus. *gag*, *pol*, *env* and *src* are structural genes for group specific antigen, reverse transcriptase, envelope proteins and transformation specific proteins respectively. See text for details.

The 5' terminus is capped by the structure $5'-m^7G_{ppp}Gm^{11,12}$ followed by a repeating sequence shown as R (found at either end of the genome). Following the R sequence at the 5' end is a sequence unique to that end (U5) and the primer binding site for the synthesis of the minus strand of DNA (see below). Genes designated as *gag*, *pol*, *env* and *src* are found in that order in the 5'—3' direction following the (−) primer binding site; *gag* (group specific antigen) codes for structural proteins of the viral core, *pol* (polymerase) for reverse transcriptase, *env* (envelope) for the glycoprotines of the viral envelope and *src* (sarcoma) for the transformation specific protein. The genes *gag*, *pol* and *env* are required for the replication of infectious virus, and are common to all viruses that can replicate in the absence of a helper virus. The deletion of *src* has no effect on viral replication and the transformation defective viruses and leukemia viruses in general do not contain this gene in the viral genome. Following the *gag*, *pol*, *env* and *src* (for genomes containing the *src* gene) are the unique sequences specific to the 3' end (U3), the R sequence and then a poly A stretch, the exact length of which varies from virus to virus. The primer site of the (+) strand of DNA is believed to be located before U3 as shown in Figure 1.

Replication

The synthesis of (−) DNA starts from the primer binding site and proceeds to R (Figure 2). At this stage the new DNA [(−) strong stop DNA] is detached and 'jumps' from its original position at the 5' end of the RNA to a new position at the 3' end of the RNA' (Figure 2). (The poly A stretch beyond the R region at the 3' end is not shown). Notice that the direct repeat (R) sequence in the viral genome makes this attachment possible. Synthesis now proceeds upto and including the P region (D in Figure 2). Plus strand DNA synthesis is initiated at the beginning of the U3 region and proceeds to complete the plus strong stop DNA containing the U3, R, U5 and P regions (E in Figure 2).

The molecule can now circularise because of the complementary nature

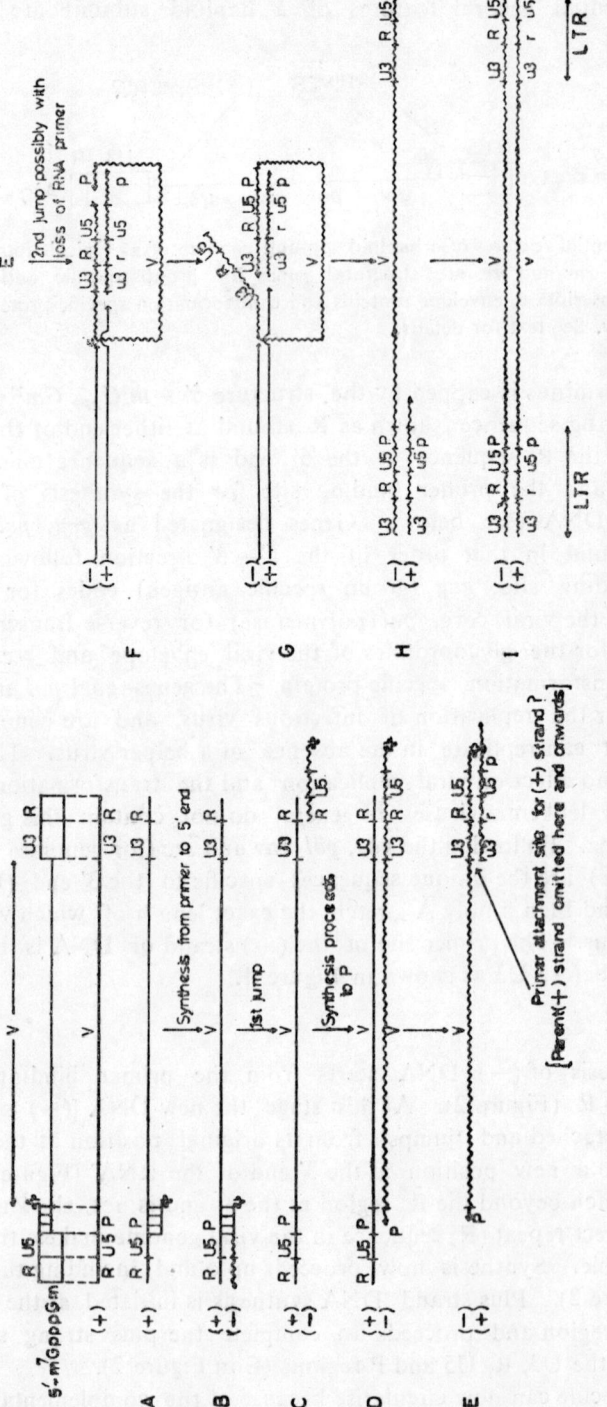

Fig. 2. Schematic representation of synthesis of a linear duplex DNA molecule from a single stranded DNA genome of a typical retrovirus. See text for details.

of the P region in the (−) strand and the p region in the (+) strand (F in Figure). From this configuration it is easy to visualise the completion of the double stranded DNA as shown in G, H and I in Figure 2. Notice that in this process the double stranded DNA has acquired the U3RU5 sequence (the long terminal repeat or LTR) at either end of the molecule[13].

src genes

There is compelling evidence from recovery experiments mode in a number of laboratories that the genomes of transforming viruses contain sequences acquired from the host cell that have the capability of transformation [14-30]. These sequences when present in the viral genome[14, 31-34] is referred to as the *src* gene (Figure 1). They are designated as *sarc* genes when located in the host-DNA. It also appears that more than one transforming gene may be present in the normal cells of a single species[14, 31-34]. Some of the retroviral specific sequences, not necessarily the *sarc* gene alone, found in normal cells emerged quite early in evolution, and they have been conserved over extensive periods during the course of evolution[35]. Most such genes are unique loci situated at constant positions on the chromosomes in a given species. Several of them also have intervening sequences.

Role in Evolution

Let us now briefly examine the significance of the above observations in a wider context. It is generally assumed that the expression of tumour specific host genes or proto oncogenes present in normal cells may not be deleterious to the cell. Further, it has also been argued that they may even be vital for normal growth and development providing tissue specific regulatory mechanisms[36]. Oncogenesis probably results when the protooncogenes are joined to regulatory elements which increase their expression[37]. On the other hand, despite earlier speculations and some tenuous experimental observations, there is very little concrete evidence for any gene amplificative function for RT in normal cells.

With regard to the evolution of retroviruses from cells, as early as 1970, Temin put forward the protovirus hypothesis[38] in which he suggested that retroviruses evolved from moveable genetic elements in cells. More recently he has analysed[39] all available data on sequences in cells and has attempted to provide a fair rationale for the origin of retroviruses. This idea still remains a possibility. So far, experimental support is available for the assimilation of retroviruses with protooncogenes by recombination. But, nothing is known about mechanisms that could eliminate intervening sequences of protooncogenes and generate the uninterrupted coding-units of viral oncogenes[40].

There has also been considerable interest in the similarity of the structure of the provirus and other transposable genetic elements like the *copia* like elements in Drosophila melanogaster[41, 42]. The similarity between *copia* like elements and the *Ty 1*, transposable elements discovered in yeast, Saccharomyas cervisiae, has also been pointed out. The striking similarity of the LTRs and the flanking regions of *copia* like elements is unmistakable[43]. Other than common mechanisms for circularisation and integration the structural similarity does not at present clarify any possible elements and retroviruses.

Nevertheless, even in the absence of full details of the sequence of moveable genetic elements or 'jumping genes' in all instances, the notion of a fluid genome in which there exists the possibility of a constant flux of sequences is being generally accepted[44]. The implication of these ideas with respect to evolutionary leaps assume considerable importance.

Do retroviruses or retroviral replication mechanism plays any role in evolution, in general? If tumour viruses, with the in-built capability of copying an RNA to DNA, are capable integrating their information in host chromosomes and in addition bring in foreign information by transduction, could they bring in fairly large chunk of foreign DNA into the host cell genome? In 1974, in fact, Zhdanov and Tickchonebko have suggested[45] the possibility of a tumour viral mechanism playing a role in evolution in addition to the generally accepted concept of point mutations and recombinations as factors responsible for the evolutionary process at a molecular level. Probably direct heterologous exchange of large units of genetic information between different species and even bigger taxonomic groups showing bigger differences is conceivable. Essentially, one is looking either at the neutral and asymptomatic or symbiotic relationship between a virus and its host in this situation.

There could be an evolutionary equilibrium balancing the acquisition of viral or virally brought information. In other words, viruses are being inserted into the germ line at very low frequency, after which they require many thousands of millions of years to be mutated away because they have little or no detrimental effect on the animals in which they are resident. Viruses that did have a detrimental effect would be lost rapidly, along with their hosts. Useful sequences could be selected.

These considerations assume importance at least with regard to the life cycle and evolution of retroviruses if not evolution in general. Among other things, a study of molecular mechanisms leading to efficient reverse transcription are of importance in that they will have a cumulative effect in any biological process dependent on RT function.

DNA Synthesis by Reverse Transcriptase: Role in Evolution 259

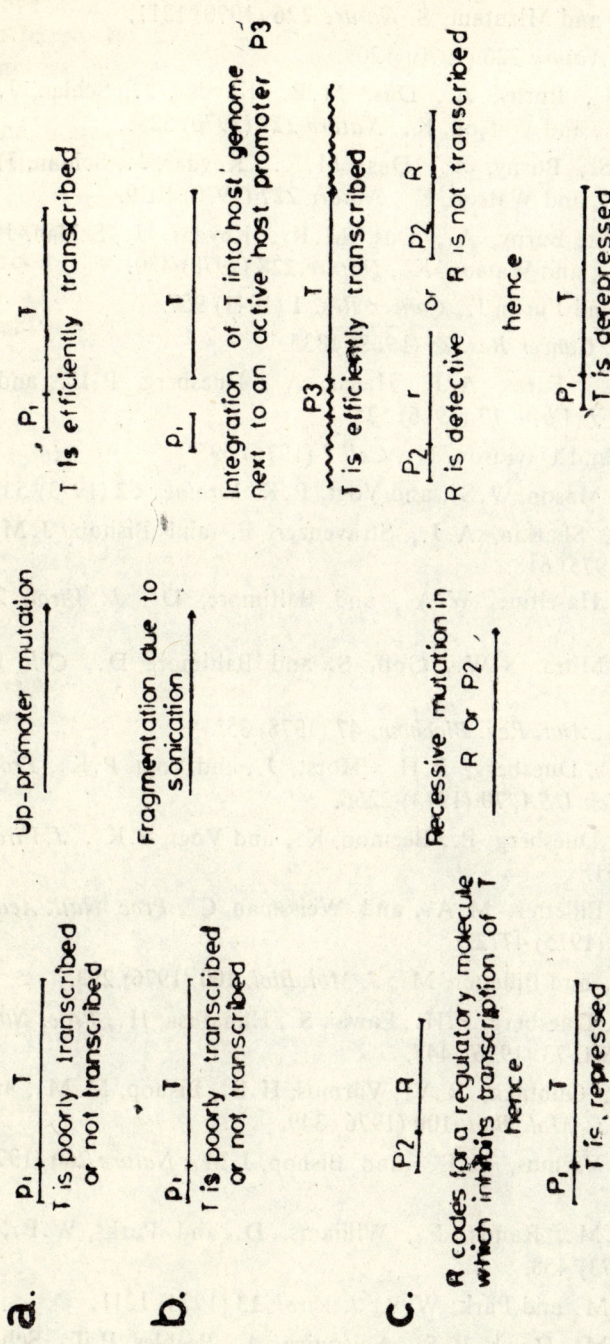

Fig. 3 Diagramatic representation of three possible modes of (over) expression of the transforming gene T. $p_1 P_1$ are inactive and active promoters of T respectively. R and P_2 are the structural gene and promoter for the diffusable regulatory molecule, r and P_2 are their recessive variants. P_2 is an active promoter in the 'host' genome. 'Host' refers to the cell into which DNA is transfected. See text for detailed discussion.

References

1. Temin, H.M., and Mizutani, S. *Nature* **226** (1970) 1211.
2. Baltimore, D. *Nature* **226** (1970) 1209.
3. Spiegelman, S., Burny, A., Das, M.R., Keydar, J., Schlan, J., Travnicek, M., and Watson, K., *Nature* **227** (1970) 5258.
4. Spiegelman, S., Burny, A., Das, M.R., Keydar, J., Schlan, J., Travnicek, M., and Watson, K., *Nature* **227** (1970) 1029.
5. Spiegelman, S., Burny, A., Das, M.R., Keydar, J., Schlan, J., Travnicek, M., and Watson, K., *Nature* **228** (1978) 430.
6. Cole, R.K., and Furth, J., *Cancer Res.* **1** (1941) 957.
7. Temin, H.M., *Cancer Res.* **28** (1968) 1835.
8. Beeman, K.L., Faras, A.J., Haase, A., Duesberg, P.H., and Maisel, J.E., *J. Virol.* **17** (1976) 525.
9. Bender, W., and Davidson, N., *Cell* **7** (1976) 595.
10. Weiss, R.A., Mason, W.S., and Vogt, P.K. *Virology* **52** (1973) 535.
11. Furuichi, Y., Shatkin, A.J., Stravenzer, E. and Bishop J.M., *Nature* **257** (1975) 618.
12. Rose, J.K., Haseltine, W.A., and Baltimore, D., *J. Virol.* **20** (1976) 324.
13. Gilboa, E., Mitra, S.W., Goff, S. and Baltimore D., *Cell* **18** (1979) 93.
14. Bishop, J.M., *Ann. Rev. Biochem.* **47** (1978) 35.
15. Lai, M.M.C., Duesberg, P.H., Horst, J., and Vogt, P.K., *Proc. Natl. Acad. Sci. USA.* **70** (1973) 2266.
16. Wang, L.H., Duesberg, P., Beemon, K., and Vogt, P.K., *J.Virol.* 16 (1975) 1051.
17. Joho, R.H., Billetter, M.A., and Weissman, C., *Proc. Natl. Acad. Sci. USA.* **72** (1975) 4772.
18. Coffin, J.M., and Billetter, M., *J. Mol. Biol.* **100** (1976) 293.
19. Wang, L.H., Duesberg, P.H., Eawai, S., Hanafusa, H., *Proc. Natl. Acad. Sci. USA.* 73 (1976) 447.
20. Stebelin, D., Guntaka, R.V., Varmus, H.E., Bishop, M.M., and Vogt, P.K., *J. Mol. Biol.* **100** (1976) 349.
21. Stehelin, D., Varmus, H.E., and Bishop, J.M., *Nature* **260** (1976) 170.
22. Scolmick, E.M., Rands, F., Williams, D., and Parks, W.P., *J. Virol.* **12** (1973) 458.
23. Scolmick, E.M. and Park, W.P., *J. Virol.* **13** (1974) 1211.
24. Scolmick, E.M., Howk, R.S., Anisouica, A., Peebles, P.T., Scher, C.D., and Parks, W.P., *Proc. Natl. Acad. Sci. USA.* **72** (1975) 4650.

25. Dina, D., Beeman, K., and Duesberg, P., *Cell* **9** (1976) 299.
26. Fraenkel, A.E., and Fischinger, P.J., *Proc. Natl. Acad. Sci. USA.* **73** (1976) 370.
27. Fraenkel, A.E., Neubaner, R.L. and Fischinger, P.J., *J. Virol.* **18** (1976) 481.
28. Hu, S., Davidson, N., and Verma, I.M., *Cell* **10** (1977) 469.
29. Vogt, P.K., in *"Comprehensive Virology"*, Frankel-Conrat, H. and Wagner, R.R., Eds. (Plenum, New York, 1977) *Vol.* 9, p. 341.
30. Hanafusa, H., in *"Comprehensive Virology"*, Frankel-Conrat, H. and Wagner, R.R., Eds. (Plenum, New York, 1977) *Vol.* 9, p. 401.
31. Roussel, M., Sanle, S., Lagran, C., Rommens, C., Beng, H., Graf, T., and Stehelin, D., Nature 281 (1979) 452.
32. Feldman, R.A., Hansfusa, T., and Hanafusa, H. *Cell* **22** (1980) 767.
33. Favera, R.D., Gelman, E.P., Gallo, R.C., and Wong-Staal, F., *Nature* **292** (1981) 31.
34. Van Beveran, C., Gallenshaw, J.A., Jonas, V., Berns, A.J.M., Doolittle, R.F., Donoghue, D.J., and Verma, I.M., *Nature* **289** (1981) 258.
35. Todaro, G.J., Callahan, R., Rapp, U.R., De Farco, J.F., *Proc. Boy. Soc. B.* **210** (1980) 367.
36. Chen, J., *Virol.* **36** (1980) 162.
37. Oskarsson, M., McClements, W.L., Blair, D.G., Maizel, J.V., and Vande, Woude, G.F., *Science* **207** (1980) 1222.
38. Temin, H.M., *Persp. Biol. Med.* **14** (1970) 11.
39. Temin, H.M., *Cell* **21** (1980) 599.
40. Bishop, J.M., *Cell* **23** (1981) 5.
41. Porter, S.S., Brorien, W.J., Dunsmur, P., and Rubin, G.M., *Cell* **17** (1979) 415.
42. Technkov, N.A., Ilyin, Y.V., Ananive, E.V., and Georgiew, G.P., *Nucleic Acids Res.* **6** (1978) 2169.
43. Finnegan, D.J., *Nature* **292** (1981) 800.
44. Lewin, R., *Science* **213** (1981) 634.
45. Zhdanov, V.M., and Tickchonebkov, Adv. Virus. Res. **19** (1974) 361.

Protein Information and the Living State

K. Sundaram and V. N. Viswanadhan

Department of Crystallography and Biophysics
University of Madras, Guindy Campus, Madras-600025

Introduction

A living system is an information sink in its immediate neighbourhood. While individual beings age and die, the cumulative total information of the population increases over several generations of the individuals. Over long enough time periods, evolution of a population would exhibit the same pattern of variation as the individual being. In terms of such an informational analysis, therefore, events such as chemical evolution, planetary evolution and the evolution of the universe are unified into a single kind of phenomenon (a cycle of rising and falling information) with different life spans. The continued increase of information in the growth phase is aided by the influx of energy (e.g., solar radiation). Fascinating features about evolution are the continued sustenance of information build up and the existence of a feed forward mechanism whereby the product of information growth at one stage acts as a substrate or a catalyst involved in the information jump of a later stage—as if ordained. To exemplify this statement we would like to point out the preevolution of water, clay matter, etc., in anticipation, as it were, of the prebiotic chemical evolution.

In the biosphere, proteins are important carriers of information. Proteins are mainly responsible for the gross structural features of organisms. Moreover, proteins also perform specific biochemical functions of regulation and control. The specificity of a protein is widely believed to be coded in some way in its three dimensional structure.

A quantitative measure of information carried by a protein through its specific three dimensional shape is likely to be helpful for understanding the role of proteins in biological evolution in the long run. Such measures can also be used for the quantitative characterization of protein specificities. Here we briefly review some methods we have used for the evaluation of information contained in protein shapes.

Protein Information from Sequential Internal Parameters

Consider a general protein chain, n units long, a section of which is shown in Fig. 1. The total shape of the protein can be expressed as a function of the torsion angles ϕ_i, ψ_i, $\varkappa_i$ and ω_i.

$$\text{Total shape} = f(\phi_i, \psi_i, \varkappa_i, \omega_i)_{i=1, 2, \cdots n}$$

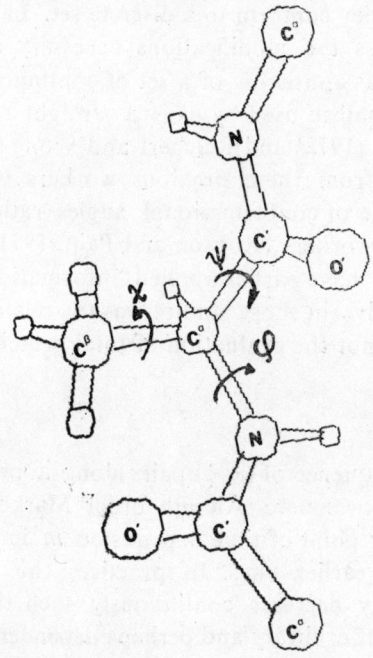

Fig. 1   The molecular formula of a segment of a general protein.

Further, the flow of the backbone in space is determined by the angle ϕ_i and ψ_i, the disposition of the side chains along being dependent on the angles $\varkappa_i$. Further, if we assume all peptides to be in the planar trans configuration,

$$\text{Total shape} = f(\phi_i, \psi_i)_{i=1, 2, \cdots n}$$

Let us denote by the single letter c_i the pair of conformational angles ϕ_i and ψ_i. The successive residues of c_i along a protein chain can be considered the output of a generalized Markov Source. The output at each point is an alphabet of a protein conformational language. One can work with alphabets c_i defined to different degree of fineness. Based on the empirically observed distribution of ϕ and ψ conformational angles in a large number of globular proteins a three letter model appears appropriate for most analysis. In this model the alphabet set of the language is

$$A\,(-180 < \phi \leqslant 0\,;\ \ -135 < \psi < 45),$$

$B(-180 < \phi \leqslant 0\,;\ -135 < \psi \leqslant 45)$ and

$C(-180 < \phi \leqslant 0)$

In the following we will first discuss the information theory of protein shapes based on this three letter language. The formulas are written in a generalized form, so that they can be readily used also for more refined alphabets so long as they conform to a discrete set. In a later part of this section we also discuss the modifications necessary for treating the conformational language as consisting of a set of continuous letters.

The theoretical formalism used by us is a straight forward extension of the methods of Gatlin (1972) and Reichert and Wong (1971). However, in significant departure from these previous workers we have applied the formulas to a sequence of conformational angles rather than to residue types. Robson and coworkers (Robson and Pain, 1971; 1974a & b; 1976; Garnier et al., 1978) have earlier applied information theory to protein shapes, but the objective in those studies was to relate backbone shape to residue sequence and not the evaluation of total protein shape information per se.

Discrete Model

Let us consider the sequence of (ϕ, ψ) pairs along a protein as the output of a generalized Markov source. An mth order Markov source is one of which the state at any point of time depends on m immediately precedent states and not on any earlier ones. In practice, the dependence on the previous symbols may decrease continuously such that the cut off point defining m is somewhat arbitrary and perhaps dependent on observational limitations.

Consider a generalized Markov source emitting a sequence of letters from among the set

$$S = \{x_i : i = 1, a\}$$

Let us denote by $p(i_1, i_2 \ldots i_n)$ the probability of occurrence of the n-tuple $(x_{i_1}, x_{i_2}, \ldots x_{i_n})$ and by $p_{i_1, i_2 \ldots i_n}$ the conditional probability of occurrence of x_n following the sequence $x_{i_1}, x_{i_2} \ldots x_{i_{n-1}}$. The probabilities satisfy the following relationships:

$$p(i_1 i_2 \ldots i_n) = p_{i_1}\, p_{i_1 i_2} \ldots p_{i_1 i_2 \ldots i_n} \qquad (1)$$

$$\sum_{i_1=1}^{a} \ldots \sum_{i_n=1}^{a} p(i_1 i_2 \ldots i_n) = 1 \qquad (2)$$

$$\sum_{i_m=1}^{a} \ldots \sum_{i_n=1}^{a} p(i_1 i_2 \ldots i_n) = p(i_1 i_2 \ldots i_{m-1}) \qquad (3)$$

$$\sum_{i_n=1}^{a} p_{i_1 i_2 \ldots i_n} = 1 \qquad (3)$$

For an mth order Markov source

$$p_{i_1 i_2 \ldots i_m i_{m+1} i_{m+2}} = p_{i_2 i_3 \ldots i_{m+2}} \tag{5}$$

There is a Shannon entropy associated with the distribution of n-tuples, given by,

$$H_n = -\sum_{i_1=1}^{a} \cdots \sum_{i_n=1}^{a} p(i_1 \ldots i_n) \log p(i_1 \ldots i_n) \tag{6}$$

Using equations (1), (4) and (5) we get, for the mth order Markov Source

$$H_n = H(0) + H(1) + H(2) + \ldots + H(m-1) + (n-m)H(m) \tag{7}$$

where $H(m)$'s are the Markov entropies of various orders defined by

$$H_m(k) = -\sum_{i_1=1}^{a} \cdots \sum_{i_{k+1}=1}^{a} p_{i_1} p_{i_1 i_2} \cdots p_{i_1 i_2 \ldots i_{k+1}} \log p_{i_1 i_2 \ldots i_{k+1}} \tag{8}$$

The maximum value of H_n is reached (see eqn. (6)) when all n-tuples have equal probabilities ($=1/a^n$) and we get $H_n^{max} = n \log a$. Further it can be shown that the differential entropy $H_m(k) - H_m(k-1)$ is a measure of the dependence of any symbol on the string of k preceding symbols; this differential entropy is a component of the information density characteristics of the source and has been given the notation D_K. The total information density considering n-tuples for a source of memory $n-1$, is,

$$I_d = D(0) + D(1) + \ldots + D(n) = \log a - H_{m-1} \tag{9}$$

As Reichert and Wong (1971) have pointed out the information density I_d and other related parameters like D_K and H_m are measures per character appropriate for long sequences like the string of nucleotides in a genome. For shorter sequences (sentences) such as a protein, more appropriate indices are the self-information per character, the total self-information for the sentence and related Markov information contents (entropies H_m), introduced by Reichert and Wong (1971).

Generalizing from the expression of Reichert and Wong (1971), we have, considering all n-tuples emitted by an $(n-1)$th order Markov source,

$$TI_{\text{self}_n} = \sum_{k=0}^{n-1} TI_{\text{self}_M}(k) \tag{10}$$

$$TI_{\text{self}_M}(k) = -\sum_{i_1=1}^{a} \cdots \sum_{i_{k+1}=1}^{a} n_{i_1 i_2 \ldots i_{k+1}} \log p_{i_1 i_2 \ldots i_{k+1}} \tag{11}$$

where $n_{i_1 i_2 \ldots i_{k+1}}$ is the number of occurrences of the directed string $x_{i_1} x_{i_2} \ldots x_{i_{k+1}}$ in the sequence

$$I_{\text{self}_n} = \sum_{k=0}^{n-1} I_{\text{self}_M}(k) \tag{12}$$

where

$$I_{\text{self}_M}(k) = TI_{\text{self}_M}(k)/N^{(k)} \tag{13}$$

$N^{(k)}$ being the sum,

$$\sum_{i_1=1}^{a} \cdots \sum_{i_{k+1}=1}^{a} n_{i_1 i_2 \ldots i_{k+1}}$$

The I's are the information per letter for the short sequence considered. TI's are the total information carried by the short sequence. While the numbers refer to the distribution in the short sequence under consideration, the probabilities $p(i_1 \ldots i_{k+1})$ refer to the distribution in the super source.

Continuous Letter Model

Extension of the above theory to the case of a continuous representation of the alphabet set is straight forward. We replace summations by integrations and introduce probability densities in place of discrete probabilities. Thus,

$$H_1 = -k \int p(c) \log p(c) \, dc$$

$$= -k \int_0^{2\pi} \int_0^{2\pi} p(\phi, \psi) \log p(\phi, \psi) \, d\phi \, d\psi$$

$$H^{\max} = 2 \log (2\pi)$$

$$H_M^{(m)} = -k \int \cdots \int p(c_i) \, p(c_i c_j) \ldots p(c_i c_j \ldots c_{m+1})$$

$$\log p(c_i c_j \ldots c_{m+1}) \, dc_i \, dc_j \ldots dc_{m+1}$$

where the p's are now probability densities around the argument conformations.

For the case of $m = 1$, we have,

$$H^D = H_1 - (n-1) \iint p(c_i) \, p(c_i c_j) \log p(c_i c_j) \, dc_i \, dc_j$$

This implementation of the continuous letter language resolves the problem associated with the uncertainties concerning the exact location of demarcations between the letters of a discrete letter language. However, it introduces another problem resulting from the equal treatment given to transitions between neighbouring letters and those between letters far separated. This bias can be corrected by working in terms of weighted transition probabilities

$$P'_{ij} = P_{ij} R_{ij} / \lambda_i$$

where R_{ij} is the distance on the (ϕ, ψ) space between the letters c_i and c_j and λ_i can be called the mean traversal distance of conformations out of c_i. More general such weighting schemes will have to be worked out for the higher order transitions.

Applications

1. Sequentially coded information of Globular Proteins:

The three letter model described above was applied to calculate informational parameters for a number of proteins. For the evaluation of the super source characteristics we pooled together the information available on 83 globular proteins whose three dimensional structures have been determined by X-ray crystallography. The data was obtained from the Protein Data Bank (Bernstein et al., 1977). Of the 83 protein structures, only 45 were independent, the others being closely related to those 45. Frequencies of various conformational letter strings were first averaged over groups of related proteins. Then the various letter string probabilities for the supersource were evaluated by averaging over all the 45 groups. The progression of selected letter string probabilities as the supersource data was built up is shown in Fig. 2. The arbitrary order in which the proteins were added to form the super source is indicated along the abscissa of Fig. 2. The name of the protein corresponding to the number listed on the abscissa can be found by reference to Table 1. As expected the probabilities fluctuate initially and stabilize as the super source size develops. It has been observed that for a super source if 'Hn/n' is plotted against n, it will have an asymptotic behaviour tending towards the limiting value of H_M for large n if the source is ergodic. We have computed Hn/n upto order 5 for the present super source and the values (see Fig. 3) seem to satisfy this criterion. In Table 1 we have listed self-Markov entropies per letter of the first (SHL1), second (SHL2) and third order (SHL3), their sum, the total self information per letter (TIL3) and the total self information of the whole protein (TI3) of several proteins. The higher the value of the total self information per letter (TIL3) the larger the occurrrence of less likely conformations and folds. The proteins that stand out with the highest total self information per letter are proteinase B, wheat germ agglutinin, alcohol dehydrogenase, cytochrome c from bonninto heart, high potential iron protein (HPIP), superoxide-dismutase and virus coat protein. In most cases, a little more than a third (37%) of TIL3 is contributed by the single letter part, SHL1, and approximately half the remainder comes from the two and three letter components each. HPIP alone is rather unique among globular proteins in having a relatively high triplet information content (37%). The sequence of local conformations in HPIP is shown in Fig. 4. On comparison with similar plots for other proteins (not shown here) the reason for the high triplet information density of HPIP becomes clear; HPIP has unusually large numbers of the rare triplet sequences, AAB, BAA, ABB and BBA. This is a subtle structural feature (in contrast to the prominent α-helix, β-sheet and reverse turn) discerned through this information theoretic analysis.

Information content in this formalism is also expected to correlate with

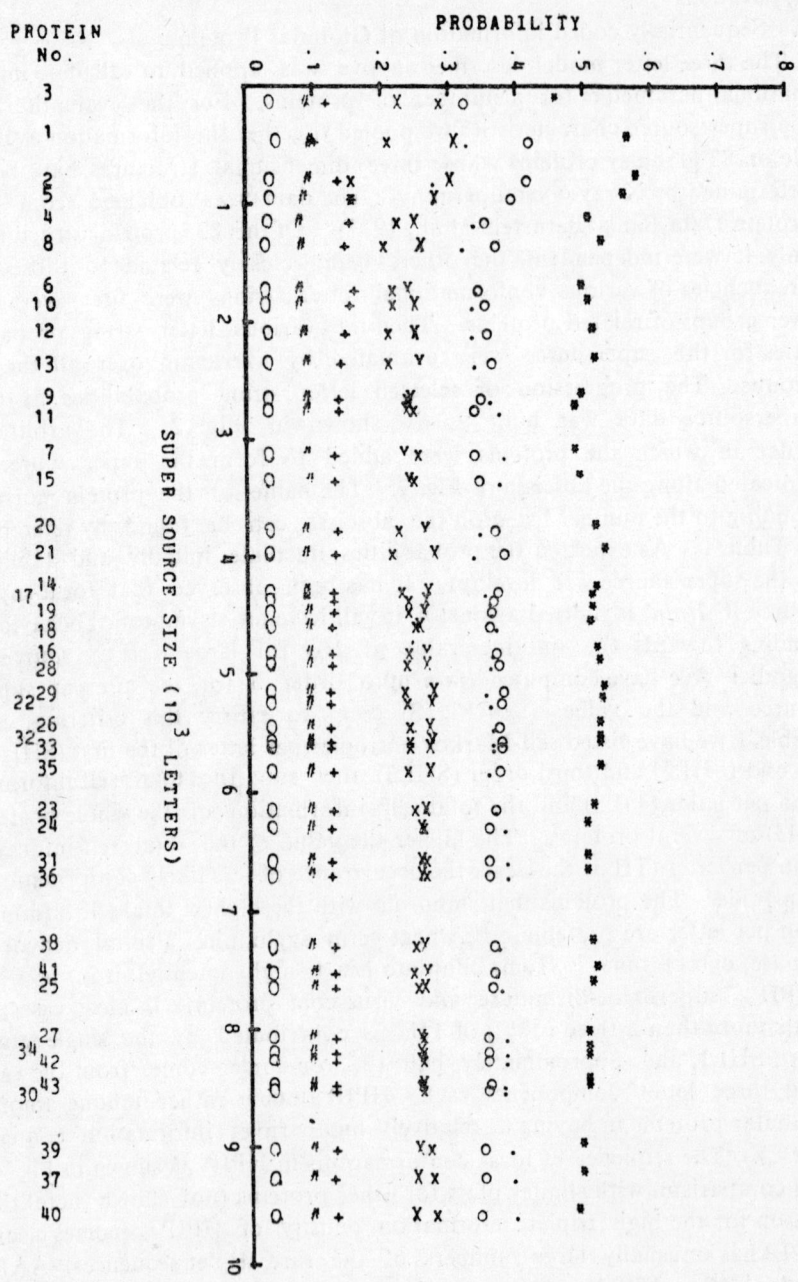

Fig. 2 Variation of selected individual letter and letter string probabilities with super source size.

Table—1 Self markov entropies per letter of the first (SHL1), second (SHL2), and third order (SHL3), total self information per letter (TIL3), and total self information for whole sequence (TI3) of several proteins and related structures.

NO.	PROTEIN	SHL1	SHL2	SHL3	TIL3	TI3 (×1000)
1.1	Acid Proteinase (Penicellum Janthinellum)	1.31	1.31	1.14	3.58	1.14
2.1	Acid Proteinase (Rhisopus Chinensis)	1.49	1.41	1.36	4.26	1.37
3.1	Actinidin (Sulfhydryl Proteinase)	1.41	1.21	1.17	3.79	0.81
4.1	Adenylate Kinase (Porcine Muscle)	1.50	1.03	0.96	3.49	0.67
5.1	Agglutinin (Wheat Germ)	1.64	1.50	1.53	4.67	0.75
6.1	Alcohol Dehydrogenase+ADP+RIB	1.57	1.39	1.34	4.30	1.59
6.2	Alcohol Dehydrogenase+Orthophen	1.57	1.39	1.34	4.30	1.59
7.1	Apo-Liver Alcohol Dehydrogenase	1.35	1.20	1.16	3.71	1.38
8.1	Alpha Lytic Protease	1.36	1.28	1.25	3.89	0.76
9.1	L-Arabinose-Binding Protein	1.44	1.14	1.06	3.63	1.10
10.1	Azurin	1.28	1.06	1.05	3.39	0.42
11.1	Calcium-Binding Parvalbumin	1.42	1.11	1.11	3.65	0.39
12.1	Carbonic Anhydrase B (Human)	1.34	1.30	1.26	3.90	0.96
13.1	Carbonic Anhydrase C (Human)	1.31	1.25	1.22	3.78	0.94
14.1	Carboxypeptidase A (Bovine)	1.38	1.19	1.12	3.70	1.12
15.1	Alpha Chymotrypsin (Tosyl)	1.30	1.21	1.20	3.71	0.85
15.2	Alpha Chymotrypsin	1.45	1.34	1.30	4.09	0.94
15.3	Gamma Chymotrypsin	1.30	1.20	1.20	3.70	0.85
15.4	Chymotrypsinogen	1.30	1.17	1.19	3.67	0.81
16.1	Concannavalin A	1.28	1.17	1.17	3.63	0.85
16.2	Concanavalin A	1.34	1.26	1.25	3.85	0.90
17.1	Cytochrome B5 (Oxidized)	1.37	1.08	1.08	3.53	0.29

Table—1: (Contd.)

NO.	PROTEIN	SHL1	SHL2	SHL3	TIL3	TI3 (×1000)
17.2	Cytrochrome B562 (B. Coli, Oxidized)	1.40	0.82	0.72	2.93	0.29
18.1	Cytochrome C (Albachore, Oxidized)	1.37	1.18	1.12	3.66	0.74
18.2	Cytochrome C (Albachore, Reduced)	1.37	1.17	1.11	3.65	0.37
18.3	Cytochrome C (Bonito Heart)	1.61	1.44	1.30	4.35	0.44
19.1	High Potential Iron Protein (Oxidized)	1.30	1.32	1.46	4.07	0.33
20.1	Immunoglobulin FAB	1.42	1.16	1.13	3.72	1.57
21.1	Immunoglobulin B-J Fragment (V-Dimer) REI	1.26	1.13	1.10	3.49	0.73
22.1	Insulin (Porcine, 2-ZINC)	1.35	0.93	0.95	3.23	0.30
23.1	Lactate Dehydrogenase	1.46	1.19	1.11	3.76	1.23
23.2	Lactate Dehydrogenase + NAD + PYRUVATE	1.44	1.12	1.06	3.62	1.18
23.3	Lactate Dehydrogenase (Mouse Testes)	1.51	1.22	1.16	3.89	1.27
24.1	Leghemoglobin	1.40	0.86	0.76	3.02	0.45
24.2	Myoglobin (Sperm Whale, Met)	1.37	0.85	0.77	3.00	0.45
24.3	Myoglobin (Sperm Whale, Met)	1.37	0.85	0.77	3.00	0.45
24.4	Myoglobin (Sperm Whale, Deoxy)	1.37	0.90	0.80	3.07	0.46
24.5	Myoglobin (Seal. Met)	1.41	1.03	0.95	3.38	0.55
25.1	Lysozyme (Bacteriophage)	1.45	1.23	1.23	3.91	0.48
25.2	Lysozyme (Hen Egg White, Set W2)	1.44	1.21	1.22	3.88	0.49
25.3	Lysozyme (Hen Egg White, Res 5D)	1.44	1.21	1.22	3.88	0.49
25.4	Lysozyme (Hen Egg White, Res 6A)	1.46	1.25	1.25	3.95	0.50
25.5	Lysozyme (Hen Egg White, Res 9A)	1.46	1.25	1.25	3.95	0.50
25.6	Lysozyme (Hen Egg White, Res 12A)	1.46	1.25	1.25	3.95	0.50
25.7	Lysozyme (Hen Egg White, Res 16A)	1.46	1.25	1.25	3.95	0.50

Table—1: (Contd.)

NO.	PROTEIN	SHL1	SHL2	SHL3	TIL3	TI3 (×1000)
25.8	Lysozyme (Hen Egg White, Triclinic)	1.45	1.20	1.20	3.86	0.49
25.9	Lysozyme (Iodine–Inactivated)	1.44	1.21	1.22	3.88	0.49
26.1	Papain (Native)	1.42	1.24	1.21	3.87	0.81
26.2	Papain (Ace-Ala-Ala-Phe-Ala, Cys–25)	1.41	1.23	1.20	3.84	0.81
26.3	Papain (Cysteinyl Der of Cys–25)	1.42	1.24	1.21	3.87	0.81
26.4	Papain (Oxidized Cys–25)	1.42	1.24	1.21	3.87	0.81
26.5	Papain (Tos–Cys, Cys–25)	1.42	1.24	1.21	3.87	0.81
26.6	Papain (Benzoxy-Gly-Phe-Gly, Cys–25)	1.42	1.23	1.21	3.86	0.81
26.7	Papain (Benzoxy-Phe-Ala, Cys–25)	1.42	1.24	1.21	3.87	0.81
27.1	Phosphoglycerate Kinase (Horse Muscle)	1.46	1.16	1.10	3.72	1.51
28.1	Plastocyanin	1.27	1.08	1.10	3.45	0.33
29.1	Prealbumin (Human Plasma)	1.26	1.08	1.06	3.41	0.76
30.1	Relaxin (Model, Confn. A, Unrefined)	1.38	1.11	1.06	3.56	0.17
30.2	Relaxin (Model, Confn. B, Unrefined)	1.38	1.11	1.06	3.56	0.17
30.3	Relaxin (Model, Confn. A, Refined)	1.40	1.08	1.03	3.51	0.17
30.4	Relaxin (Model, Confn. B, Refined)	1.40	1.08	1.03	3.51	0.17
31.1	Rhodanese	1.36	1.13	1.13	3.62	1.05
32.1	Ribonuclease A	1.28	1.02	1.01	3.30	0.40
32.2	Ribonuclease S	1.34	1.15	1.10	3.59	0.43
33.1	Rubredoxin (Clostridium Pasteurianum)	1.35	1.26	1.35	3.97	0.20
33.2	Rubredoxin (Desulfovibrio Vulgaris)	1.43	1.24	1.36	4.02	0.20
34.1	Staphylococcal Nuclease	1.45	1.17	1.13	3.75	0.52
35.1	Proteinase A (Streptomyces Griseus)	1.49	1.36	1.32	4.17	0.74

Table—1: (Contd.)

NO.	PROTEIN	SHL1	SHL2	SHL3	TIL3	TI3 (×1000)
35.2	Proteinase B (Streptomyces Griseus)	1.67	1.54	1.51	4.72	0.86
36.1	Suptilisin Inhibitor (Streptomyces)	1.35	1.28	1.21	3.84	0.40
37.1	Subtilisin (BPN')	1.39	1.23	1.23	3.84	1.04
37.2	Subtilisin NOVO	1.45	1.26	1.23	3.94	1.06
38.1	Superoxide Dismutase	1.43	1.39	1.35	4.17	2.50
39.1	Thermolysin (Unrefined)	1.46	1.15	1.11	3.71	1.16
39.2	Thermolysin (Refined)	1.49	1.21	1.16	3.86	1.21
40.1	Triose Phosphate Isomerase	1.37	1.05	1.00	3.42	1.67
41.1	Trypsin (Native, PH 8)	1.34	1.26	1.23	3.83	0.85
41.2	Trypsin (Benzamidine Inhibited, PH 7)	1.34	1.26	1.23	3.83	0.85
41.3	Trypsin+Trypsin Inhibitor	1.32	1.24	1.19	3.75	1.03
41.4	Trypsin (DIP Inhibited)	1.33	1.25	1.21	3.79	0.83
41.5	Trypsinogen+Trypsin Inhibitor	1.33	1.23	1.20	3.76	1.03
41.6	Trypsinogen+Trypsin Inhibitor+ILE-VAL	1.32	1.23	1.20	3.76	1.03
41.7	Trypsinogen (MGSO4, Without CA)	1.35	1.27	1.25	3.88	0.86
41.8	Trypsinogen (PEG, With CA)	1.35	1.27	1.25	3.88	0.86
41.9	Trypsinogen	1.35	1.30	1.28	3.93	0.86
42.1	Trypsin Inhibitor (Bovine Pancreas)	1.33	1.24	1.17	3.75	0.21
43.1	Virus Coat Protein (Southern Bean Mosaic)	1.54	1.38	1.33	4.25	0.89
44.1	Murein Lipoprotein (Hypothetical)	1.35	0.53	0.34	2.22	0.25
45.1	Valinomycin (Cyclic LD Depsipeptide)	2.10	2.42	2.09	6.61	
46.1	Pure Alpha Helix (Hypothetical)	1.35	0.53	0.34	2.22	
47.1	Pure Beta Sheet Strand (Hypothetical)	1.04	0.51	0.49	2.04	

Protein Information and the Living State   273

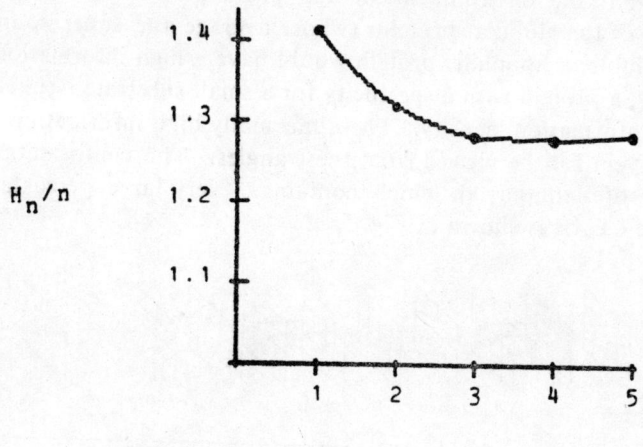

Fig. 3 Variation of H_n/n with source order.

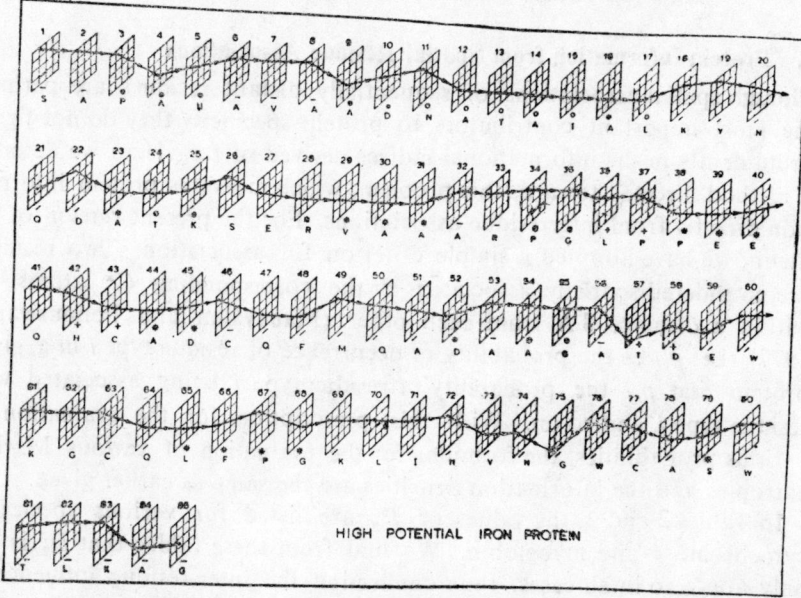

Fig. 4. The conformational trajectory of high potential iron protein (chromatium vinosum D), shown as a sequential array of the $2-D$ plots connecting the $(\phi-\psi)$ pairs at each α-carbon. Amino acid residues are indicated by their one-letter code abbreviated according to IUPAC–IUB Commission on Biochemical Nomenclature (1968). J. Biol. Chem., 243, 3557-3559. The secondary structure is represented by the graphic symbols given by Levitt and Greer (1977) ('O' for helix, '/' β-strand, '*' for right turn, '+' for left turn and '—' for coil conformations).

the nature of the environment of the protein. For instance, since the majority of the globular proteins (whose average the super source is) are water soluble, a lipophilic protein would have a high information content. Similarly, a protein with a specificity for a small substrate is likely to have a high information density. The abnormally high information density of Valinomycin can be viewed from these angles. The conformational letter sequence of Valinomycin which contains a very large percentage of rare sequence $CCAA$ is shown in Fig. 5.

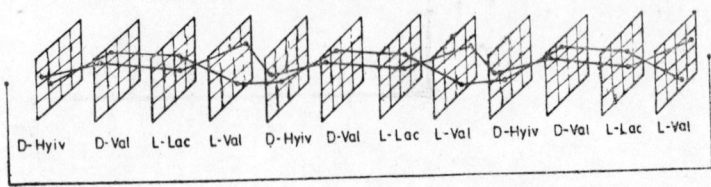

VALINOMYCIN

Fig. 5 The trajectories of two conformations of valinomycin (Sundaram, 1980). The connecting line indicates cyclicity.

2. Protein Information from Spatial Residue Associations

Though spatial associations of sequentially distant residues are perhaps the most important contributors to protein specificity they do not figure prominently in the informational indices derived so far. Here we describe a method directed towards this problem, which measures protein information directly from interresidue associations. For the present version of the theory we have adopted a simple criterion for association. Two residues are considered spatially associated if the corresponding C^α atoms are within a distance 8 Å from each other (Manavalan and Ponnuswamy, 1977). Let p_{ij} be the probability of occurrence of residue type i in a given protein and p_{ij} the probability or residue type i being associated with residue type j, and so on. With this prescription for the evaluation of various probabilities, the formulas for the evaluation of various Markov entropies and the information densities are the same as earlier given.

In Tables 2 and 3, the values of D_2 are listed for various species of cytochrome c and myoglobin. We find from these tables that D_2 is not only non-zero in all cases, thus confirming the inter-residue influence in three dimensions, but that D_2 is actually on the increase along the taxonomic hierarchy. There are, however, a few exceptions, we notice in particular the high value of sperm whale myoglobin. The tendency of D_2 to increase along the hierarchy is remarkably well preserved in both classes studied, indicating that non-bonded coupling effects are sharpened during protein evolution. The divergence from equiprobability, D_1 which is also evaluated (difference between the maximum possible compositional entropy and actual compositional entropy) and the information density, I_d (the sum of D_1 and D_2) do not significantly correlate with the taxonomic order and hence are not listed.

Table—2 Values of D_2 for Different Species of Cytochrome c

Sl. No.	Species	D_2 (in bits)
1	Human	0.331
2	Pig, Bovine and Sheep	0.297
3	Chicken	0.276
4	Snapping turtle	0.269
5	Bull frog	0.279
6	Bonito	0.290
7	Pacific lamprey	0.269

Table—3 Values of D_2 for Different Species of Myoglobin

Sl. No.	Species	D_2 (in bits)
1	Human	0.167
2	European hedge dog	0.136
3	Horse	0.132
4	Bovine	0.135
5	Badger	0.126
6	Slow loris	0.124
7	Sperm whale	0.175
8	Bottlenose Dolphin	0.140
9	Platypus	0.125
10	Chicken	0.117

A further point of investigation here is the kind of spatial coupling involved. An inspection of the conditional probability matrices showed that, the conditional probabilities between pairs of the same residue type are, in general high.

It may be observed that this sharpening of the inter-residue clustering during the course of evolution, is a simple way of enhancing the information density of the molecule. Early in evolution when polypeptides were formed by thermal copolymerization or via a primitive genetic code, specific preferences among residues were limited in nature. In the later stages, when there was a demand for increased specificity in structure, this was presumably achieved through sharp inter-residue associations and this seems to be a continuing trend in evolution.

Changes in amino acid composition are known to be the result of 'mutation pressure'. The observed increase in D_2 leading to enhanced specificity in inter-residue association cannot be readily attributed to 'genetic drift' or 'mutation pressure' and is possibly a direct indicator of 'selection pressure'.

Conclusions

What we have attempted to demonstrate through the above theoretical developments and actual applications to selected groups of proteins can be summarized as follows. Information theory can be applied to quantitate three dimensional features of proteins. By utilizing the concepts of super source and self information the parameters obtained are reduced to a "common denominator" enabling one to compare indices obtained for one protein with those of another. Information is stored in proteins through various forms of structural order. By studying these various forms of stored information vis-a-vis phylogenetic hierarchy, biochemical specificity, etc., it appears possible to understand the specific structures of proteins and the patterns of evolutionary residue replacements as goal directed rather than random.

Acknowledgement

The diagrams and tables presented here were obtained using a graphic package the development of which was aided by a grant from the Department of Science and Technology (India) and computer facilities extended by Dr. R. D. MacElroy of NASA Ames Research Center, U S.A. Department of Atomic Energy (India) provided a fellowship grant to V. N. Viswanadhan.

References

1. Bernstein, V.C., Koetzle, T.G., Williams, G.J.B., Meyer, E.F.Jr., Brice, M. D., Rogers, J. R., Kennard, O., Shimanouchi, T. and Tasumi, M. (1977). *J. Mol. Biol.*, **112**, 535–542.
2. Garnier, J., Osguthrope, D.J., and Robson, B. (1978), *J. Mol. Biol.*, **120**, 97–120.
3. Gatlin, L.L. *Information Theory and Living System*, Columbia University Press, New York (1972).
4. Levitt, M. and Greer, J. (1977). *J. Mol. Biol.*, **114**, 181–239.
5. Manavalan, P. and Ponnuswamy, P.K. (1977). *Arch. Biochem. Biophys.*, **184**, 476–487.
6. Reichert, T. and Wong, A.K.C. (1971). *J. Molec. Evol.*, **1**, 97–111.
7. Robson, B. and Pain, R.H. (1971). *J. Mol. Biol.*, **581**, 237–259.
8. Robson, B. and Pain, R.H. (1974a). *Biochem. J.*, **141**, 869–882.
9. Robson, B. and Pain, R.H. (1974b). *Biochem. J.*, **141**, 899–904.
10. Robson, B. and Pain, R.H. (1976). *Biochem. J.*, **155**, 331–344.
11. Sundaram, K. (1980). In Biomolecular Structure Conformation, Function and Evolution (Srinivasan, R., ed.), Pergamon Press, New York.

Biological Fields, Morphogenesis and Phase Transitions

B.C. Goodwin and J.L. Rius*

School of Biological Sciences,
University of Sussex, Falmer,
Brighton BN1 9QG, Sussex.

Introduction

Biological morphogenesis involves the appearance of localized demains of increased macromolecular and cellular order, giving rise to characteristic spatial patterns which transform one into the other in a regular and repeatable manner. The dimensions over which this order manifests are characteristically in the range of microns to millimeters ($10^{-6}-10^{-3}$ m), with large structures ($10^{-2}-1$ m) usually arising by growth of previously established spatial patterns. Since variety of form is what most distinctively characterises the biological realm, there is no need to do other than point to living organisms to recognise the distinctive morphologies which define different species. These morphologies, such as the shape of the lotus flower or the structure of the human arm, arise during embryogenesis from spatially homogeneous masses of cells, and so it is necessary to conduct analytical enquiries into the nature of the process which generate these patterns by investigations of the embryonic stages of the life-cycle.

Research in this area during the past century or so has given rise to two somewhat divergent types of description of the causal basis of biological pattern formation. On the one hand, studies in genetics and molecular biology have resulted in an "atomistic" description of organisms, their embryogenesis, and their evolution, which emphasizes their particulate nature as interacting aggregates of genes and macromolecules. The macromolecular products of genes are considered to give rise to spatial patterns as a result of self-assembly or crystallisation-type aggregation processes, which depend upon short-range forms of molecular interaction (electrostatic, van der Waals and Heitler-London forces). On the other hand, a less dominant tradition of analysis in embryology uses

*Present address: Department of Physics, UNAM.

the concept of fields of long-range order which organise molecules and cells into spatial patterns, specific morphologies arising from a combination of general field constraints together with specific "boundary conditions" which are partly determined by the proportions of specific macromolecules (e.g. gene products) and partly by other features such as electrical potentials across membranes.

Physicists have long learned to recognise the complementarity of particle and field descriptions, and in this paper we adopt the same view-point with respect to morphogenesis. Thus we accept that the spatial organisation that characterises living organisms is a manifestation of long-range and short-range field phenomena which both arise from and organise the molecules of which they are composed. Within such a perspective it is perfectly consistent to consider that at particular phases of its development the whole organism or extensive domains within it are predominantly field-like, in accordance with empirical observation, so that the appropriate descriptive language for morphogenesis stresses the wholeness of the field and its spatio-temporal transformations. This view has been advanced by a number of outstanding embryologists, among them Driesch, Needham and Waddington, and it is one which has been elaborated at length elsewhere (Webster and Goodwin, 1982; Goodwin, 1982). The problem then becomes that of characterising the properties of morphogenetic fields in mathematical terms, showing the nature of the transformation which characterise morphogenetic processes, and drawing some informed conclusions about the nature of these biological fields. Organisms could then be understood as self-organising entities conforming to the general physical principle of field-particle complementarity, though it may well turn out that the fields involved as well as the "particles" (i.e. macromolecules) are distinctive to the biological domain.

Order Parameters and Morphogenesis

The local domains of order arising during morphogenesis can be diversely characterised in terms of such processes as alignment of macromolecules (e.g. microfilaments, microtubules, collagen, etc.) condensation of cells as in chondrogenesis (i.e. cartilage formation, preceding bone formation), or specific deformations of cell sheets as in gastrulation (the transformation of a hollow one-layered spherical embryo into an elongated three-layered structure). It is therefore natural to seek a description of such processes in terms of the general concept of an order parameter, as was suggested by Haken (1978) Thus we assume that morphogenesis involves the passage of the developing organism through critical states which behave like second-order phase transitions, second-order because the new organisation arising from broken symmetry increases continuously rather than discontinuously during the morphogenetic process. Thus during cell division, chondrogenesis, or somitogenesis (formation of somites, the segmental muscles of the body axis in vertebrates), for example, new types of order gradually increase in "amplitude" as microtubles and microfilaments condense or cells aggregate.

Using the Landau assumption that phase transitions can be described in the neighbourhood of critical points solely in terms of a macroscopic order parameter u a potential function, F, analogous to the free energy, can be expanded in a power series for small values of u viz.

$$F = Fo + au + bu^2 + cu^3 + du^4 + \cdots$$

where a, b, c, etc. are analytic functions of $(\sigma - \sigma_c)$, σ being the energy supply to the system and σ_c its value at the critical point. The equilibrium phase is characterised by minimum F, hence

$$\frac{\partial F}{\partial u} = 0$$

Since the minimum must, by definition, correspond to $u = 0$, defining the critical point, we find that $a = 0$. Symmetry considerations result in the odd terms of higher order vanishing, so we are left with

$$F = Fo + bu^2 + du^4 + \cdots$$

Ignoring terms higher than fourth order, this gives for the minimum, the expression

$$bu_0 + du_0^3 = 0$$

The solutions are then $u_0 = 0$ and $u_0^2 = \dfrac{-b}{d}$. Since we want $u_0 = 0$ to be the only solution for $\sigma < \sigma_c$, whereas for $\sigma > \sigma_c$ we want a solution $u_0 = 0$, we take the σ-dependence of b and d such that $-\dfrac{b}{d}$ is positive for $\sigma > \sigma_c$ and negative for $\sigma < \sigma_c$. In addition, d is positive for all σ. Hence b must be negative for $\sigma > \sigma_c$ and positive for $\sigma < \sigma_c$. The simplest relations satisfying these conditions are

$$b(\sigma) = B(\sigma_c - \sigma)$$

$$d(\sigma) = d(\sigma_c) = d > 0$$

The familiar behaviour of F now follows, giving the picture of Figure 1 showing the way F changes in going through the phase transition.

When the order parameter is not a constant in space, the "free energy density", f, will depend not only on u but also on its spatial derivative. Assessing that u varies slowly, only the lowest derivative need be included in the expansion, giving,

$$f(u) = bu^2 + du^4 + \alpha(\nabla u)^2$$

where f is the value of F per unit volume. The expression for F is then given by the spatial integral over the domain D, viz,

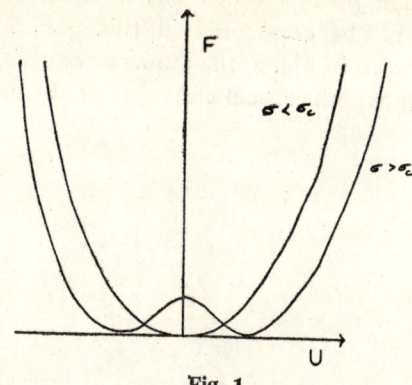

Fig. 1

$$F = \int_D f\,dV = \int_D [bu^2 + du^4 + \alpha(\nabla u)^2]\,dV \qquad (1)$$

If now the stable states of the system are taken to correspond to minima of the total "free energy", then the variational calculus applied to F gives the Euler-Lagrange equation for u, namely

$$\alpha \nabla^2 u + bu + du^3 = 0$$

Let us now consider how this general approach to morphogenesis can be applied to specific problems, and what the order parameter may described.

Cleavage

The first stage of morphogenesis in the majority of organisms whose development is initiated by the fertilization of an egg cell by a spermatozoan is a series of cleavage divisions (cell divisions without growth) which transforms the solid sphere of the egg into a hollow spherical ball of cells, the blastula (Fig. 2). This process shows a well-defined spatial and temporal organisation throughout a number of cleavages which differs between species, global order gradually decaying while local order persists. As the cleavage field decays, another global field increases in strength and gives

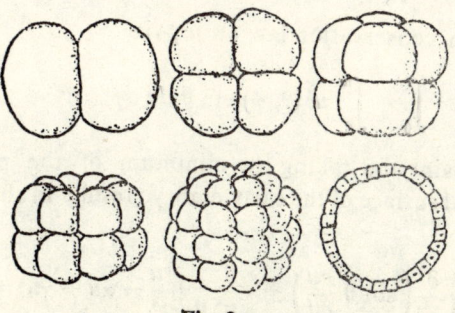

Fig. 2

rise to the gastrulation process which has its own characteristic spatial pattern, but this will not be considered in this paper. A typical initial cleavage pattern is shown in Fig. 3, the sequence of division planes showing orthogonal relationships with vertical cleavages preceeding horizontal ones.

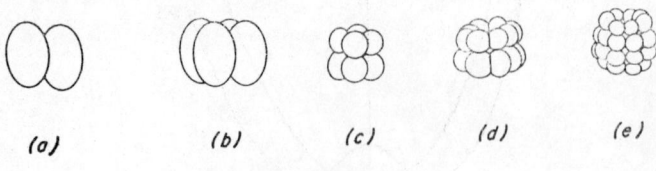

(a)   (b)   (c)   (d)   (e)

Fig. 3

The order shown in the lines of cleavages is a surface or 2-dimensional aspect of a three-dimensional organisation which involves the spatial orientation of the spindle-shaped mitotic apparatus consisting of centrioles and microtubules. Among the components involved in the surface order are the microfilaments which are located in the surface cytoplasmic layer or the cortex of the cell, and the cell membrane itself with its electrical potential of some 20–50 mV, which varies with the species of embryo and the stage of development. A projection of the presumptive cleavage planes back onto the spherical egg gives the pattern shown in Fig. 4.

The fact that there is a well-defined cleavage pattern over the whole of the developing embryo shows that there are global constraints operating, hence a global field, and that each cell is not behaving simply as an autonomous unit in the process. If the latter were the case, then cleavage planes could be at any relative angles in successive divisions rather than the orthogonal relationships observed. Restricting our description of the cleavage field to the cell surface, we can describe a potential function, F, in terms of an order parameter, u in accordance with expression (1), where we now use spherical coordinates to define position on the surface:

$$F(\theta, \phi) = A \int_0^{2\pi} \cdot \int_0^{\pi} \left[\beta u^2 + \delta u^4 + \left(\frac{\partial u}{\partial \theta}\right)^2 + \frac{1}{\sin^2 \theta} \left(\frac{\partial u}{\partial \phi}\right)^2 \right] \sin \theta \, d\theta \, d\phi$$

We assume also a conservation law on u^2, viz,

$$\int_0^{2\pi} \int_0^{\pi} u^2(\theta, \phi) \sin \theta \, d\theta \, d\phi = 1$$

Using this constraint and taking the minimum of the potential function over the spherical surface of the early embryo results in the Euler-Lagrange equation

$$\frac{1}{\sin \theta} \frac{\partial}{\partial \theta} \left(\sin \theta \frac{\partial u}{\partial \theta} \right) + \frac{1}{\sin^2 \theta} \frac{\partial^2 u}{\partial \phi^2} - \alpha u - \gamma u^3 = 0 \qquad (2)$$

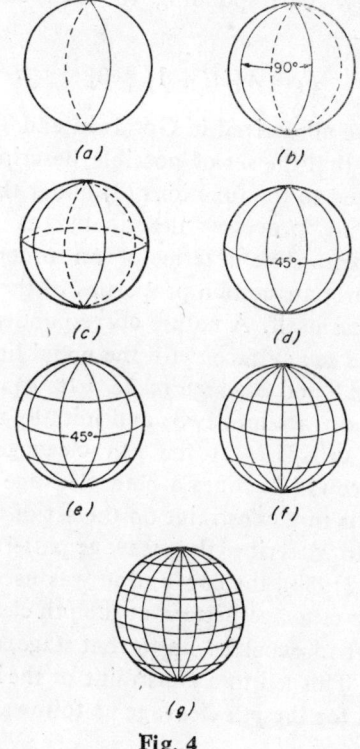

Fig. 4

where α incorporates the parameter β and on undetermined multiplier of the conservation constraint on u^2. Since the order parameter u takes arbitrarily small values near the transition point, in the neighbourhood of this point equation (2) has the linearised approximation

$$\frac{1}{\sin \theta} \frac{\partial}{\partial \theta} \left(\sin \theta \frac{\partial u}{\partial \theta} \right) + \frac{1}{\sin^2 \theta} \frac{\partial^2 u}{\partial \phi^2} - \alpha u = 0$$

which is the well-known Helmholtz equation. Assuming the order parameter to be finite, single-valued, and continuous over the sphere, the possible solutions are restricted to a characteristic (eigenfunction) set, viz., the spherical harmonics

$$u(\theta, \phi) = Y_l^m(\theta, \phi) = \sqrt{2} \, N_{lm} \, P_l^m (\cos \theta) \cos m\phi \qquad (3)$$

where l takes on the integral values, 0, 1, 2, etc., and for given l, the m values are integers ranging from $-l$ to $+l$. The parameter α is restricted to the corresponding characteristic values (eigenvalues) $l(l+1)$. In equation (3), the $P_l^m (\cos \theta)$ are associated Legendre polynomials and the N_{lm} are the normalisation constants given by

$$N_{lm} = \left[\frac{2l+1}{4\pi} \frac{(l-|m|)!}{(l+|m|)!} \right]^{1/2} \qquad (4)$$

The "surface free energy" corresponding to each characteristic cleavage state is given by

$$F_l = A[l(l+1) + \beta] \tag{5}$$

These relationships were all derived in Goodwin and Trainor (1980).

This analysis tells us that the set of possible descriptors of the ordered cleavage field is restricted to the functions (3) under the assumptions made about cleavage involving processes describable as second-order phase transitions which are expressible in terms of an order parameter $u(\theta, \phi)$. But we need now to correlate certain properties of the spherical harmonics with the cleavage process itself. A nature correspondence is to identify the cleavage furrows on the cell surface, with the nodal lines of the harmonics, i.e. with those values of θ and ϕ which make $u(\theta, \phi) = 0$. Since cleavage is binary (i.e., one cell divides into two) and initially synchronous, at each division the number of cells is 2^p at the pth cleavage. Identifying nodal lines with cleavage furrows (the lines where cleavage planes intersect the cell surface) then results in a constraint on the set of spherical harmonics which can be used to describe the cleavage pattern. In the paper by Goodwin and Trainor (1980), the convention was used that the harmonic function describing the cleavage pattern at the pth cleavage defined all the cleavage planes which had occurred up to that stage, so that a cumulative description was given. This led to a constraint or the integers (l, m) defining the harmonic function for the pth cleavage as follows:

$$2^p = l + 1 \quad \text{if } m = 0$$

$$2^p = 2m(l - m + 1) \quad \text{if } m \neq 0$$

If we now use a slightly different convention, that the harmonic function $Y_l^m(\theta, \phi)$ describes only the new planes occurring at the pth synchronous division, then we have a different constraint on the (l, m) pairs: if p is even, so that $p = 2n$, then $l = m = 2^{n-1}$; but if p is odd ($p = 2n + 1$), then $l = 2^{n-1}$, $m = 0$, except for the case $p = 1$ ($n = 0$), in which case the solution is $Y_1^1(\theta, \phi) = \sin\theta \sin\phi$, where $\sin\phi$ is taken instead of $\cos\phi$. These solutions, up to the 8th cleavage, are shown in Table 1.

Since the "free energy" involved in cleavage is solely a function of l as shown in (5), there is degeneracy of the pattern for $p = 2n$ and $2n + 1$ for any n. This is resolved by selecting the maximum value of m as the solution which is expressed first for given l, a selection rule which, as previously described (Goodwin and Trainor, 1980), may be interpreted biologically as the action of a weak field in the direction of the animal-vegetal axis which selects vertical over horizontal cleavages when there is a choice due to equal values of the "free energy". This field is evident in most holoblastically-cleaving eggs as a pigment and/or yolk gradient, revealing the animal-vegetal polarity. Thus the cleavage pattern is interpreted to arise from four factors: a global field constraint, "free energy" minimisation,

Table 1

Cleavage number (p)	Number of new cleavage planes	(l, m) pair	Solution: $Y_l^m(\theta, \phi)$
1	1	(1, 1)	$Y_1^1 = C_1 \sin\theta \sin\phi$
2	1	(1, 1)	$Y_1^1 = C_2 \sin\theta \cos\phi$
3	1	(1, 0)	$Y_1^0 = C_3 \cos\theta$
4	2	(2, 2)	$Y_2^2 = C_4 \sin^2\theta \cos 2\phi$
5	2	(2, 0)	$Y_2^0 = C_5(3\cos^2\theta - 1)$
6	4	(4, 4)	$Y_4^4 = C_6 \sin^4\theta \cos 4\phi$
7	4	(4, 0)	$Y_4^0 = C_7(35\cos^4\theta - 30\cos^2\theta + 3)$
8	8	(8, 8)	$Y_8^8 = C_8 \sin^8\theta \cos 8\phi$

The constants C_i are defined in equations (3) and (4).

binary cell division, and a weak polar field which resolves degeneracies. The description given here is a linearisation of a complex problem in which the selection rules express the non-linearities responsible for the stabilisation of particular field solutions during the passage of the developing embryo through the critical points of a series of phase transitions. What drives the embryo through this sequence of critical points is the periodic metabolism of the cell cycle and the systematic change in boundary conditions expressed in the selection rules, the whole defining an integrated four-dimensional process.

The Nature of the Order Parameter

A question of considerable interest relates to the microscopic origin of the order parameter in the cleavage process. It is often assumed in discussions of morphogenesis that spatial order comes about as a result of a systematic concentration difference in some biochemical species over the domain in which the pattern appears. Thus pattern formation in *Hydra* has been described (Wolpert, 1971; Gierer and Meinhardt, 1972) in terms of monotonically-varying concentrations of "morphogens" governed by equations of the type first studied by Turing (1952), whose demonstration of spatial symmetry-breaking by processes involving chemical reactions and diffusion led him to believe that he had discovered the chemical basis of morphogenesis. While this mechanism may well play an important role in some morphogenetic process, there are good reasons for doubting that it is a universal mechanism (Frankel, 1979). Certainly one needs much more than monotonic gradients to account causally for complex pattern formation such as vertebrate limb morphogenesis. Furthermore, the invariance of pattern observed in holoblastic cleavage over large variation in size, either between species or within a single species such as Chaetopterus (Wilson, 1928) is not readily accounted for by diffusion-reaction as the primary determinative process. This is because constancy of the diffusion parameter means constancy of chemical wavelength and hence a change of pattern for a change of size, rather than the observed invariance. Hunding's (1981) analysis of mitosis in Aulacantha shows that parameters other than cell size can be used to control spatial pattern in diffusion-reaction processes, but what seems to be emerging in all such studies is the effect of different non-linearities on the stabilisation or selection of solutions of Laplace's equation at bifurcation points. The treatment of the problem of spatial organisation in developing organisms presented here is therefore not only the more general one, but it also leads one to enquire into alternative types of field which could provide both the requisite spatial pattern and the forces needed to organise macromolecules into the spatial arrays which are such a dramatic accompaniment of cell division in particular, and morphogenesis in general.

One very interesting candidate for such a field comes from Frohlich's (1968) demonstration of the possibility of coherently-oscillating dipoles in biological systems, giving long-range forces of order $1/R^3$, where R is the distance from the dipole. Such a process was originally conceived of as being intimately linked to the electrical potential across the cell membrane, but can also arise within and between polarisable macromolecules such as proteins and nucleic acids. A discussion of the process in relation to the cell cycle, showing the correspondence of different aspects of the phonon and the pseudo-ferro-electric phase transitions of the theory to different phases of the cell cycle, has been presented by Cooper (1979), while a detailed study of this mechanism in relation to holoblastic cleavage has been carried out by Rius (1981). The latter analysis demonstrated how the harmonic functions described in Table 1 can be interpreted in terms of the order parameter of Fröhlich's model, which is the amplitude of the coherent electromagnetic vibrations. The nodal lines of the harmonic function then correspond to regions of zero amplitude of the order parameter, while the centrioles or microtuble organising centres of the mitotic spindle are oriented towards the regions of maximum amplitude by virtue of the attractive forces exerted by these excited regions, as shown in Figure 5. A consistent correspondence may thus be set up between the formal cleavage model described in terms of a field defined by an order parameter, and the specific theory of long-range coherence given by Fröhlich. Because this mechanism can organise polarisable molecules (e.g. microtubles, microfilaments) into spatial patterns with the symmetries and spatial order observed during cleavage, it provides a plausible basis for the cleavage field. There is, furthermore, experimental evidence now accumulating that coherent electromagnetic oscillations in the microwave range are a necessary accompaniment of cell division (Webb, 1980; Pohl, 1980; Zimmerman et al., 1981).

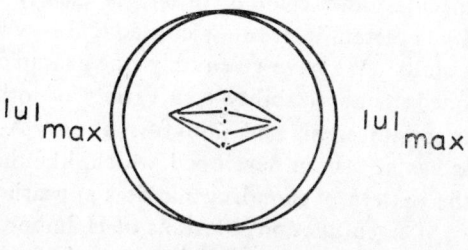

Fig. 5.

There is thus the exciting possibility that a field theory may now be available which can account for the cleavage process in terms which satisfy the requirements of a complete causal description involving spatial organising forces capable of generating the observed macroscopic patterns. This is in contrast to theories based upon diffusion-reaction processes and biochemical gradients in which the nature of the forces which produce the molecular and cellular patterns remain obscure and undefined. Additional assumptions are required about mechanisms which turn continuous gradients into discontinuous states, for which switch-like genetic control processes are commonly used (Davidson, 1968; Wolpert and Lewis, 1976). Such genetic discontinuities, however, fail to describe actual spatial patterns between cells of the type observed in the cleavage process, since these are concerned with the formation and the orientation of cleavage planes in cells which are in the same state; and they cannot be applied to spatial discontinuities of pattern within single cells as observed throughout the kingdom of the protists (single-celled organisms), so that one is really no closer to an understanding of a vast range of morphogenetic processes by the use of a purely biochemical mechanism for the generation of spatial order. This is not to say that such mechanisms play no role in morphogenesis. On the contrary, they are clearly of significance, including as they do the very important category of process in which the diffusing and transported species are actually charged ions, so that the electrical currents and potentials described by Jaffe and others (Jaffe et al. (1974), Borgens et al. (1977), in developing systems fall within this class of mechanism. The point to recognise is that finally one needs a way of going from continuity to ordered discontinuity, within and between cells, and one needs to suggest what the forces are which orientate macromolecules and organelles in space, again within and between cells. It is for these reasons that the theoretical approach based upon fields of long-range order and phase transitions is analytically useful and descriptively attractive.

The Field Approach to Biological Form

The approach to morphogenesis given above, wherein pattern formation results from global fields defined by an order parameter which describes the relevant molecular and cellular order, is clearly of quite general applicability and has certain interesting consequences with respect to one's view of the living state. We have given only one example of this approach, but have mentioned its applicability to a variety of other morphogenetic processes such as somitogenesis and limb formation. A detailed model of the latter process has now been developed which, like the theory presented here, describes the pattern of chondrogenic sites appearing in the developing limb as a sequence of eigenfunction solutions of Helmholtz's equation which are "selected" by specific boundary conditions and "energy" values which vary systematically along the proximodistal axis of the limb. Variation in this "energy" profile generates the variety of bone patterns that is observed in tetrapods. But what is equally important, the model provides a basis for understanding the regularity of the tetrapod limb pattern, that basic invariance of form which makes possible the recognition that each particular type of limb, whether that of the bird or the horse or the human, are all transformations of one another under changes in the generative parameters. In the same way, the great variety of pattern shown by species with holoblastic cleavage, whether radial, biradial, bilaterally symmetric, or spiral and with various degrees of size difference between dorsal and ventral blastomeres (cells), can all be described as members of the same set of field solutions selected by particular parameter values.

A research programme which arises from this view of biological form is to identify the properties of the groups to which biological morphologies belong, and within which they are equivalent under transformation. It has been found, for example, that the holoblastic cleavage pattern parametrised by the (l, m) pairs as originally defined by Goodwin and Trainor (1980) (La Croix and Goodwin, unpublished), belong to an infinite cyclic group. This approach to the classification or organisms of specific form (species) represents an attempt to discover a rational taxonomy in biology, the analogue of the periodic table of the elements in physics. The basic conception underlying such a research programme is the belief that the living state is one which is governed by laws of organisation or order, giving rise to a deep regularity or unity behind the diversity of its manifestations. And the particular form of this hypothesis expressed here is that it is the field properties of organisms which give rise to the underlying unity of the biological realm as expressed in morphology or form. The rich variety of pattern and form observed in organisms, reflected in the hierarchical classification scheme of species, genera, orders, classes, kingdoms etc., may then be systematically described by a hierarchy of constraints on the group structure defining degrees of relatedness or equivalence under transformation of morphologies. If such a scheme can be developed in biology, then this science will take on a structure much more similar to that of physics than its present empirical, historically-based systematics would suggest. From such a perspective, it is possible to imagine quite profound changes in the subject, some of which have been discussed elsewhere (Goodwin, 1982). Not the least of these would be the demonstration of the rational order and the principles of unity and harmony which many biologists have long felt to characterise the living realm.

References

1. Borgens, R.B., Venable, J.W. & Jaffe, L.F. (1977). Bioelectricity and regeneration. Proc. Nat. Acad. Sci. U.S.A. **74**, 4528–4532.
2. Cooper, M.S. (1979). Long-range dielectric aspects of the eukaryotic cell cycle. Physiol. Chem. and Phys. **11**, 435–443.
3. Davidson, E.H. (1976). Gene Activity in Early Development. Academic Press, N. Y. and London.
4. Frankel, J. (1979). An analysis of cell-surface patterning in *Tetrahymena*, pp. 215–246. In: "Determinants of Spatial Organisation", ed. S. Subtelny and I.R. Konigsberg. Academic Press, N. Y.
5. Frohlich, H. (1968). Long-range coherence and energy storage in biological systems. Int. J. Quantum Chem. **2**, 641–649.
6. Gierer, A. and Meinhardt, H. (1972). A theory of biological pattern formation. Kybernetik, **12**, 30–39.
7. Goodwin, B C. (1982). Development and evolution. J. theor. Biol., in the Press.
8. Goodwin, B.C. and Trainor, L.E.H. (1980). A field description of the cleavage process in embryogenesis. J. theor. Biol. **85**, 757–770.
9. Haken, H. (1978). Synergetics. Springer-Verlag, Berlin.
10. Hunding, A. (1981). Possible patterns governing mitosis: the mechanism of spindle-free chromosome movement in *Aulacantha Scolymatha*. J. theor. Biol. **89**, 353–385.
11. Jaffe, L.F., Robinson, K.R. and Nuccitelli, R. (1974). Local cation entry and self-electrophoresis as on intracellular localization mechanism. Ann. N. Y. Acad. Sci. **238**, 372–389.
12. Pohl, H.A. (1980). Oscillating fields about growing cells. Int. J. Quantum Chem.: Quantum Biology Symposium, **7**, 411–431.
13. Rius, J.L. (1981). A field description of cleavage and early morphogenesis. D. Phil. thesis, Sussex University.
14. Turing, A.M. (1952). The chemical basis of morphogenesis. Phil. Trans. Roy. Soc. B. **237**, 37–72.

15. Webb, S.J. (1980). Physics Reports, 60(4), 201–224.
16. Webster, G.C. and Goodwin, B.C. (1982). The orgin of species: a structuralist approach. J. Soc. Biol. Struct. In the press.
17. Wilson, E.B. (1929). The development of egg-fragments in annelids. Roux Archiv. 117, 179–210.
18. Wolpert, L. Positional information and pattern formation. Curr. Top. Devel. Biol. 6, 183–211. (1971).
19. Wolpert, L. and Lewis, J.H. (1975). Towards a theory of development. Fed. Proc. 34, 14–20.
20. Zimmerman, U., Vienken, J. and Pilwat, G. (1981). Z. Naturforsch, 36c, 173–177.

The Self-Organization of Dynamical Order in the Evolution of Metazoan Gene Regulation

Stuart A. Kauffman

Department of Biochemistry and Biophysics
University of Pennsylvania
School of Medicine
Philadelphia, Pennsylvania 19104

Introduction: Cell Differentiation

A higher metazoan such as man contains sufficient DNA per nucleus to encode on the order of 2,000,000 average size proteins [1–3]. A very large fraction of that total may not be codonic, nevertheless, the complexity of RNA transcript sequences now detectable in the nuclei of higher plant and animal cells is of the order of 20,000 to 100,000 distinct sequences. A central dogma of developmental biology is that, with unusual exceptions, the genetic constitution of each cell in a higher plant or animal is essentially identical, and therefore that differentiation of the zygote into an array of distinct cell types is associated with alternative, coordinated patterns of gene expression; each cell type is thought of as associated with a delimited set of expressed genes. Increasingly refined data are coming available both with respect to the regulation of gene expression and the overlap in expression in different cell types of one organism. Expression appears to be modulated at a variety of levels: transcription itself, processing the primary transcript by splicing, capping, and polyadenylation, transporting the mature message from the nucleus, translation to protein, subsequent secondary modifications of the protein [1]. With respect to the overlap of gene expression patterns in different cell types the observations depend somewhat upon whether consideration is restricted to cytoplasmic RNA, or to total nuclear plus cytoplasmic transcripts. For total transcript complexity, the central observations are that a very large fraction of sequences appear to be ubiquitously transcribed in all cell types of the organism. The numbers range from 60%–85% of total transcribed sequence complexity. Different cell types, therefore, tend to differ from one another in the transcription of a fairly small fraction of the total transcribed complexity, typically ranging from a few to 10%-15% [3-6]. Two typical cell types in one higher plant or animal might share a common 20,000 sequences transcribed in both, and differ in the transcription of 2000.

Orchestration of differentiation during development therefore depends upon mechanisms which coordinate the expression of very large numbers of distinct genes, and the evolution of differentiation must have been associated not only with the evolution of novel structural genes, but of novel arrangements of regulatory mechanisms to coordinate the expression of useful combinations of genes. A substantial amount of work has been devoted to the study of protein evolution. Almost nothing is known of the principles, means, mechanisms, obstacles, and solutions to the evolution of an adequate regulatory system. My purpose in this article is to show briefly three things: First, chromosomal mutations which duplicate and disperse genetic regulatory loci throughout the genome provide mechanisms which not only generate novel regulatory loci arrangements, but which drive toward a well defined "average" regulatory architecture. Second, even without further selection, such "average" regulatory systems spontaneously crystallize highly ordered dynamical behavior with sharply alternative, highly coordinated patterns of gene activities around a shared set of ubiquitously active genes. Third, fundamental features of this highly coordinated dynamical behavior provide just the proper preconditions for successful stepwise selection of favorable patterns of gene activities in evolution.

Ensemble Constraints: Evolutionarily "Average" Gene Regulatory Networks
While the mechanisms regulating gene expression are not fully characterized, it now seems clear that these include *cis* and *trans* acting genetic loci. Cis acting loci act on an adjacent domain of genes on the same chromosome. Transacting genes act on genes which may be different on different chromosomes, presumably through diffusible products. Families of such genetic loci have now been found in yeast, mouse, maize, and *Drosophila* [7-12]. Furthermore, it is now widely appreciated that tandem gene duplications have arisen in evolution [13-16], while recent evidence demonstrates that fairly rapid dispersion of some genetic elements occurs through chromosomal mutations [17] including transpositions, translocations, inversions, and recombination [17-20]. Dispersal of loci by processes such as these provide the potential to move cis acting sites to new positions and thereby create novel regulatory connections, opening novel evolutionary possibilities [8, 12, 17, 21]. This raises a general question: If duplication and dispersion occur with characterizable probabilities per site, then in the absence of selection, is it possible to build a statistical theory of the expected control structure of a genetic regulatory system after many such transformations?

An initial extremely simple approach to this complex question is illustrated in Fig. 1b. Here I have assumed that the genome has only four kinds of genetic elements, Cis acting (Cx), trans-acting (Tx), structural (Sx) and empty (—). All types of elements are indexed. Any cis acting site is assumed to act in polar fashion on all trans acting and structural genes in a domain extending to its right to the first blank locus. Each indexed trans acting gene, Tx, regulates all copies of the corresponding cis acting gene, Cx, wherever they may exist on the chromosome set. Structural genes, Sx, play no regulatory roles. In Fig. 1b I have arrayed

CHROMOSOME 1   C1 T2 S1—C2 T3 S2—C3 T4 S3—C4 T5 S4—

CHROMOSOME 2   C5 T6 S5—C6 T7 S6—C7 T8 S7—C8 T9 S8—

CHROMOSOME 3   C9 T10 S9—C10 T11 S10—C11 T12 S11—C12 T13 S12—

CHROMOSOME 4   C13 T14 S13—C14 T15 T14—C15 T16 S15—C16 T1 S16—

Fig. 1a. Hypothetical set of 4 haploid chromosomes with 16 kinds of cis acting (C1, C2, ...), trans acting (T1, T2, ...), structural (S1, S2, ...) and "blank" genes arranged in sets of 4 Cx. Tx, Sx—. See text.

CHROMOSOME 1   C1 T1 S1—C2 T2 S2—C3 T3 S3—C4 T4 S4—

CHROMOSOME 2   C5 T5 S5—C6 T6 S6—C7 T7 S7—C8 T8 S8—

CHROMOSOME 3   C9 T9 S9—C10 T10 S10—C11 T11 S11—C12 T12 S12—

CHROMOSOME 4   C13 T13 S13—C14 T14 S14—C15 T15 S15—C16 T16 S16—

Fig. 1b. Similar to 1a, except the triads are $C(x)$, $T(x+1)$, $S(x)$—. See text.

16 sets of triads of cis-acting, trans-acting and structural genes separated by blanks on four "chromosomes". A graphical representation of the control interactions among these hypothetical genes is shown in Fig. 2a, in which an arrow is directed from each labeled gene, shown as a dot, to each gene which it affects. Thus, in Fig. 2a, C1 sends an arrow to T1 and an arrow to S1, while T1 sends an arrow to C1. A similar simple architecture occurs for each of the 16 triads of genes, creating 16 separate genetic feedback loops. By contrast, in Fig. 1a each triad carries the indexes (Cx, $Tx+1$, Sx), while the 16th is (C16, T1, S16). This permutation yields a control architecture containing one long feedback loop, Fig. 2b.

To begin to study the effects of duplication and dispersion of loci on such simple networks, I ignored questions of recombination, inversion, deletion, translocation, and point mutations completely, and modelled dispersion by using transposition alone. I used a simple program which

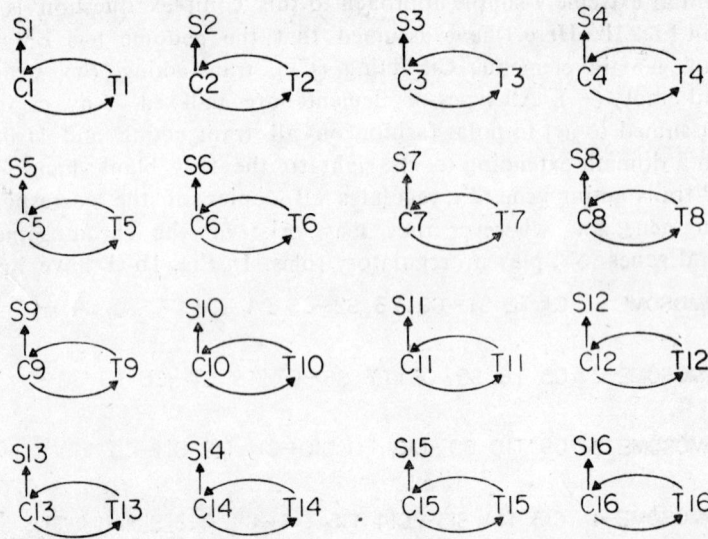

Fig. 2a. Representation of regulatory interactions according to rules in text, among genes in chromosome set of 1b.

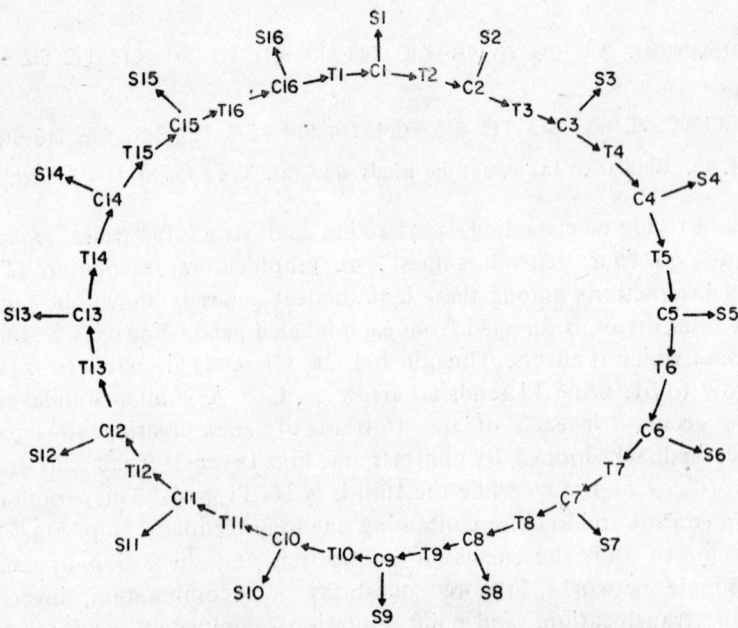

Fig. 2b. Regulatory interactions of chromosome set in 1a.

decided at random for the haploid chromosome set whether a duplication or transposition occurred at each iteration, over how large a linear range of loci, between which loci duplication occurred, and into which position transposition occurred.

Even with these enormous simplifications, the kinetics of this system is complex and scantily explored. Since, to simplify the model, loci cannot be destroyed, the rate of formation of a locus depends upon the number of copies already in existence, and approximately stochastic exponential growth of each locus is expected. This is modified by the spatial range of duplication, which affords a positive correlation of duplication of neighboring loci, and further modified by the frequency ratio of duplication to transposition. The further assumption of first order destruction of loci would decrease the exponential growth rates. However, the kinetics are not further discussed here, since the major purpose of this simple model is to examine the regulatory *architecture* after many instances of duplication and transposition have occurred. I show the results for the two distinct initial networks in Fig. 3a and 3b for conditions in which transposition occurs much more frequently than duplication, 90:10, and 2000 iterations have occurred. The effect of transpositions is to randomize the regulatory connections in the system. Consequently, while the placement of individual genes differs, the overall architectures of the two resulting networks look fairly similar after adequate transpositions have occurred.

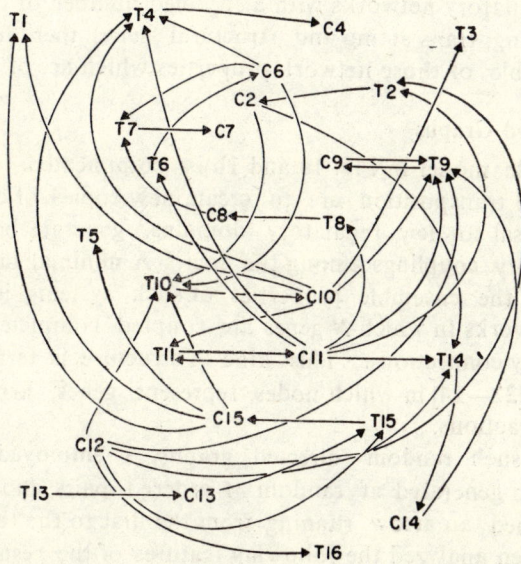

Fig. 3a. Regulatory interactions from chromosome set in 1a after 2000 transpositions and duplications have occurred, in ratio 90:10, each event including 1 to 5 adjacent loci.

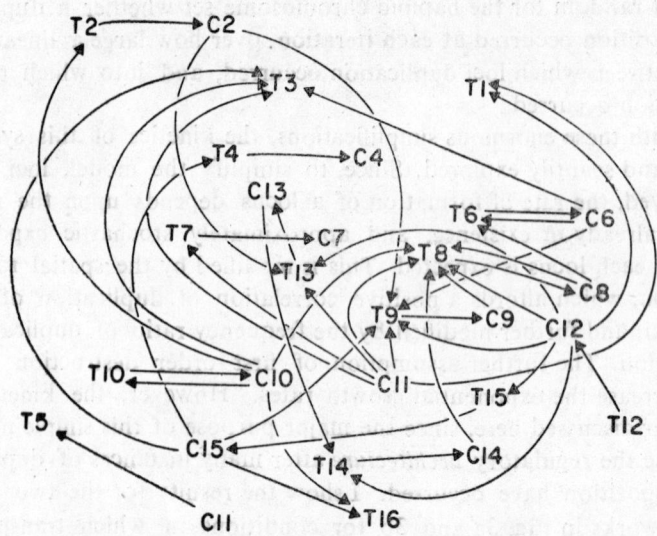

Fig. 3b. Similar to 3a, after random transposition and duplication in chromosome set from 1b.

This result raises the possibility of studying more formally the structurla properties of such randomized regulatory networks to seek their statistically stable features. Conceptually, this study would construct the ensemble of all possible regulatory networks with a specified number of copies of each type of cis acting, trans acting and structural gene, then form averages, over the ensemble, of those network properties which are of interest.

Random Directed Graphs

Even the simple model in Fig. 1a and 1b is complicated. The effects of duplication and transposition are to create new copies of old genes, and by their dispersal to new regulatory domains, generate both more, and novel, regulatory couplings among the loci. A minimal initial approach to the study of the ensemble properties of such systems is to study the features of networks in which N genes are coupled completely at random by M regulatory connections. This kind of structure is termed a random directed graph [22—24] in which nodes represent genes, arrows represent regulatory interactions.

To analyze such random directed graphs, I employed a computer algorithm which generated at random M ordered pairs chosen among N "genes", assigned an arrow running from the first to the second member of each pair, then analyzed the following features of the resulting directed graph: (1) The number of genes directly or indirectly influenced by each single gene, termed the *descendents* from each gene; (2) The *radius* from each gene, defined as the minimum number of steps for influence to progate to all its descendents; (3) The fraction of genes lying on genetic feedback loops among the total N genes; (4) The length of the smallest feedback loops for any gene which lies on a feedback loop.

As would be expected, these properties depend upon both the number of genes, N, and the number of regulatory arrows, M, which connect them. Figures 4a–d show the results for networks with 200 genes and regulatory connections, M, ranging from 0 up to 720. Figure 4a shows the mean number of descendent genes in a network of 200 genes. As M increases past N, a crystallization of large descendent structures occurs, and some genes begin directly or indirectly to influence a large number of genes. The curve is sigmoidal, and by $M=3N$, on average any gene can directly or indirectly influence .87 of the genes. Figure 4b shows the mean radius as a function of M. It is a non-monotonic function of the number of regulatory connections, low when connections are few and each gene can influence all of its descendents in a few steps, maximum when the number of connections is between $1.5N$ and $2N$, and each gene can influence a large number of other genes through a still relatively sparse network, then declining as additional regulatory connections provide shorter routes from each gene to influence all its descendents. As would be intuitively expected, the fraction of genes lying on feedback loops parallels the mean number of genes descendent from each gene. As M increases, the chance of forming closed loops goes up, Fig. 4c. Similarly, the lengths of the smallest feedback loops on which genes lie parallels the radius distribution, reaching a maximum for M between $1.5N$ and $2N$, then declining as M increases, Fig. 4d.

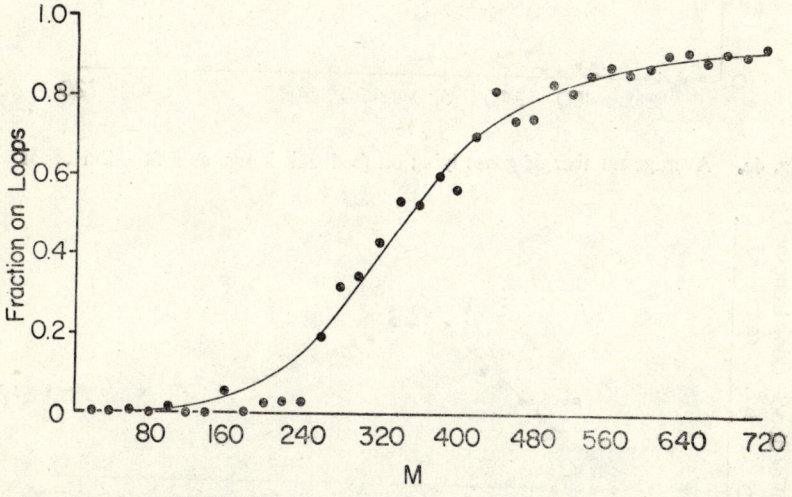

Fig. 4a. Descendent distribution, showing the average number of genes each gene directly or indirectly influences as a function of the number of genes (200) and regulatory interactions, M.

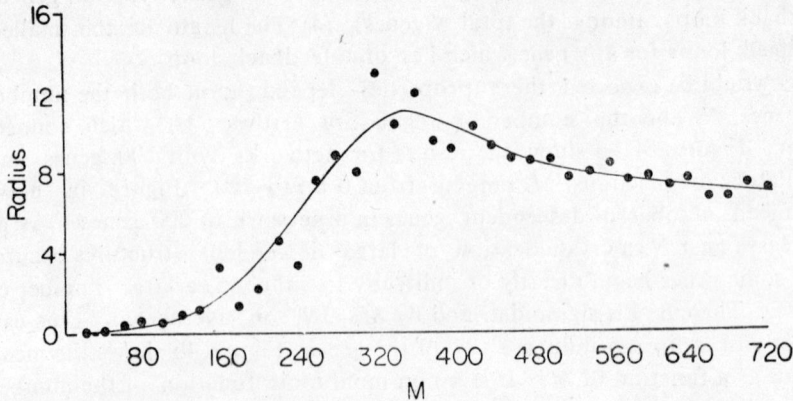

Fig. 4b. Radius distribution, showing the mean number of steps for influence to propagate to all descendents of a gene, as a function of M.

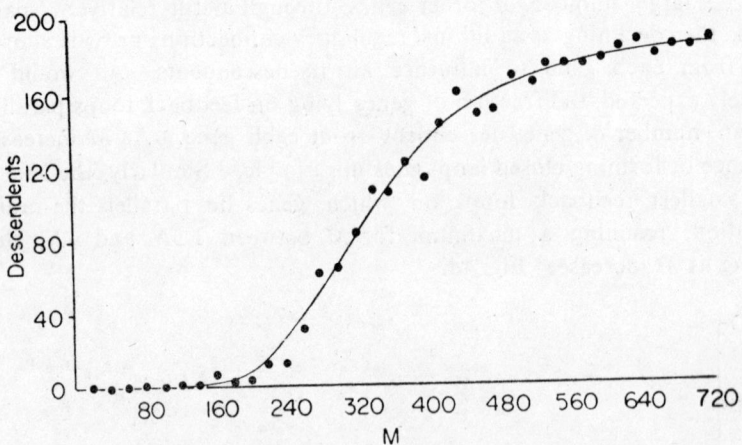

Fig. 4c. Average number of genes lying on feedback loops, as a function of M.

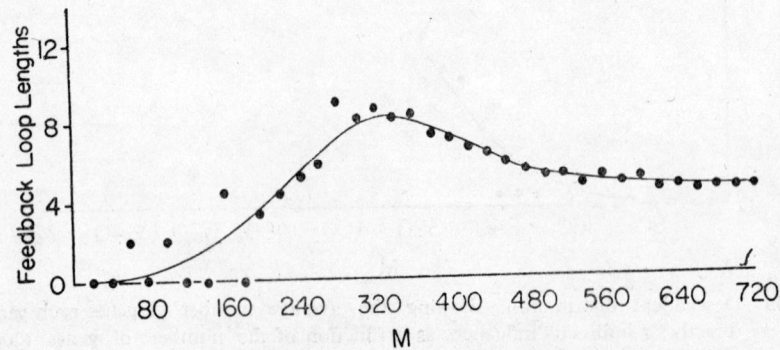

Fig. 4d. Average length of shortest feedback loops genes lie on as a function of M.

This brief analysis of the structural properties of random genetic networks suggests how robust such statistically typical features are expected to be. These first, simplest models of the expected regulatory architecture among the genetic elements in a eukaryote under the actions of duplication, transposition, translocation, recombination, deletion, point mutations, of course, are no more than the schemata of a theory. Their purpose, at this stage, is to indicate that such a theory should be constructable. Such a theory would characterize the statistically robust structural properties of large, evolving genetic regulatory networks in the absence of selection. Simultaneously, such a theory would demonstrate the form of the genetic network towards which those organisms would fall in the absence of selection, and perhaps allow an assessment of how strong selection would have to be to differ by a given degree from the average. In short, the mean statistical features of such networks are constraints in the evolution of the genomic regulatory system which selection must continuously overcome.

Crystallization of Coordinated Dynamical Behavior

Genetic regulatory networks are of importance because their organized dynamical behavior coordinates the activities of genes and their products in the regulation of cell differentiation. If it is possible to envision a theory which characterizes the expected *architecture* of gene regulatory systems under the actions of duplication, transposition, recombination, deletion, point mutations, in genomic evolution then it becomes of interest to assess the expected dynamical behavior of such regulatory systems. My own previous work [25-27] as well as that of a number of other authors [28-40] has already begun to shed some light on this problem. Again, analysis has largely been carried out only for the simplest kinds of dynamical models of regulatory systems, in which each gene is assumed to be a simple, binary, on-off device, regulated by input arrows from other binary genes in a randomized genetic network, in which M arrows ($M=2N$) connect N genes. The behavior of each binary gene is governed by randomly assigning to it one of the possible binary Boolean "switching" rules. For example, a given gene might be active at the next moment if any of its cis and trans acting regulatory genes are currently active, or if any is inactive, etc. Although formulated in terms of transcriptional regulation, the approach can include regulation of transcript processing and transport, by cis acting sequences on transcripts, and trans acting intranuclear signals [3].

Figures 5a-c exhibit the formal properties of such Boolean genetic networks. In Fig. 5a, three genes are shown, each receives regulatory input arrows from the other two. Two genes are activated at the next time moment if either input is active at the present moment (the OR function) while one is activated only if both inputs are active (the AND

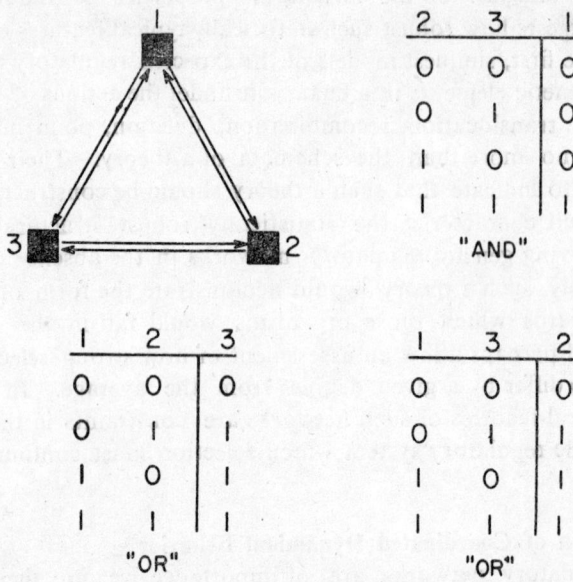

Fig. 5a. 3 genes, each regulated by the other two, according to the "OR" and "AND" Boolean function.

	T			T+1	
1	2	3	1	2	3
0	0	0	0	0	0
0	0	1	0	1	0
0	1	0	0	0	1
0	1	1	1	1	1
1	0	0	0	1	1
1	0	1	0	1	1
1	1	0	0	1	1
1	1	1	1	1	1

Fig. 5b. All $2^3=8$ states of binary activity of the 3 genes at time T, and the state at time T+1 into which each transforms.

function). Figure 5b rewrites this in terms of the current $2^N=2^3$ states of the small network, and the next state into which each state is transformed. The resulting sequence of state transitions is shown in Fig. 4c. Since there is a finite number of states, as the system follows a sequence of state transitions, it must arrive at a state previously encountered. Since the network is deterministic, it thereafter cycles repeatedly. If started at

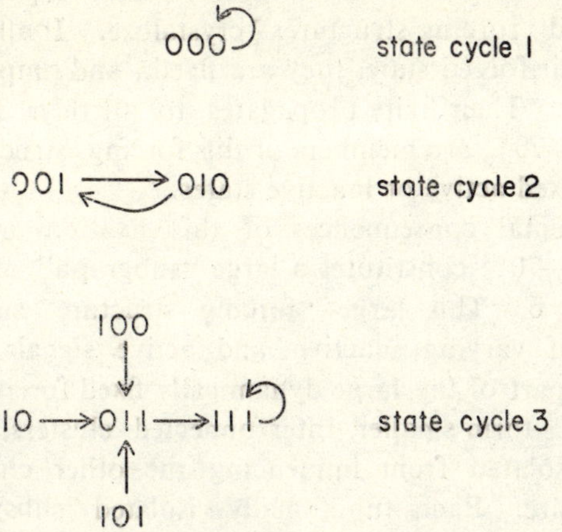

```
          000↺                  state cycle 1


       001 ⇌ 010                state cycle 2


           100
            ↓
       110 → 011 → 111↺         state cycle 3
            ↑
           101
```

Fig. 5c. The sequence of state transitions from 5b, showing 3 state cycles, see text.

a different initial state, the system may follow a sequence which runs into the first *state cycle*, or some different state cycle. Because each state either runs into or lies on a state cycle, each state cycle is an *attractor* lying in a *basin of attraction*, and the basins of attraction partition the entire space of 2^N states. The existence of at least one state cycle is a trivial property of these sequential finite automata. The interesting questions concern the number of states lying on one state cycle, which might range from 1 to 2^N, how similar the states lying on one state cycle are to one another, how many different state cycles are in the dynamical repertoire of the system, how stable any state cycle is to minimal perturbations reversing the activity of any single binary gene, how many genes are ubiquitously active (or inactive) on all state cycles, how different patterns of gene activities are on different state cycles, and a variety of further questions.

The quite surprising result of studies on large networks of binary genes, having up to 10,000 binary elements and $2^{10,000} = 10^{3000}$ possible combinations of gene activities, is that such randomized networks spontaneously exhibit extraordinarily constrained, ordered dynamical properties reminiscent in many respects to those found in even contemporary cells.

First, among the N genes, approximately 60%-70% settle into fixed "on" or fixed "off" states [25–27]. This fixation is due to the crystallization of a particularly powerful subsidiary control structure, termed a *forcing structure*, in these arbitrary binary dynamical systems [26, 27]. Although discussed only briefly here, forcing structures are characterized by regulatory connections having the property that one state (active or inactive) of a given gene suffices to force its descendent to a single state of activity, regardless of the activities of other genes influencing that descendent gene. If the fraction of regulatory connections which are

302 Stuart A. Kauffman

"forcing" surpasses the number of genes, then as expected from Fig. 4a-d, large connected "forcing structures" crystallize. If all genes on a forcing loop are in their forced state, they are fixed, and impervious to influence by other genes. Their fixity propagates to all their forced descendents. Typically 60%-70% are members of this forcing structure and are those which fall to fixed active or inactive states.

The fundamental consequences of this fixation are simply pictured. This fixed 60%-70% constitutes a large "subgraph" of the entire genetic network, Fig. 6. This large "forcing structure" subgraph blocks the propagation of varying inactive and active signals from genetic loci which are not part of the large dynamically fixed forcing structure. Those remaining genes form smaller interconnected clusters, each of which is functionally isolated from influencing the other clusters by the large forcing structure. Each functionally isolated subsystem has a small number of alternative dynamical modes of behavior into which it settles: alternative steady states, or oscillatory patterns of gene activities among the genes in that cluster. Since the clusters are functionally isolated, the alternative dynamical properties of the entire genetic system include the ubiquitously fixed active (and inactive) genes, and the *possible combinations* of the J alternative modes of behavior in the first cluster, the K of the second,... That is, the dynamical behavior is inherently combinatorial,

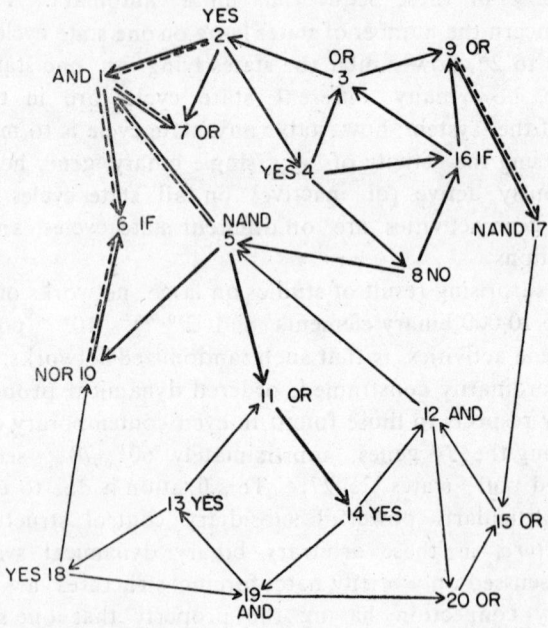

Fig. 6. A network of binary genes regulated by Boolean functions YES, NO, AND, OR, NAND, NOR. Dark arrows and dark arrows with parallel dashed lines comprise forcing structure (see text).

comprised by the ubiquitous set plus the possible combinations of the alternate modes of the functionally isolated subsystems, each of which provides a highly coordinated pattern of activities among a rather small subset of the nonfixed genes.

The observed simulation results [25] show that the entire large network settles into extremely small localized state cycle attractors, typically cycling among about $\sqrt{N}$ highly similar states. Thus model networks with 10,000 binary variables typically settle into attractors comprised of only 100 of the 10^{3000} possible combinations of gene activities. The number of possible alternative state cycle modes of dynamical behavior of the entire network is also roughly $\sqrt{N}$. Thus a model genome with 10,000 binary genes settles into one of about 100 alternative coordinated patterns of gene activity, each pattern comprised by a recurring set of about 100 similar states. These alternatives reflect the combinatorial behavior of the functionally isolated subsystems described above.

This allows initial construction of a picture of the coordinated gene activities corresponding to a cell type. The genes fixed in the active state would correspond to and satisfy the need in genetic regulatory systems for "housekeeping" genes ubiquitously active in all cell types. Each different cell type would be comprised in addition to the ubiquitous set by a unique combination of the alternative modes of behavior of the functionally isolated subsystems. The extent to which these model cell types differ from one another in patterns of gene expression (typically 5%-10%) and share a ubiquitously active set (typically 60%-80%) is surprisingly similar to the numbers observed in contemporary cells. Simulation results also show that each model cell type is stable to transient reversal of activity of most of the genes, one at a time, but that most cells can be stimulated to differentiate into one of a few "neighboring" cell types by transiently altering the activity of a few critical genes, mirroring the restricted pathways of differentiation open to contemporary cells. A particularly interesting feature of these model gene regulatory systems is that removal of a gene which is fixed *inactive* in all cell types does not alter those cell types, but may alter the response to perturbations away from the cell type attractor, hence can alter the *pathways* of *differentiation* between cell types induced by signals to key genes, while preserving the cell types themselves.

Space precludes a complete discussion, but the results presented suffice to indicate that even in genetic regulatory systems whose architecture is scrambled by duplication and dispersion of regulatory loci throughout the genome, and in which this scrambling leads to entirely *arbitrary* assignment of the control rule regulating the response of each gene to those controlling its activity, nevertheless, highly coordinated dynamical behavior spontaneously crystallizes. Even without selection, many of the dynamical properties of these model genomes are strongly reminiscent of those seen in contemporary cells. At the least, this suggests that our intuitions are wrong. Crystallization of orderly alternative coordinated patterns of gene activities about a set of ubiquitously active genes appears to be a *typical* property of large genomic systems which does not itself require substantial selection. The task of selection instead is to pick the particularly favorable combinations of genes which are to be coordinately expressed.

Preconditions for Selection

The next point I want to make is that the spontaneously crystallized orderly dynamics already provides the preconditions requisite for selection to act upon. In order for selective evolution to have succeeded it must be relatively simple to achieve at least a modestly workable starting system, then it is critical that *partial* successes can accumulate. This implies that a local rearrangement of regulatory architecture or modification of the control rule governing some single gene will not grossly alter the dynamical behavior of the entire system. Just this requirement is provided by the functional isolation of different clusters of genes. Each can be modified without often affecting the other clusters, allowing modifications in the patterns of activities of a small number of genes in one or a few types at a time.

Insufficient information is available to begin to test this hypothesis critically. However, it is of interest that among the Hawaiian picture wing *Drosophilae*, closely related species differ in quixotic ways in the array of tissues in which a battery of different enzymes are synthesized [10, 11], suggesting that piecemeal alterations to new combinations of gene activities in a few cell types at a time may be a common mode of evolution in gene regulation. Further, the capacity of systems with functionally isolated subsystems to accumulate partial successes suggests that maintenance of functionally isolated subsystems may have been advantageous and therefore that current differentiation may be fundamentally combinatorial. Such combinatorial features have been suggested on entirely independent grounds in *Drosophila* [41, 42], and mouse development [43].

A second aspect of the capacity to accumulate partial successes is also provided by the spontaneously crystallized dynamical order in these model genomic systems. As noted, deletion or addition of a gene which is *inactive* on all cell types can alter the *pathways* of differentiation among cell types by signals to critical genes. Thus such mutants can preserve useful cell types but alter the time and conditions of their occurrence. To evolve by piecewise alterations of developmental pathways, it is necessary that any such mutant affects only a *few* of the pathways of differentiation among cell types. Just this property is found in these model systems. Hence the capacity for restricted selectively advantageous modifications of pathways of development appears to be typical even of arbitrarily scrambled genomic regulatory systems. Since this capacity for piecewise modification of pathways itself seems to be highly useful, it is plausible that it may also have been maintained during the course of metazoan evolution.

Caveats

The results described apply to highly simplified models of gene regulation and its evolution by chromosomal rearrangements. The dynamical models are binary idealizations of gene regulation. Some grounds exist to think that the very powerful properties of forcing structures carry over to a class of continuous strongly sigmoidal rate equations of the forms:

Continuous	Boolean Homologue	
$\dot{x} = \dfrac{y^2}{\theta + y^2} - Kx$	(YES)	(1)
$\dot{x} = \dfrac{1}{\theta + y^2} - Kx$	(NO)	(2)
$\dot{x} = \dfrac{(y+z)^2}{\theta + (y+z)^2} - Kx$	(OR)	(3)
$\dot{x} = \dfrac{(yz)^2}{\theta + (yz)^2} - Kx$	(AND)	(4)
$\dot{x} = \dfrac{1}{\theta + (y+z)^2} - Kx$	(NOR)	(5)
$\dot{x} = \dfrac{1}{\theta + (yz)^2} - Kx$	(NAND)	(6)

For example, substitution of these continuous equations into the binary model in Fig. 6 yielded a system in which the "forcing structure" members assumed extreme active or inactive steady states expected from the binary idealization, while the remaining subsystem exhibited alternative steady states homologous to the binary system. Finding analytic results for these systems has hardly begun. The homologies, however, raise the hope that the powerful ordering properties seen in the binary system are robust enough to carry over to continuous systems. If so, a substantial principle of dynamical order is likely to be found in forcing structures. Beyond mathematical interest, however, if, as it appears, the ordering is not an artifact of the binary idealization, it will strongly support the suggestion from binary models that many dynamical features of contemporary gene regulation crystallize spontaneously in large systems of coupled biochemical reactions far from thermodynamic equilibrium.

References

1. Brown, D.D. *Science* **211,** 667 (1981).
2. Britten, R.J. and E.H. Davidson. *Science* **165,** 349 (1969).
3. Davidson, E.H. and R.J. Britten. *Science* **204,** 1052 (1979).
4. Chikaraishi, D.M., S.S. Deeb and N. Sueoka. *Cell* **13,** 111 (1978).
5. Goldberg, R. Personal Communication (1981).
6. Kleene, K.C. and T. Humphreys. *Cell* **12,** 143 (1977).
7. McClintock, B. *Cold Spring Harbor Symposium on Quantitative Biology* **21,** 197 (1957).
8. Sherman, F. and C. Helms. *Genetics* **88,** 689 (1978).
9. Paigen, K. In: *Physiological Genetics,* ed. J.G. Scandalias (Academic Press, New York, 1979).
10. Dickenson, W.J. *Science* **207,** 995 (1980a).
11. Dickenson, W.J. *Genetics* **1,** 229 (1980b).
12. Errede, B., T.S. Cardillo, F. Sherman, E. Dubas, J. Deshuchamps, J-M. Waine. *Cell* **22,** 427 (1980).
13. Rubin, G.M., D.J. Finnegan, and D.S. Hogness. *Prog. in Nucl. Acid Res. and Biol.* **19,** 221 (1976).
14. Wilson, A.C., S.S. Carlson, T.J. White. *Ann. Rev. Biochem.* **46,** 573 (1977a).
15. Wilson, A.C., T.J. White, S.S. Carlson and L.M. Cherry. In: *Molecular Human Cytogenetics,* ed. R.S. Sparks and D.E. Comings (Academic Press, New York, 1977b).
16. Livak, K.J., R. Freund, E.R. Schweber, P.C. Wensink, and M. Meselson. *Proc. Natl. Acad. Sci.* **75,** 5613 (1978).
17. Bush, G.L. In: *Essays on Evolution and Speciation* (Cambridge University Press, Cambridge, 1980).
18. Dooner, H.K. and O.E. Nelson. *Proc. Natl. Acad. Sci.* **74,** 5623 (1977).
19. Dover, G.A. In: *Insect Cytogenetics,* ed. R.L. Blackman, G.M. Hewett and M. Ashburner. Royal Entomological Society London Symposium 9 (Blackwell, Oxford, 1979).
20. Young, M.W. *Proc. Natl. Acad. Sci.* **76,** 6274 (1979).
21. Krone, W. and V. Wolf. *Hereditas* **86,** 31 (1977).
22. Erdos, P. and A. Renyi. In: *On the Random Graphs* **1,** vol. 6 (Inst. Math. Univ. DeBreceniensis, Debrecar, Hungary, 1959).
23. Erdos, P. and A. Renyi. In: *On the Evolution of Random Graphs,* publ. n. 5 (Math. Inst. Hung. Acad. Sci. 1960).
24. Berge, C. *The Theory of Graphs and its Applications* (Methusena, London. 1962).

25. Kauffman, S.A. *J. Theor. Biol.* **22**, 437 (1969).
26. Kauffman, S.A. In: *Current Topics in Developmental Biology* **6**, 145 (1971).
27. Kauffman, S.A. *J. Theor. Biol.* **44**, 167 (1974).
28. Walker, C.C. and W.R. Ashby. *Kybernetik* **3**, 100 (1966).
29. Newman, S.A., S.A. Rice. *Proc. Natl. Acad. Sci.* **68**, 92 (1971).
30. Glass, L., and S.A. Kauffman. *J. Theor. Biol.* **34**, 219 (1972).
31. Aleksander, I. *Int. Journ. Man/Machine Studies* **5**, 115 (1973).
32. Babcock, A.K. Ph.D. thesis, State University of New York, Buffalo (1976).
33. Cavender, J.A. *J. Theor. Biol.* **65**, 791 (1977).
34. Thomas, R. In: *Lecture Notes in Biomathematics* **29**, ed. by R. Thomas (Springer-Verlag New York 1979).
35. Sherlock, R.A. *Bull. Math. Biol.* **41**, 687 (1979a).
36. Sherlock, R.A. *Bull. Math. Biol.* **41**, 707 (1979b).
37. Walker, C.C. and A.E. Gelfand. *Behavioral Science* **24**, 112 (1979).
38. Gelfand, A.E. Submitted (1980).
39. Fogelman-Soulie, F., E. Goles-Chacc and G. Weisbuch. Submitted (1981).
40. Gelfand, A.E. and C.C. Walker. Submitted (1981).
41. Kauffman, S.A. *Phil. Trans. Roy. Soc. London* **B295**, 567 (1981).
42. Kauffman, S.A. *J. Theor. Biol.* **44**, 167 (1974).
43. Gardner, R.L. In: *Results and Problems in Cell Differentiation*, ed. W. Gehring (Springer-Verlag, Berlin, Heidelberg, New York, 1978).

Biological Patterns and the Problem of Regulation

Vidyanand Nanjundiah and Balakrishna L. Lokeshwar

Tata Institute of Fundamental Research,
Bombay 400 005
and
Indian Institute of Science,
Bangalore 560 012

Introduction

The word *regulation* means many different things in biology; here, we use it in the sense of scale-invariance of spatial patterning. Namely, a system is said to regulate if the relative sizes of differentiated structures in the adult are independent of the total size of the embryo. This is admittedly a somewhat loose formulation, particularly with regard to the meaning of "size" (which could mean any one of number, mass, volume, or linear dimension, for instance), but trying to be more precise at this stage will lead to an unnecessary restriction in the range of phenomena to be examined. Besides, experimental results with "regulative" systems are somewhat variable, and depend both on the system and on the pattern that is examined. For a discussion related to this point, see Sander (1971). Developmental biologists usually contrast regulative organisms with mosaic ones, in which local damage to the embryo typically results in a missing part in the adult (Davidson; 1976, Chap. 7). In this article we wish to draw attention to the phenomenon of regulation particularly from the point of view of its implications for theories of pattern formation. Much of what we will say has been said before by other workers, and E.B. Wilson's (1900, 1925) classic work, which contains references to (among other things) Driesch's original formulation of the problem, is even today a very useful reference.

The Problem

In biological systems it is meaningful to distinguish between an embryonic state in which a future pattern is determined but not expressed, and the adult state in which the pattern is visibly manifested (Balinsky (1970),

Chap. 10). The questions we address ourselves to here refer to the specification of the determined state. We are talking of events occurring over one to a few hours in an embryo consisting of about 10^3 cells and roughly 1 mm in linear dimension (e.g. the pre-gastrula amphibian embryo). These 10^3 cells give rise, after further division, growth and tissue rearrangements, to an adult with 10^{11} to 10^{12} cells made up of about 10^2 distinct cell types. Though the process of pattern determination is usually accompanied by other events, like those mentioned above, there are reasons to believe that one can think of it in isolation in many cases (see e.g. Cooke (1973)); at any rate, we shall do so here. Once one knows the position of a group of cells in the embryo, one can predict its future fate in the adult quite reliably (Balinsky, 1970). In a regulative system, the fate must depend, not on position as such, but on position relative to the size of the entire embryo. The case of identical twins, where each half-embryo gives rise to a normal foetus, illustrates this point very nicely. Such instances, where a reduction (or for that matter, increase) in overall size does not affect form, lead to two inferences. One is that there must be mutual communication among the cells of a developing embryo. The other, and more interesting, inference is that the form of communication must be such that a cell (or cell-group) is characterised by its relative position in the embryo. If we simplify matters and consider only one spatial axis (x) to be of importance, and the length of the embryo along that axis is L, the fate of a cell at position x depends on x/L. Put thus, the problem is to postulate plausible means of intercellular communication so as to generate function forms $f = f(x/L)$. Why is this a problem at all? A qualitative answer is that any dynamical model for signalling between cells would be expected to contain within it at least one natural scale of length. Then the number of units that made up a pattern would change with the total size of the system, and therefore the pattern itself would vary with size.

Experimental evidence

When the right kind of experiment is performed, regulation can be demonstrated in organisms ranging from unicellular ciliates to mammals. Here we cite three examples chosen as much for their diversity as for illustrative purposes. In all three cases we assume for the sake of convenience that the patterns are either one-dimensional, or that they can be decomposed into non-interacting patterns in matually orthogonal directions (Wolpert, 1971).

(i) Cortical pattern in ciliates

Ciliates are single-celled organisms which nevertheless exhibit striking patterns of spatial differentiation on their surface (or cortex). Their theoretical interest lies in the fact that such patterns, normally the outcome of cell-cell communication, are found here in a single cell with no internal compartments. To paraphrase Nanney (1971) slightly, the arithmetical and geometrical abilities of organisms reside here within one cell. The information given below relies heavily on two reviews by Frankel (1974, 1975); those and the article by Bakowska and Jerka-Dziadosz (1980) should be consulted for further references.

There are two sorts of evidence for regulation, one based on a study of normally developing populations, and the other based on observations following surgical intervention. The first is typified by the spacing pattern of the dorsal ciliary units in *Euplotes minuta*. The cilia form a rectangular array of units arranged in parallel vertical rows. The number of units per row ($u \div r$) is not constant; besides, the number of rows r varies from one individual to another. Frankel made a study of the distribution of units among rows in *Euplotes* from different strains, with r ranging from 7 ($u \approx 84$) to 10 ($u \approx 112$). After compensating for the fact that the percentage of units per row decreased with r, he found that at a given geometrical location this percentage was the same in all individuals. We may formally describe these results thus: let the width of the dorsal surface be L (this varies, but is not correlated with the number of rows), and let row i, at geometrical location x; when measured from one of the dorsal boundaries, have n: units; $u = \sum_{i=1}^{r} n:$. Then $r.n:/u$ differs from unity in a systematic manner, but only as a function of $x:/L$. Interestingly, this result also holds for genetically altered (mutant) strains of *Euplotes*, but the functional form of the dependence changes.

The surgical experiments result in a reorganization of pattern in the single cell. Though the data are not as quantitative as those discussed above, the basic observation is highly suggestive: the larger ciliates can adjust to the right sort of surgical fragmentation by re-adjusting their cortical pattern so that it looks normal. To take a specific instance, during development in *Paraurostyla weissei* out of the six cirri (bunches of cilia, used for crawling) near the anterior end, numbers 4, 5 and 6 reorientate and form ciliary streaks; these eventually give rise to new cirri. Numbers 1, 2 and 3 are anteriormost and get resorbed. However, if a transverse amputation removes 1, 2 and 3 from the cell, the *remaining* three cirri get ultimately resorbed. The ciliary streaks which ought to have developed from them now do so from a somewhat posterior location.

(ii) **Regeneration in the cellular slime molds**

These slime molds are soil amoebae which if well-fed live as singly growing and dividing cells, and if starved enter a gregarious phase and form an aggregation of (typically) 10^4 to 10^5 cells (Bonner, 1967). The rest of the development of this organism occurs without any further growth and very little cell division, if any. The aggregate migrates for some time as a polarised, slug-shaped mass (say $1\,\text{mm} \times .1\,\text{mm}$) characterised by a smooth tip at its anterior end. Here we have a situation in which the embryonic structure (the slug) is formed, not by multiple divisions of a single cell at one place, but by the coming together of a number of spatially separated identical units (but this is not unique to the slime molds; see Wourms, cited in Berrill (1971), Chapt. 22; Wireman & Dworkin, 1975). Normally the front of the slug differentiates into a thin, erect cellular stalk and the back into a mass of spore cells supported on top of the stalk. The ratio of spore and stalk sizes (as judged by a variety of criteria; see Bonner (1967)) varies very little even when the total number of cells in the aggregate varies over a factor of 1000. Also, the slug can stand repeated fragmentations perpendicular to its length; each fragment re-forms a polarised structure with a tip at its front and subsequently differentiates into a normal-looking fruiting body with stalk and spores (Raper, 1941; our own studies (unpublished) suggest that systematically cutting all the slugs on a plate into two halves does not affect the total yield of spores from that plate).

We are currently engaged in a study of pattern formation in the slime mold *Dictyostelium discoideum*. In particular, we are examining the time-distance relationships involved in the process of determination in the slug; this is with the hope that such studies will enable us to abstract empirical laws which might be used later for theorising. The most interesting result we have so far is that we have been able to confirm that there is a system of communication in this embryo which yields a scale-invariant pattern (Lokeshwar & Nanjundiah, 1982). The basic experiment involves amputating the slug (total length L) at various distances x from the front, and monitoring the time τ_R needed for a new tip to regenerate in the posterior fragment. We find that

$$\tau_R = C + \tau.x/L$$

with $C = -3.2$ min and $\tau = 152.3$ min. Therefore the time needed for tip regeneration is the same at the same percentage level of the cut (τ_R has a coefficient of variation of about 6% when x varies about five-fold, x/L being held fixed).

(iii) **Retina-tectum connexions in amphibia and fish**

When stimulated by light, photoreceptors in the retina pass on their message to a region of the brain (the optic tectum in lower animals) specialised to process visual information. The correspondence between the positions of receptor cells in the eye and associated tectal regions in the brain is both one-to-one (within the limits of experimental accuracy) and invariant from one individual to another (Gaze, 1970; Jacobson, 1970). Also, the retinal neurons project onto the tectum in an orderly fashion: for instance, if a visual stimulus is moved from the front of a frog towards its back (anterior to posterior), the subset of neurons that get stimulated in the tectum also moves from anterior to posterior. Experiments of interest to us here involve ablations of parts of the retina or the tectum before normal connexions have formed, the aim being to see how half a retina maps onto a normal tectum, or how a normal retina connects with half a tectum. The results are not as quantitative as one might like, and there are experiments suggesting a lack of regulation under certain circumstances (retinal axons being forced to connect with ipsilateral tectum; see the discussion in Straznicky et al. (1981)). The interested reader should consult the original literature for an appreciation of the complexities of the system. We shall take a somewhat optimistic view of the situation and consider those cases which indicate that regulation does occur. If half of the tectum is removed in a goldfish, the remaining half receives orderly but compressed projections from the entire retina (the operation does not create any blind spots; Yoon, 1976). Similarly, if two half-retinas, one from each eye, are juxtaposed within the same orbit, each half projects onto the entire tectum (Gaze, 1970; Jacobson and Hunt, 1973). Thus the retina-tectum map regulates. A parenthetical remark: one might wonder how these animals respond behaviourally. In the sunfish, the answer is that inspite of the visual field getting compressed onto half the tectum the animal can orientate accurately (Northmore, 1981). Therefore the map of tectal neurons onto motor pathways also regulates.

To sum up, positions x in the retina (linear dimension L_R, $0 \leqslant x \leqslant L_R$) are in register with positions y in the tectum (linear dimension L_R, $0 \leqslant y \leqslant L_T$). The dependence of y on x suggests a functional relationship of the form $y/L_T = f(x/L_R)$.

Implications for models of pattern formation

We will discuss these implications in terms of the concept of *positional information* (Wolpert, 1969, 1971); other useful words are *pre pattern* (Stern, 1968) and *map* (Robertson and Cohen, 1972). Positional information is equivalent to the abstract function $f(x/L)$ that we have been using so far; it is that property, varying with position, according to which a cell is committed to one pathway of differentiation as opposed to another. To give a popular example, it could reside in the concentration of a molecular species (the morphogen), and all cells in which the concentration was above a certain threshold would differentiate in one fashion while the rest of the cells would do so in a different fashion. Put this way, the problem is to generate a scale-invariant system of positional information over a field a few hundred cells (≈ 1 mm) in length and within a time of 1 hour or so (Wolpert, 1969). Before we consider some of the ways one might do so, a few remarks are in order about what we will not do.

There are classes of models that we will not discuss here; these include models in which the mechanical properties of cells and tissues are important (Gierer, 1977). The reason for this is that forces of deformation and hydrodynamical flows do not appear to be central to the problem we have mentioned. Also, in what follows we will not refer to catastrophe theory as applied to biological development (Zeeman; 1975a, 1975b) because we are unable to assess its significance. Finally, we will not attempt to make a distinction between local, "short-range" or "inductive" methods of pattern generation and global, "long-range" or "field" ones (Vogel, 1978). This is not to say that patterns cannot be generated by nearest-neighbour interactions between cells (or between regions on a cell) as opposed to a scheme in which most cells play the role of passive responders and only a few act as sources (or sinks). However, it is our feeling that the experimental evidence at hand does not allow us to make a clear distinction between local and global theories for regulative patterns. Besides, the formal structure of the theory may not be sensitive to such a distinction. After all, the prototype of a field equation—Laplace's equation—has as its analogue in a discrete system something which depends only on local differences. Consequently, even when we use—for instance—the diffusion equation we do not mean to imply that matter is necessarily being transported through the tissue in the manner it would be if a concentration gradient were set up in a fluid (see below).

Before listing the models, we note that it is sufficient to consider smoothly varying system of positional information, since it is not difficult to translate these into discrete patterns (Edelstein, 1972). We will consider two classes of models, the first involving chemical morphogens and the second with positional information being specified in some other manner. Our main reason for doing so is that chemical models for morphogenesis have been studied extensively, and they involve quantities (e.g. diffusion constants, rates of reaction, receptor affinities) whose magnitudes are familiar—something which one cannot say in general in the other cases. We will not attempt any detailed comparison of theory and experiment, but will restrict ourselves to examining how various assumptions could lead to scale invariance. For reasons already given, we consider only one spatial dimension.

(a) Model with chemical morphogens

Positional information is specified by a set of chemical (morphogen) concentrations. There are many reasons for taking these models seriously, among them being the number of experimental results which have been fairly satisfactorily explained (Wolpert et al., 1971; Lawrence et al., 1972; Meinhardt, 1977). For one thing, the time—and distance-scales agree well with what might be expected on the basis of diffusion of a small molecule ($L = 1$ mm, $t = 1$ hr, $D \sim L^2/t \approx 3.10^{-6}$ cm^2 sec^{-1}) from cell to cell (Crick, 1970). The simplest model one has under this class is a:

(i) Source-Sink model

The field extends from point A (the "source") to point B (the "sink"). Morphogen concentrations at A and B are held fixed at C_A and C_B respectively, and the morphogen differs freely inbetween. Then its concentration is given by

$$C(x) = C_A + (C_B - C_A)x/L$$

where L is the distance AB and x is measured from A (Fig. 1(1)). This simple scheme shows perfect scale-invariance, but has one serious problem, and that is the requirement of fixed boundary conditions (Wolpert, 1969); it is not at all clear why fragmenting the field ought to result in the same boundary conditions as before. For this reason, one prefers to consider equations with either no-flow $\left(\hat{n} \cdot \dfrac{\partial c}{\partial x} = 0\right)$ or mixed, but symmetric $\left(\hat{n} \dfrac{\partial c}{\partial x} = aC\right)$ boundary conditions. A moment's consideration shows that the simple diffusion equation is no longer very good. This leads us to

(ii) Reaction-Diffusion models

The standard form is

$$\frac{\partial c}{\partial t} = \nabla \cdot (D\nabla C) + R(C),\ 0 < x < L,\ \hat{n} \cdot \nabla C = hC,\ x = 0, L,$$

where C is the vector of chemical concentrations and R that of reactions; D is in general a matrix of diffusion coefficients. There are two ways of approaching this problem. The first, due to Turing (1952), examines instabilities (if any) that occur in the system when it is perturbed about a spatially uniform steady state. The wavelength of the most rapidly growing instability is assumed to set the scale of pattern, at least for short times. The second approach is to directly look for stable solutions of the time-independent equations. The basic results are as follows. First of all, one needs at least two chemical components if no-flow boundary conditions are used; a one-component system does not have any stable spatially non-uniform solutions (Casten and Holland (1978), Fife (1979)). With a two component system

$$\left.\begin{aligned}\frac{\partial C_1}{\partial t} &= D_1 \frac{\partial^2 C_1}{\partial x^2} + R_1(C_1, C_2) \\ \frac{\partial C_2}{\partial t} &= D_2 \frac{\partial^2 C_2}{\partial x^2} + R_2(C_1, C_2)\end{aligned}\right\} 0 \leqslant x \leqslant L$$

$$\frac{\partial C_1}{\partial x} = \frac{\partial C_2}{\partial x} = 0 \text{ at } x = 0, L,$$

it is easy to show that for the linearised (i.e. perturbed) equations to have an unstable non-uniform mode, one needs

$$D_1 \neq D_2, \frac{\partial R_1}{\partial C_2} \cdot \frac{\partial R_2}{\partial C_1} < 0,$$

$$\frac{\partial R_1}{\partial C_1} \cdot \frac{\partial R_2}{\partial C_2} < 0, \text{ and } \frac{\partial R_1}{\partial C_1} + \frac{\partial R_2}{\partial C_2} < 0$$

(Turing, 1952, Maynard Smith, 1968). The conditions are the same if R_1 and R_2 are linear and we look for a stable time-independent solution (Casten and Holland (1978)). Gierer and Meinhardt (1972) have made a detailed study of this sort of scheme (with additions involving inhomogenous source terms), which they have called on activator-inhibitor model. The reason for the name becomes clear on considering the specific case

$$D_1 < D_2, \frac{\partial R_1}{\partial C_1} > 0, \frac{\partial R_2}{\partial C_2} < 0,$$

$$\frac{\partial R_1}{\partial C_2} > 0, \frac{\partial R_2}{\partial C_1} < 0, \quad \text{illustrated in Fig. 1(2).}$$

Do such systems exhibit scale-invariance? Except in a restricted range of lengths, no, because for perfect scale invariance one would need D_1, D_2 to be proportional to L^2. The exceptional case arises whenever the range of lengths, over which a perturbation corresponding to one of the normal modes of the system is unstable, does not overlap with the range of "unstable lengths" for any other mode (see Lacalli and Harrison (1978) for details). Limited scale-invariance can also be demonstrated in the schemes proposed by Edelstein (1972) and Gierer and Meinhardt (1972), though these involve non-uniform initial distributions of morphogen, inhomogeneities in the system, or both. Cohen (1972) has obtained limited scale-invariance by using the assumption that morphogen transport depends on both passive diffusion and—for low concentration gradients—polarised active transport. The one case of *exact* scale-invariance in a reaction diffusion scheme is that proposed by Othmer and Pate (1980). In addition to the morphogens, they postulate the existence of a controlling chemical which is synthesised at a constant rate everywhere and influences the diffusion of the morphogens. If the diffusion constants of the morphogens vanish in the absence of the controlling species, and if the boundaries of the system are infinitely permeable to the controlling chemical, exact scale invariance results.

(iii) Cell-contact models

As indicated earlier, the diffusion equation

$$\frac{\partial C}{\partial t} = \frac{\partial}{\partial x}\left(D \frac{\partial C}{\partial x}(x, t)\right)$$

can be written for a set of discrete compartments in the form

$$\frac{\partial C}{\partial t}(i, t) = \frac{D}{\Delta^2}[C(i+1) + C(i-1) - 2C(i)]$$

and in a more general case, as

$$\frac{\partial C}{\partial t}(i, t) = \alpha_{i+1}C(i+1) + \alpha_i C(i) + \alpha_{i-1}C(i-1)$$

If one so chooses one can interpret these equations as referring to the influence of the concentrations in its neighbouring cells on the concentration of a chemical in cell i; there is no need to think in terms of transport at all. The influences could be transmitted by cell-cell contact (McMahon, 1973; Babloyantz, 1977). McMahon has applied a variant of such a model to the slime mold slug. As might be expected from our remarks about reaction-diffusion systems, this scheme too can result in stable, non-uniform patterns which show approximate regulation.

(b) **Models without chemical morphogens**

We begin with the

(i) *Wave propagation model* of Goodwin and Cohen (1969). Developing cells are assumed to function like nonlinear oscillators, and positional information is specified in terms of the phase difference between two waves of periodic activity, originating from the same source but spreading with different velocities. The source has the highest intrinsic frequency of oscillation, and therefore acts as a pacemaker. This is sufficient to set up stable spatial patterns. Unfortunately, for the system to regulate, there must be an initial gradient of autonomous frequencies which is itself scaled with respect to total length. Probably more attention will need to be paid to wave propagation models in the future, because there is increasing evidence that periodic cellular activities are common in development (Robertson, 1979; Campbell, 1979). Fig. 1(5) shows one way in which the propagation of two signals along a line of cells can lead to a size-independent ratio if the faster signal gets reflected at the distal boundary (P is where the two signals meet). Next we consider:

(ii) *Probabilistic models.* These have not been studied in any detail; Kauffman's (1975) work on transdetermination in *Drosophila* comes close but refers to a somewhat different problem. The hypothesis here would be that a cell of any given type has a constant probability per unit time of switching to any other specified type; therefore the *ratios* of the numbers of cells constituting different types would be a constant. In order for this to generate a regulating spatial pattern, one would require that cells of a kind sort out (Steinberg, 1970) and concommitantly get stabilised with respect to their type. The germ of this idea, that constant ratios of cell types are a reflection of the constant ratios resulting from "mass action" kinetics, can be found in an important article by Rose (1952; Rose also invokes diffusion to account for bilaterally symmetric patterns and distance-effects). Durston (1981) has suggested something similar for the slime mold slug pattern (Fig. 1(4)). An alternative probabilistic model, also

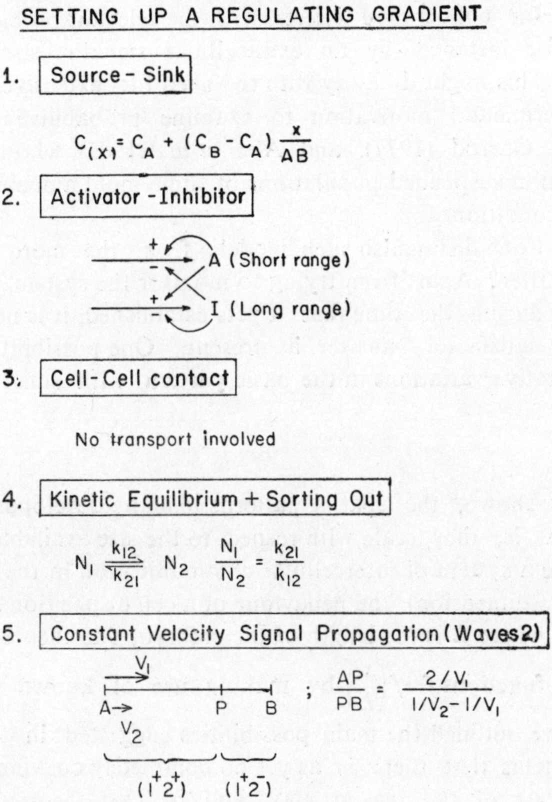

SETTING UP A REGULATING GRADIENT

1. Source–Sink

$$C_{(x)} = C_A + (C_B - C_A)\frac{x}{AB}$$

2. Activator–Inhibitor

A (Short range)
I (Long range)

3. Cell–Cell contact

No transport involved

4. Kinetic Equilibrium + Sorting Out

$$N_1 \underset{k_{21}}{\overset{k_{12}}{\rightleftharpoons}} N_2 \qquad \frac{N_1}{N_2} = \frac{k_{21}}{k_{12}}$$

5. Constant Velocity Signal Propagation (Waves 2)

$$A \xrightarrow{v_1} P \quad B \;;\; \frac{AP}{PB} = \frac{2/v_1}{1/v_2 - 1/v_1}$$
$$\underbrace{}_{(1^+2^+)} \underbrace{}_{(1^+2^-)}$$

Fig. 1   This illustrates the different ways in which a regulating system of positional information might be set up in a line of cells. The word 'gradient' is used here in the classical embryological sense of a pattern made up of graded spatial differences. In case 2. A stands for activator (concentration C_1) and I for inhibitor (concentration C_2). 'Short range' and 'long range' refer to the fact that the activator has a diffusion coefficient which is smaller than that of the inhibitor. The plus and minus symbols refer to the signs of self- or cross-activation (or inhibition); for instance, the plus sign at the end of the arrow leading from A to I indicates that $\partial R_1/\partial C_2 > 0$. In case 4, N_1, N_2, are the numbers constituting the two progenitor cell types and the ks stand for the forward and reverse rate-constants for transition from one type to the other. In case 5, the velocity of the first signal (v_1) is greater than that of the second (v_2). The region AP has therefore had both signals pass through it (symbolically indicated as 1+2+), where PB (1+2−) has experienced only the first one. The sizes AP, PB automatically scale with respect to the total length AB. The purely probabilistic ('coin-flipping') model has not been illustrated.

needing sorting-out, would be that undifferentiated cells flip a many-sided coin, so to speak, and act accordingly. These models can be improved by allowing for cooperative effects, with the relevant probabilities being influenced—for instance—by an extracellular signal released by the cells themselves. This might do away with the need for extensive sorting-out. For an experimental motivation to examine probabilistic models, see Forman and Garrod (1977), and Abe *et al.* (1981), who have obtained differentiation in suspended populations of slime mold amoebae kept in a well-shaken condition.

How would one distinguish such models from the more conventional ones listed earlier? Apart from trying to monitor the system of positional information during the time that it gets established, it is not possible to give a fully satisfactory answer at present. One possibility might be to study quantitative variations in the basic pattern with small numbers of cells.

Conclusions

As we have shown, the spatial patterns in many developmental systems regulate, that is, they scale with respect to the size available. Therefore there must be a system of intercellular communication in the embryo such that (in one dimension) the behaviour of a cell at position x depends on the ratio x/L where L is the length of the embryo. The theoretical problem is to generate functions $f = f\left(\dfrac{x}{L}\right)$ by making use of known properties of cells. We have outlined the main possibilities suggested in the past, but the fact remains that there is as yet no completely convincing theory of regulation. Part of the reason may well be that theories have tried to accomplish more than what was necessary. A basic question which has remained unanswered (except by the committed) concerns the necessity or otherwise of constructing theories of pattern formation in homogeneous systems. To put it somewhat differently, how much of the spatial inhomogeneity in the egg is essential for making the chicken? Other data which are urgently needed will have to come from careful experimental measurements on the degree of precision of regulation and on associated rates and sizes. It is our belief that future work on three classes of systems, typified by the single-celled ciliates, the genetically and developmentally identical slime mold amoebae, and the classical mosaic organisms like ascidians, will help in clarifying the situation.

Acknowledgements

We are thankful for support from the Department of Science and Technology and the Indian National Science Academy.

References

1. Abe, K., Saga, Y., Okada, H. and Yanagisawa, K. (1981). Differentiation of *Dictyostelium discoideum* mutant cells in a shaken suspension culture and the effect of cyclic AMP. J. Cell Sci. **51**, 131–142.
2. Babloyantz, A. (1977). Self organization phenomena resulting from cell-cell contact. J. Theor. Biol. **68**, 551–561.
3. Bakowska, J. and Jerka-Dziadosz, M. (1980). Ultrastructural aspect of size dependent regulation of surface pattern of complex ciliary organelle in a protozoan ciliate. J. Embryol. exp. Morph. **59**, 355–375.
4. Balinsky, B.I. (1970). An introduction to embryology. 3rd edn. Philadelphia: W.B. Saunders.
5. Berrill, N.J. (1971). Developmental Biology. New York: McGraw-Hill.
6. Bonner, J.T. (1967). The Cellular Slime Molds. 2nd ed. New Jersey: Princeton University Press.
7. Campbell, R.D. (1979). Development of hydra lacking interstitial and nerve cells ("Epithelial Hydra"). In "Determinants of Spatial Organization". Subtelny, S. ed. 267–293. New York: Academic Press Inc.
8. Casten, R.G. and Holland, C.J. (1978). Instability results for reaction diffusion equations with Neumann Boundary Conditions. J. Diffl. Eqn. **27**, 266–273.
9. Cohen, M.H. (1972). Some mathematical questions in biology. (J.D. Cowan ed.), Providence, R.I.: Amer. Math. Soc.
10. Crick, F. (1970). Diffusion in embryogenesis. Nature, Lond. **225**, 420–422.
11. Edelstein, B.B. (1972). The dynamics of cellular differentiation and associated pattern formation. J. Theor. Biol. **37**, 221–243.
12. Fife, P.C. (1979). In *Lecture Notes in Biomathematics*. Providence, R.I.: Am. Math. Soc.
13. Forman, D. and Garrod, D.R. (1977). Pattern formation in *Dictyostelium discoideum* II. Differentiation and pattern formation in non-polar aggregates. J. Embryol. exp. Morph. **40**, 229–243.
14. Frankel, J. (1974). Positional information in unicellular organisms. J. Theor. Biol. **47**, 439–481.
15. Frankel, J. (1975). Pattern formation in ciliary organelle systems of ciliated protozoa. In *Cell Patterning*, Ciba Foundation Symposium, **29** (new series) 25–49.
16. Gaze, R.M. (1970). The formation of Nerve Connections. London: Academic Press.

17. Gierer, A. (1977). Physical aspects of tissue evagination and biological form. Quart. Rev. Biophys. **10**, 529–593.
18. Gierer, A. and Meinhardt, H. (1972). A theory of biological pattern formation. Kybernetik **12**, 30–39.
19. Goodwin, B.C. and Cohen, M.H. (1969). A phase-shift model for the spatial and temporal organization of developing systems. J. Theor. Biol. **25**, 49–107.
20. Jacobson, M. and Hunt, R.K. (1973). The origins of nerve-cell specificity. Sci. Am. **228**, 26–35.
21. Jacobson, M. (1976). Developmental Neurobiology. 2nd ed. New York: Holt.
22. Kauffman, S. (1975). Control circuits for determination and transdetermination: interpreting positional information in a binary epigenetic code. In *Cell Patterning*, Ciba Foundation Symposium **29** (new series), 201–215.
23. Lacalli, T.C. and Harrison, L.G. (1978). The regulatory capacity of Turing's model for morphogenesis, with application to slime molds. J. Theor. Biol. **70**, 273–295.
24. Lawrence, P.A., Crick, F.H.C. and Munro, M. (1972). A gradient of positional information in an insect, *Rhodnius*. J. Cell. Sci. **11**, 815–853.
25. Lokeshwar, B.L. and Nanjundiah, V. The scale invariance of spatial patterning in a developing system. Wilhelm Roux's Arch. (in press).
26. Maynard Smith, J. (1968). Mathematical Ideas in Biology. Cambridge: Cambridge University Press.
27. Meinhardt, H. (1977). A model of pattern formation in insect embryogenesis. J. Cell Sci. **23**, 117–139.
28. McMahon, D. (1973). A cell contact model for cellular position determination in development. Proc. Nat. Acad. Sci. USA **70**, 2396–2400.
29. Nanny, D.L. (1971). The constancy of cortical units in *Tetrahymena* with varying numbers of ciliary rows. J. Exptl. Zool. **178**, 177–182.
30. Northmore, B.P.M. (1981). Visual localization after rearrangement of retinotectal maps in fish. Nature, Lond. **293**, 142–144.
31. Othmer, H.G. and Pate, E. (1980). Scale-invariance in reaction-diffusion models of spatial pattern formation. Proc. Nat. Acad. Sci. USA **77**, 4180–4184.

32. Raper, K.B. (1940). Pseudoplasmodium formation and organization in *Dictyostelium discoideum*. J. Elisha Mitchell Sci. Soc. **56**, 241-282.
33. Robertson, A. (1979). Waves propagated during vertebrate development: observations and comments. J. Embryol. exp. Morph. **50**, 155-167.
34. Robertson, A.J. and Cohen, M.H. (1972). Control of developing fields. Ann. Rev. Biophys. Bioeng. **1**, 409-464.
35. Rose, S.M. (1952). A hierarchy of self-limiting reactions as the basis of cellular differentiation and growth control. Am. Natur. **86**, 337-354.
36. Sander, K. (1971). Pattern formation in longitudinal halves of leaf hopper eggs (Homoptera) and some remarks on the definition of "Embryonic Regulation". Wilhelm Roux's Archives **167**, 336-352.
37. Steinberg, M.S. (1970). Does differential adhesion govern self-assembly process in histogenesis? Equilibrium configurations and the emergence of a hierarchy among populations of embryonic cells. J. Exp. Zool. **173** (4), 395-434.
38. Stern, C. (1968). "Genetic Mosaics and Other Essays". Cambridge, Mass: Harvard University Press.
39. Straznicky, C., Gaze, R.M. and Kealing, M.J. (1981). The development of the retinotectal projections from compound eyes in *Xenopus*. J. Embryol. exp. Morph. **62**, 13-35.
40. Turing, A. (1952). The chemical basis of morphogenesis. Phil. Trans. R. Soc. **B237**, 32-72.
41. Vogel, O. (1978). Pattern formation in the egg of the leafhopper *Euscelis plebejus* fall (Homoptera): Developmental capacities of fragments isolated from the polar egg regions. Dev. Biol. **67**, 357-370.
42. Wilson, E.B. (1900). The cell in development and inheritance. 2nd ed. New York: Macmillan Publishing Co.
43. Wilson, E.B. (1925). The cell in development and heredity. 3rd ed. New York: Macmillan Publishing Co.
44. Wireman, J.W. and Dwarkin, M. (1975). Morphogenesis and developmental interactions in Myxobacteria. Science **189**, 516-523.
45. Wolpert, I. (1969). Positional information and the spatial pattern of cellular differentiation. J. Theor. Biol. **25**, 1-47.
46. Wolpert, L. (1971). Positional information and pattern formation. Curr. Top. Dev. Biol. **6**, 183-224.
47. Wourms, J.P. (1967). Methods in Developmental Biology. cited in Berrill, N.J. (1971) "Developmental Biology". New York: McGraw-Hill.

48. Yoon, M.G. (1976). Progress of topographic regulation of the visual projection in the halved optic tectum of adult goldfish. J. Physiol. **257**, 62-643.
49. Zeeman, E.C. (1975a). In "Lectures on Mathematics in the life Sciences", Ch. 7. Providence, R.I.: Am. Amth. Soc.
50. Zeeman, E.C. (1975b). Catastrophe theory and biological patterns: appendix. Ann. Rev. Biophys. Bioeng. **4**, 210-213.

Addendum

Lokeshwar and Nanjundiah have established that the time taken by the slug of *Dictyostelium discoideum* to regenerate a new tip after the old tip has been removed by axial fragmentation can be used as a measure of positional information in the slug (Wilhelm Roux' Arch. Devl. Biol. **190**, 361-364; J. Embryol. exp. Morph. (1983) **73**, 151-162); significantly, this time is a function only of the relative level of fragmentation. Papageorgiou (Biophys. Chem. (1980) **11**, 183-190) has pointed out that scale-invariance, that is, x/L-dependence, can result from a mechanism which measures x and L separately and then takes their ratio. Bard (J. theor. Biol. (1982) **94**, 689-708) has developed a membrane-carrier mechanism which accounts for spontaneous regeneration of sources and sinks, and so has removed one possible objection to gradient models which need fixed boundary conditions.

Embryonic Development as a Mode of Activity of the Living State: The Case of the Vertebrate Limb

Stuart A. Newman

Department of Anatomy
New York Medical College
Valhalla, New York 10595

The embryonic development of metazoan organisms is among the most mysterious phenomena contemplated by biological science. The early embryo, and each of the organ rudiments comprising it, begin as small, relatively homogeneous masses of cells which, by means still poorly understood at the physical and chemical levels, become increasingly heterogeneous with regard to cell type and overall structure. In a world where most active processes "run down", and nonuniformities tend to even out with time, the rapid and unremitting growth of complexity during embryogenesis is a hallmark of the special nature of the living state.

The morphogenesis of the vertebrate limb, and its skeleton in particular, has long fascinated students of embryology as a singularly obvious and experimentally accessible example of the developmental process. The wide spectrum of adult limb forms exhibited by different classes and orders of vertebrates, which nonetheless always represent variations on a common structural theme, intrigued Darwin, for one (1), and has stimulated a search for embryological mechanisms consistent with such constraints and degrees of freedom (reviewed in ref 2). The resulting advances over the last half century in our descriptive knowledge of the morphogenesis of the limb, and of the biochemistry and cell biology of the limb forming-tissues, have provided opportunities for, and limitations on our attempts to provide a physicochemical understanding of this process.

In what follows, I will indicate some landmarks in the evolution of our modern understanding of limb development. No attempt will be made to be comprehensive, as excellent reviews of this field are available (2, 3). I will instead emphasize connections between the various levels of analysis, and theoretical considerations that have entered into a provisional explanatory model for this process.

The Establishment of the Limb Axes

The orientation of the limb as a whole can be described in terms of three orthogonal axes of differing degrees of polarity or asymmetry (Fig. 1). Any model of limb development must address those asymmetries, but more importantly, must account for the basic pattern on which these asymmetries are superimposed. The *proximodistal axis*, pointing outward from the body wall, has the most marked polarity of the three. With respect to the limb skeleton, the proximodistal axis is characterized by a generally increasing number of parallel bone elements in successive rows. The asymmetry of the *anteroposterior axis* (traversed, in the case of the hand, in moving from the thumb to the little finger) is less extreme than that of the proximodistal axis, but is easily recognizable in differences in shape of parallel elements (such as fingers) in a given row and the lack of a perpendicular plane of reflection in any such row. The *dorsoventral*, or front to back, asymmetry of the limb is functionally essential in that it reflects a pattern of tendon and muscle insertions leading to the ability of the hand, for example, to grasp. However, the polarity in this case is quite subtle: it takes more than a casual glance to

Fig. 1. Drawings of the skeletal patterns of the forelimbs of three vertebrate species: (left to right) human, axolotl (an amphibian), chicken. Not to scale. Axolotl limb adapted from Hinchliffe and Johnson (2).

tell whether the skeleton of a hand in a photograph is palm upward or downward.

The three axes are experimentally as well as conceptually separable. In a classic series of experiments, R.G. Harrison rotated or inverted prospective limb tissues in urodele amphibian embryos at different stages of development to determine the time and order of the establishment of the limb axes (reviewed by Balinsky (4)). Similar experiments were also performed on developing avian limbs, with results essentially identical to those in amphibians (5, 6). These studies concluded that the anteroposterior axis is the first to be irreversibly determined (i.e., not respecified by disharmonious host tissues) followed in succession by the dorsoventral and proximodistal axes. An early speculation on the physiochemical basis of this independent, sequential specification of the three limb axes, should be of particular interest to viewers of Professor R. K. Mishra's beautiful photographs of liquid crystal configurations during this Seminar. Liquid crystals are substances in which the melting transition from solid to liquid occurs not directly, but via a sequence of *para-crystalline* or *mesomorphic* states. Joseph Needham suggested that the succession of mesoforms of just such materials could underlie the developmental phenomena described. "A solid is rigid in three dimensions" he states, "a smectic mesoform in two, a nematic mesoform in one, an isotropic liquid in none...Just as the liquid crystal passes through stages of rigidity in one, two, and three dimensions, so the limb bud passes through stages of determination in one, two and three dimensions" (7).

Whatever the underlying basis may be of the axial *asymmetries*, it may be worthwhile considering these as the result of perturbations of essentially *symmetrical* processes leading to the establishment of skeletal structures. Certainly it is not difficult to take such an approach in discussing the dorsoventral polarity; the difference between the front and back of a limb is undoubtedly a second-order phenomenon from the point of view of developmental mechanisms. The anteroposterior axis can be similarly conceptualized; an ideal version of an arm with identically shaped fingers and a radius equivalent to the ulna is still recognizable as an arm, and one can readily hypothesize that the differences among bones in a given row is secondary to their similarities. Although the applicability of this logic to the differences along the limb's proximodistal axis is less obvious, it will be argued in a later section that even the differently appearing rows of the skeletal elements may best be understood by considering them to be generated by uniformly acting physicochemical process under a succession of slightly different conditions.

The next three sections will therefore outline the current understanding of the nature of the material that forms itself into the limb: that is, the cells and their phenotypic potentials, interactions, biochemistry. In the final section I will suggest how these properties may be responsible for the progressive emergence of limbs during embryonic development.

The Cell Types and Their Developmental Roles
The limb skeleton and musculature arise from a mass of loose embryonic connective tissue known as *mesenchyme*. This tissue constitutes the mesoblast, which is sheathed in a thin covering of epithelial cells known as the *ectoderm*. The mesoblast is vascularized with a fine capillary network during the course of development, and receives a supply of nerve fibers from the spinal cord and spinal ganglia at about the time overt pattern formation is initiated (8, 9). In analyzing the mechanisms of limb skeletal pattern formation it is essential to determine which limb bud cell types are necessary and which are dispensible for this process.

A healthy vascular system would seem to be necessary for normal development if only to satisfy nutritional requirements of the cells. Since it is clear that both muscle and cartilage differentiation can occur in artificial media in the absence of an organized capillary network (10, 11) it is doubtful that the vascular system *per se* plays any role in the cytodifferentiation of the two main components of the limb. A suggestion of several years ago that the spatial organization of the vasculature may play a role in the position-dependent choice of the muscle *vs* the cartilage phenotype in the differentiating mesoblast (12), is made less likely by current understanding that the premuscle and precartilage cells have already diversified by the time of emergence of the limb bud (see below).

The situation regarding the dependence of limb skeletal development on nerves is complex. A large body of experiments indicates that limbs are capable of developing and regenerating *aneurogenically*, that is, in the absence of nerves (13, 14, 15). However, a number of studies also indicate that limb tissues become "addicted" to nerve-supplied factors during normal development (16), and that if they are made to develop aneurogenically, other tissues, in particular the ectoderm (17), compensate by contributing factors otherwise supplied by the nerves. It has been suggested that nerves may exert their influence on developing limbs by supplying an *angiogenic* factor, that is, something necessary for proper capillary formation (18). Since the apical ectodermal ridge, a structure essential for limb outgrowth (see below) also appears to provide such factors (19), nerves and ectodermal cell may promote limb development by a common mechanism.

The dispensibility of muscle-forming tissue for skeletal differentiation has been demonstrated more recently (20). This finding depended on previous knowledge that the premuscle mesenchyme of the limb bud originates in and migrates from the somites, blocks of tissue along the embryonic spine (21, 22). The skeleton-forming cells of the limb bud, in contrast, originate in the flank, or body wall, from which the bud eventually emerges. When deprived of cells of somite origin, the flank-derived limb cells are capable of undergoing position-dependent differentiation into normally arranged skeletal structures (20).

The independence of the processes of skeletal and muscle formation, and the separate origins and fates of the respective precursor cells (21–23) run counter to an earlier suggestion that cell diversification in the limb occurs late in development among cells uncommitted to either of these lineages until the time of pattern formation (24). However, despite the lack of evident structural or biochemical markers that might permit us to distinguish between mesenchymal cells prospective for muscle and those prospective for skeletal and other connective tissues, these subpopulations of cells apparently abide by an early, covert determination as they participate in normal development (23, 25). Thus, the attempt to understand the mechanisms of skeletal pattern formation in the limb is now generally centered on the regulation of cell differentiation and cell death within the distinct subpopulation of mesenchyme cells that gives rise to cartilage, the tissue type that constitutes the skeletal primordia (25, 2). Although the cartilage skeleton is replaced by bone during subsequent development (26, 27) the processes involved probably have little in common with the pattern-forming mechanisms operative during the chondrogenic (cartilage-forming) phase.

It is, of course, logically possible that the limb skeletal pattern is already established, on a small scale, in the precartilage mesenchyme of the early limb bud as it emerges from the body wall. This preformationist hypothesis, while not going so far as to assert that the limb pattern is present in the fertilized egg, nonetheless consigns to irrelevance any analysis of cellular, biochemical and biophysical events subsequent to the emergence of the limb bud, with respect to an understanding of the pattern-forming process. However, though all the skeletal elements of the chick limb, for example, can be projected back to small, specific regions in the young bud (28), there is no evidence that those regions are uniquely determined to give rise to the structures into which they map. Dissociated, randomized early limb mesenchyme cells, packed into a limb ectodermal hull, usually develop into jointed cartilage structures, recognizably limb-like in their patterning (29). In Driesch's terms, the prospective *potential* of these cells is greater than their prospective *significance* (or fate) (30). But while such regulative phenomena convinced Driesch of the essential nonphysical nature of the goal-directedness of embryonic development, it would seem that they might more reasonably play the opposite role in embryological theory. Regulative capacity implies a lability in cellular behavior, suggesting that final fates can be *assigned* by some physicochemical process, such as the nonuniform distribution of a substance. In a certain sense it does not matter which cell in the labile population is chosen to play a given role, since they are presumed to be inter-changable. This is certainly more congenial to the mechanism-minded than the logically alternative concept of the embryo as a cabal of cells, each born with the secret of its ineluctable role.

Cell and Tissue Interactions During Limb Development

The development of the limb skeleton in those vertebrates in which it has been investigated depends on the establishment of aggregates or condensations of cells in the precartilage mesenchyme tissue that prefigure the positions of the cartilage primordia (3, 31, 32). The proximodistal progression of the establishment of these cellular condensations depends in turn on outgrowth of the limb during a critical period characteristic of each species. Each of these processes, outgrowth and condensation, involves cellular interactions: the first between tissues of different types (heterotypic), and the second between cells of the same type (homotypic).

The heterotypic interaction between the mesoblast and the ectoderm has been alluded to briefly in the previous section. Rimming the distal margin of the paddle-shaped limb bud during the entire phase of skeletal pattern formation is a ridge of ectodermal cells morphologically distinct from the rest of the cells of the ectodermal sheath. These taller cells constitute the apical ectodermal ridge or AER (33), and appear to be absolutely required for limb outgrowth and the consequent establishment of progressively more distal parts during development. Removal of the AER at any time during the critical period results in a truncated limb which is normal in size and appearance up to a point characteristic of the stage at which the ridge was removed. Beyond that point there is nothing (33). Although the AER provides no stage-specific (34), limb type-specific (35), or even species-specific (36, 37) information to the growing mesoblast, its action on the developing limb bud is unique, in that ectoderm from other regions of the embryo, limb or nonlimb, cannot substitute for it (38). Despite the profound influence the AER has on the mesoblast, there is no direct cellular contact between these two tissues (39).

A transient reduction in the extracellular space between precartilage mesenchyme cells has been well-documented in the chick limb (32) and appears to constitute an early step in the homotypic interaction of these cells. The dependence of cartilage differentiation in culture on the proximity of these cells has also been studied (25, 40). However it is not clear what new modes of cellular communication are established during the condensation phase, since the mesenchymal cells appear to already have extensive gap junctional contacts with one another before condensations are underway (41). One possibility is that the broad contacts that occur between the precartilage cell surfaces during condensation allow an interaction between receptors that are not located at the pre-existing focal contacts. It has been suggested that the intracellular signal molecule cyclic AMP is transiently elevated during the condensation phase, and that this might be a link in the chain of events leading to the establishment of the cartilage phenotype (42, 43). The "interacting mode" of the precartilage mesenchyme cells, established during the condensation phase (25) could well potentiate this metabolic event.

Biochemical Aspects of Limb Development

Any tenable model for limb development must encompass recent findings on the macromolecular changes accompanying this process. Indeed, the assumption of modern biology is that among these changes is a causal chain which drives the process forward. Here I will summarize the four aspects of this question which are most relevant to the theoretical analysis of limb skeletal formation: changes in the extracellular matrix during chondrogenesis; metabolic changes accompanying cellular interaction; developmentally-associated changes in the precartilage cell nucleus; and molecular correlates of AER activity.

Differentiated cartilage, such as that found in the limb skeletal primordia following pattern formation, is a tissue whose tough, rubbery consistency is a consequence of the properties of the macromolecules filling its intercellular spaces. These comprise a type of collagen (a fibrous protein) virtually unique to cartilage, and a glycosaminoglycan (a complex type of polysaccharide found in connective tissues) known as chondroitin sulfate. This last molecule is covalently bound to a cartilage-specific "core protein" which is in turn bound noncovalently to small amounts of another glycosaminoglycan known as hyaluronic acid (see reference 44 for a review).

In contrast, the mesenchymal cells that differentiate into cartilage are embedded in a matrix which contains abundant amounts of hydrated hyaluronic acid, and a glycoprotein known as *fibronectin*. The small amount of collagen in this matrix is not of the cartilage-specific type, but appears to be the same type found almost ubiquitously in the body's other connective tissues (45). During the early stages of limb chondrogenesis the enzyme *hyaluronidase* is rapidly synthesized, presumably by the mesenchyme (46). The removal of hyaluronate is accompanied by a marked increase in the synthesis of chondroitin sulfate relative to hyaluronate (46). At about the same time mesenchymal condensations appear, with fibronectin, initially uniformly distributed, accumulating between the condensing cells (47-49). The differentiating cells now stop synthesizing ubiquitous collagen and begin making the cartilage-specific variety (44, 45), as well as the chondroitin sulfate core protein (50). The fibronectin gene appears to become inactive (51).

Studies of chondrogenesis in culture have indicated that the intracellular signal molecule cyclic AMP is transiently elevated in concentration during the early phase of differentiation (42). Because chondrogenesis occurs precociously when cyclic AMP levels are artificially enhanced, it has been suggested that the molecule plays a triggering role in reprogramming the mesenchyme cell, possibly as a concomitant of cellular interaction (52, 53).

One cellular target of enhanced cyclic AMP levels is one of the two abundant precartilage nonhistone chromatin proteins that is lost as the nuclear material becomes reorganized during chondrogenesis (54, 55). The protein in question, designated PCP 35.5 is strongly bound to DNA in the precartilage cell nucleus, and is localized near transcriptionally-activated chromatin domains of that cell type (56). Recent studies have determined that this protein is modified by the addition of phosphate groups during cartilage differention: a cyclic AMP-dependent enzyme located in the same precartilage nuclei is probably responsible for this phosphorylation (57).

Work on biochemical correlates of the inductive activity of the AER is an area that has recently yielded some promising results. One investigation has demonstrated that the AER synthesizes hyaluronic acid in amounts significantly greater than other limb ectodermal cells (58). Another study, using an immunolocalization technique, has shown that fibronectin accumulates beneath the AER to a markedly greater extent than it does beneath the rest of the limb ectoderm (49). In each case it has been suggested that the subridge enhancement of the corresponding matrix macromolecule may play a role in the outgrowth-promoting function of the AER.

How it Might Work

Limb development is a complex process that unfolds simultaneously on many levels. A fashionable conceptualization of such embryonic processes takes its cue from the operation of a digital computer. The attempt is made to discern a "program" of development that is played out in response to an input of "information". The information may be an external stimulus, or as development proceeds, can be generated by the embryo itself, in a form that is separable from the cells' "interpretative" functions. In contrast to this computer analogy, we favor a view of embryogenesis that sees regulated gene expression and spatial patterning of tissue structures as an emergent property of interacting cells that behave predictably but blindly. At no stage is cellular activity presumed to be formally codified by the organism into anything resembling a program.

We have proposed a dynamical model for limb formation that is consistent with these ideas, and synthesizes much of the experimental information outlined in the previous sections (59). It is suggested that the directional outgrowth of the limb bud is promoted by extracellular matrix materials, including hyaluronate and fibronectin, that accumulate beneath the cells of the AER (49, 60). As outgrowth proceeds, a chain of events is set into motion that leads to cartilage differentiation in a pattern of rows of elements that are generally increasing in number. These events take place in precartilage cells that have developmentally passed beyond the stage where a muscle-forming option is available to them. The only options of these cells are to differentiate into cartilage or soft connective tissue, or to die off (25, 61).

There is compelling evidence that condensations of the precartilage cells are obligatory to their exercising the chondrogenic option (25, 40). We have therefore suggested that a molecule that encourages cellular adhesion could be responsible for the positioning of these condensations and the consequent cartilage elements, were it to be distributed appropriately during development (59). Fibronectin, which is abundant in the limb bud mesenchyme, is one such molecule (62), and its localization between the condensing cells makes it a candidate for the putative morphogenetic agent. However, fibronectin and hyaluronate, which is also abundant in extracellular matrix of the early limb bud, bind one another (63) and this interaction would be expected to impede the redistribution of the glycoprotein during skeletal pattern formation. We have therefore suggested that the developmentally-regulated appearance of hyaluronidase (46) could initiate pattern formation by freeing fibronectin from its interaction with hyaluronate and allowing it to diffuse through the extracellular space (59, 49). A diffusible morphogen would be subject to the *reaction-diffusion* equation first introduced by Turing as an explanation for biological morphogenesis (64). Analysis of this equation shows that under appropriate boundary conditions a substance subject to synthesis and diffusion will assume spatially nonuniform distributions of concentration (64, 59; see also Prigogine, this volume, and references therein).

Application of the reaction-diffusion law to the case of a diffusing substance in a chamber of the dimensions of the chick limb bud, yields sinusoidal solutions for the concentration distribution of the putative morphogen (59). In our representation the wave number of the solution increases as the proximodistal width of mesenchyme remaining to be organized into cartilage elements decreases. Therefore the chick wing bud, which maintains constant anteroposterior and dorsoventral dimensions during development (28), is predicted by this model to generate one, two, and then three elements, emerging proximodistally, in successive rows. The course of development and final form of the chicken forelimb agrees with these predictions. A characteristic feature of this model is a direct proportionality (other things being equal), between the number of bones formed in any row and the anteroposterior width of the mesenchyme when they are being established. Because the human limb bud expands at the tip during development, the five digits of the human limb are a natural consequence of our reaction-diffusion scheme.

The skeletal structures generated by the dynamical model described above are limb-like in their final patterns and in the order of their emergence. That is, a generally increasing number of elements is established in a proximodistal sequence over the course of a few days. In a physically natural way the stepwise reduction in the length of tissue remaining to be organized polarizes the proximodistal axis during the course of development by abruptly and symmetrically increasing the number of elements in the anteroposterior dimension (59). Carrying our interpretation to the limit, anteroposterior and dorsoventral polarities can be considered to result from asymmetries in the shape of the developing limb bud, or from nonuniformities in the absorption of the putative morphogen at the limb bud's periphery (59).

This dynamical model and the suggestion that fibronectin concentrations provide a spatial *prepattern* for skeletal differentiation must be considered tentative until critical experimental tests are performed. Whatever mechanism triggers the precartilage mesenchymal condensations, it appears likely that a subsequent link in the causal chain of developmental events is the elevation of cyclic AMP levels in the condensing cells (52, 53). Our identification of the developmentally-regulated DNA-binding protein PCP 35.5 as a target of cyclic AMP activity (57), and the association of this protein with transcriptionally-activated regions of chromatin (56), leaves open the question of which genes in the precartilage cell (if any) are affected in their expression by the modification of this protein. Are precartilage-specific genes (such as fibronectin) turned off during this process, or are cartilage-specific genes (such as cartilage-type collagen) turned on? This is currently under investigation, as is the significance of the finding that the *talpid*^{2} mutant of the chicken, which exhibits poor spatial regulation of limb chondrogenesis, is apparently deficient in PCP 35.5 (55).

A distinctive feature of this model is that the formation of cartilage elements is considered to arise from the *local*, stereo-typical behavior of precartilage cells confronted with an extracellular substance causing them to condense. The cartilage that forms is essentially the same regardless of the element into which the responding cells happen to be recruited. This contrasts with the point of view that each cell has a genetically-based ability to sense its *global* position on the basis of a local signal or an internal clock, and that the cartilages that form in response to each different positional value are *nonequivalent* from their inception (65).

It is clear that some of the pieces of the puzzle of vertebrate limb development are now at hand, and that they can be forced together into a plausible mechanistic model. However, considering the complexity of embryonic processes and the elusiveness of explanatory concepts in this field, it would be imprudent not to expect the puzzle to spring apart more than once before all the pieces are finally in place.

References

1. Darwin, C. (1872). "On the Origin of Species" (6th ed.) John Murray, London.

2. Hinchliffe, J. R. and Johnson, D. R. (1980). "The Development of the Vertebrate Limb" Oxford University Press, Oxford.
3. Hall, B. D. (1978). "Developmental and Cellular Skeletal Biology" Academic Press, New York.
4. Balinsky, B.I. (1971). An Introduction to Embryology (3rd ed.) W.B. Saunders, Philadelphia.
5. Chaube, S. (1959) *J. Exp. Zool.* **140**, 29–78.
6. MacCabe, J.A., Errick, J.E. and Saunders, J.W., Jr. (1974). Develop. Biol. **39**, 69–82.
7. Needham, J. (1968). *Order and Life,* MIT Press, Cambridge, Mass. p. 164.
8. Fouvet, B. (1973). Arch. Anat. Microsc. Morph. Exp. **62**, 269–280.
9. Lance-Jones, C. and Landmesser, L. (1981). *Proc. R. Soc. Lond* B **214**, 1–18.
10. Konigsberg, I. (1963). Science **140**, 1273–1284.
11. Coon, H. (1966). Proc. Nat. Acad. Sci. USA **55**, 66–73.
12. Caplan, A.I. and Koutroupas, S. (1973). J. Embryol. Exp. Morph. **29**, 571–583.
13. Hamburger, V. (1939). J. Exp. Zool. **80**, 347–389.
14. Hamburger, V. and Waugh, M. (1940). Physiol. Zool. **13**, 367–380.
15. Yntema, C.L. (1959). J. Exp. Zool. **140**, 101–123.
16. Singer, M. (1952). Quart. Rev. Biol. **27**, 169–200.
17. Steen, T.P. and Thornton, C.S. (1963). J. Exp. Zool. **154**, 207–221.
18. Smith, A.R. and Wolpert, L. (1975). Nature **257**, 224–225.
19. Feinberg, R. and Saunders, J.W., Jr. (1982) J. Exp. Zool. **219**, 345–354.
20. Chevallier, A., Kieny, M., Mauger, A. and Sengel, P. (1977). In Vertebrate Limb and Somite Morphogenesis (D. A. Ede, J. R. Hinchliffe, and M. Balls, eds.) Cambridge Univ. Press, Cambridge, pp. 421–432.
21. Christ, B., Jacob, H.J. and Jacob, M. (1977). Anat. Embryol. **150**, 171–186.
22. Chevallier, A., Kieny, M. and Mauger, A. (1977). J. Embryol. Exp. Morph. **41**, 245–258.
23. Newman, S.A., Pautou, M.-P. and Kieny, M. (1981). Develop. Biol. **84**, 440–448.
24. Zwilling, E. (1968). Develop. Biol. (Suppl.) **2**, 184–207.
25. Newman, S.A. (1977). In "Vertebrate Limb and Somite Morphogenesis" (D.A. Ede, J.R. Hinchliffe and M. Balls, eds.) Cambridge Univ. Press, Cambridge, pp. 181–197.

26. Ham, A.W. and Cormack, D.H. (1979). "Histology" (8th ed.) J.B. Lippincott, Philadelphia.
27. Osdoby, P. and Caplan, A.I. (1979). Develop. Biol. 73, 84–102.
28. Stark, R.J. and Searls, R.L. (1973). Develop. Biol. 33, 138–153.
29. Finch, R. and Zwilling, E. (1971). J. Exp. Zool. 176, 397–408.
30. Driesch, H. (1929). "The Science and Philosophy of the Organism" (2nd ed.) Black, London.
31. Fell, H.B. and Conti, R.G. (1924). Proc. R. Soc. Lond. B 116, 316–349.
32. Thoroughgood, P.V. and Hinchliffe, J.R. (1975). J. Embryol. Exp. Morph. 33, 581–606.
33. Saunders, J.W., Jr. (1948). *J. Exp. Zool.* 108, 363–404.
34. Rubin, L. and Saunders, J.W. Jr. (1972). *Develop. Biol.* 28, 94–110.
35. Zwilling, E. (1955). *J. Exp. Zool.* 128, 423–441.
36. Zwilling, E. (1956). *Cold Spring Harbor Symp. Quant. Biol.* 21, 349–354.
37. Jorquera, B. and Pugin, E. (1971). *Comptes rendus de l'Acad. des Science, Paris* 272, 1522–1525.
38. Errick, J. and Saunders, J.W. Jr. (1976). Develop. Biol. 50, 20–34.
39. Kelley, R.O. and Fallon, J.F. (1981). In "Morphogenesis and Pattern Formation" (Connolly, T.G., Brinkley, L.L. and Carlson, B.M., eds.) Raven Press, New York, pp. 49–85.
40. Solursh, M. and Reiter, R.S. (1980). *Develop. Biol.* 78, 141–150.
41. Kelley, R.O. and Fallon, J.F. (1978). *J. Embryol. Exp. Morph.* 46, 99–110.
42. Solursh, M., Reiter, R.S., Ahrens, P.B. and Pratt, R.M. (1979). Differentiation 15, 183–186.
43. Kosher, R.A., Savage, M. and Chan, S.C. (1979). J. Exp. Zool. 209, 221–228.
44. Levitt, D. and Dorfman, A. (1974). In "Current Topics in Developmental Biology" Vol. 8 (Moscona, A.A. and Monroy, A. eds.) Academic Press, New York, pp. 103–144.
45. von der Mark, K. and von der Mark, H. (1977). J. Cell Biol. 73, 736–747.
46. Toole, B.P. (1972). Develop. Biol. 29, 321–329.
47. Silver, M.H., Foidart, J.-M. and Pratt, R.M. (1981). **Differentiation** 18, 141–149.
48. Melnick, N., Jaskoll, T., Brownell, A.G., MacDougall, M., Bessem, C. and Slavkin, H.C. (1981). J. Embryol. Exp. Morph. 63, 193–206.

49. Tomasek, J.J., Mazurkiewicz, J.E. and Newman, S.A. (1982) Develop. Biol. **90**, 118–126.
50. Vertel, B.M. and Dorfman, A. (1978). Develop. Biol. **62**, 1–12.
51. Lewis, C.A., Pratt, R.M., Pennypacker, J.P. and Hassell, J.R. (1978). Develop. Biol. **64**, 31–47.
52. Kosher, R.A. and Savage, M.P. (1980). J. Embryol. Exp. Morph. **56**, 91–105.
53. Solursh, M., Reiter, R.S., Ahrens, P.B. and Vertel, B.M. (1981). Develop. Biol. **83**, 9–19.
54. Newman, S.A., Birnbaum, J., and Yeoh, G.C.T. (1976). Nature **259**, 417–418.
55. Perle, M.A. and Newman, S.A. (1980) Proc. Nat. Acad. Sci. USA **77**, 4828–4830.
56. Perle, M.A., Leonard, C.M. and Newman, S.A. (1982). Biochemistry, in press.
57. Newman, S.A. and Leonard, C.M., in preparation.
58. Kosher, R.A. and Savage, M.P. (1981). *Nature*, **291**, 231–232.
59. Newman, S.A. and Frisch, H.L. (1979). Science **205**, 662–668.
60. Newman, S.A., Frisch, H.L., Perle, M. and Tomasek, J.J. (1981). In "Morphogenesis and Pattern Formation" (Connelly, T.G., Brinkley, L.L. and Carlson, B.M., eds.) Raven Press, New York, pp. 163–177.
61. Newman, S.A. (1980). J. Embryol. Exp. Morph. **56**, 191–200.
62. Ruoslathti, E., Engvall, E. and Hayman, E.G. (1981). Cell Res. **1**, 95–128.
63. Yamada, K.M., Kennedy, D.W., Kimata, K., and Pratt, R.M. (1980). J. Biol. Chem. **255**, 6055–6063.
64. Turing, A.M. (1952). Phil. Trans. R. Soc. Lond. Ser B **237**, 37–72.
65. Wolpert, L. 1981), In "Morphogenesis and Pattern Formation" (Connelly, T.G., Brinkley, L.L. and Carlson, B.M. eds) Raven Press, New York, pp. 5–20.

An Evolutionary Model of a Neural Network

Jitendra C. Parikh and Ram Verma Pratap

Physical Research Laboratory
Ahmedabad–380 009

The human brain cotains around 10^{10} to 10^{12} nerve cells or neurons with a complex network of connections between them. These neurons communicate with each other by means of two types of signals. The question, how the brain processes information using these signals between the basic components (neurons), is one of the most challenging ones.

At the present time, one has a fairly clear picture of the behaviour of a single neuron [1] under a stimulus. One also has some understanding of the early processing of sensory information [1]—in particular visual—largely due to the work of Hubel and Wiesel. In order to understand, more complex functions carried out by the brain, at this stage at least, one has to resort to mathematical models. The models should naturally begin from the well-known electrical properties of single neurons (even in an idealized or schematic manner) and for a collection of them be able to simulate some of the features of the "higher" brain functions.

In this note a very general mathematical framework is proposed to study neural networks. The basic question we want to pose is, that, given a certain input stimulus at time $t = 0$, what is the output of the neural network, which takes into account past $(-\infty < t < 0)$ experiences and associations of the system which in turn evolve with time and undergo change.

We consider the network* to be made up of N idealized neurons that are all linked to each other. In the absence of an input stimulus the neurons would have a certain amount of spontaneous activity which is described by the firing frequencies r_j^0 ($j = 1, \ldots N$). Since the spontaneous activity is known to vary randomly with time, having a well defined constant time average, the firing frequency r_j^0 should be a random variable in time.

---

*The notation used here is identical to that of Cooper (see ref. 2).

In our model we take each r_j^0 to be a random variable having a distribution around the time-averaged mean. If there is an input stimulus at time $t = 0$ the network will be represented by specifying the firing frequencies f_j^0 ($j = 1, \ldots N$) for the N neurons. The frequencies f_j^0 are, more precisely, the differences between the actual firing rates r_j and the spontaneous firing rates r_j^0—i.e.,

$$f_j^0 = r_j - r_j^0 \qquad (1)$$

It is also known that the same stimulus can lead to firing patterns that vary from trial to trial, but the average (over many trials) of the firing rate is constant. Therefore, the frequencies f_j^0 ($j = 1, \ldots N$) are also random variables distributed about appropriate mean values.

In order to simulate 'memory' effects, we say, following Cooper [2], that during the time interval $-\infty$ to 0 (at instants $\beta_1, \beta_2 \ldots \beta_k$) there have been K experiences and associated correlations, defined by certain input and output firing rates in the neurons. We denote* these firing rates by f_j^ν and g_i^μ respectively, where ν and μ take values from 1 to K, and $i, j = 1, \ldots N$.

In view of these considerations it is appropriate to define the state (probabilistics) of the neural network at time $t = 0$ by a distribution function $\rho(\{g_i^\mu\}, \{f_j^\nu\}, \{f_j^0\}, t = 0)$. Our problem is to obtain the distribution function $\rho(\{g_i\}, t)$ for the output firing rates $\{g_i\}$ at time t in the presence of the stimulus $\{f_j^0\}$ and the past experiences and associations described by the firing frequencies $\{g_i^\mu\}$ and $\{f_{jj}^\nu\}$.

We now make the 'ansatz' that $\rho(\{g_i\}, t)$ is given by the evolution equation [3]

$$\rho(\{g_\alpha\}, t) = \int_0^t d\tau \int d\{g_i^\mu\} \int d\{f_j^\nu\} \, G(\{g_\alpha\} \mid \{g_i^\mu\}, \{f_j^\nu\}, t - \tau)$$
$$\times \rho(\{g_i^\mu\}, \{f_j^\nu\}, \{f_j^0\}, \tau) \qquad (2)$$

Here $G(\{g_\alpha\} \mid \{g_i^\mu\}, \{f_j^\nu\}, t - \tau)$ is the kernel in the integral equation. It is quite clear that the solution of this equation will depend on the form of $\rho(\{g_i^\mu\}, \{f_j^\nu\}, \{f_j^0\}, t = 0)$ and the kernel.

We now rederive Coopers [2] results from the present formulation by making specific assumptions about the kernel and the distribution function. We define the forms of G and ρ, by

$$G(\{g_\alpha\} \mid \{g_i^\mu\}, \{f_j^\nu\}, t - \tau) = \sum_{\mu\nu ij} G_{ij}^{\mu\nu}(\{g_\alpha\} \mid g_i^\mu, f_j^\nu, t - \tau) \qquad (3)$$

where

$$G_{ij}^{\mu\nu} = C_{\mu\nu} \, \delta(t - \tau) \, \delta(g_i - g_i^\mu) f_j^\nu \qquad (4)$$

and

$$\rho(\{g_{i'}^{\mu'}\}, \{f_{j'}^{\nu'}\}, \{f_{j'}^0\}, \tau) = \prod_{\mu'\nu'i'j'} \rho_{i'}^{\mu'}(g_{i'}^{\mu'}, \tau) \, \rho_{j'}^{\nu'}(f_{j'}^{\nu'}, \tau) f_{j'}^0 \, \delta(\tau) \qquad (5)$$

If we now multiply Eq. (2) by g_l, and integrate over all g_l we get on substituting from Eq. (3)–(5)

$$\bar{g}_l(t) = \sum_{\mu,\nu,j} \int_0^t d\tau \, C_{\mu\nu} \, \delta(t-\tau) \, \bar{g}_l^\mu(\tau) \, \bar{f}_j^\nu(\tau) \, f_j^0 \, \delta(\tau) \tag{6}$$

Integrating w.r.t. time, Eq. (6) reduces to

$$\bar{g}_l(0) = \sum_{\mu\nu} \sum_j C_{\mu\nu} \bar{g}_l^\mu(0) \, \bar{f}_j^\nu(0) \, f_j^0$$

i.e.

$$\bar{g}_l(0) = \sum_{j=1}^N A_{lj}(0) \, f_j^0 \tag{7}$$

with

$$A_{lj}(0) = \sum_{\mu\nu} C_{\mu\nu} \, \bar{g}_l^\mu(0) \, \bar{f}_j^\nu(0) \tag{8}$$

If we further impose the condition that $\bar{g}_l^\mu$ and $\bar{f}_j^\nu$ are independent of time then Eqs. (7)–(8) reduce exactly to Cooper's [2] model.

The point of obtaining Cooper's results is that Cooper and his colleagues have developed and analyzed a "class of neural models for the acquisition and storage of distributed memories that display on a primitive level, features such as recognition, association and generalization and which suggest some of the mental behaviour associated with animal memory and behaviour". These features are contained in the mapping described by Eqs. (7)–(8).

The framework proposed in this note is more general and can take care of some of the limitations of Cooper's model. Firstly, we are able to deal with the actual firing rates rather than their averages over many trials and secondly we have explicitly introduced time via the evolutionary Eq. (2). It should be clear that in writing the evolutionary equation, we want to account for the formation of memory and its development with time. Thus in the present model we allow for the possibility of memory being "sharpened" by repeated similar inputs and thereby account for the learning process. In addition memory can "decay" with time if the corresponding information is not used. It should be pointed out that in contrast to the memory in a living system the memory in a computer is static so that there is no learning or "decay".

We would further like to explore the present model to account for some of the "higher" brain functions such as abstraction.

References

1. S.W. Kuffler and J.G. Nicholls, "From Neuron to Brain", Sinauer Associates, Inc. Publishers (Sunderland, Mass.) 1976.

2. L.N. Cooper, "Distributed Memory in the Central Nervous System, Possible Test of Assumptions in Visual Cortex" (Preprint) Center for Neural Science and Deptt. of Physics, Brown University, Providence, Rhode Island 02912.
3. I. Prigogine, Nonequilibrium Statistical Mechanics, Interscience (New York) 1962.

Metastable Membrane States and the Generation of Electric Signals in Biology

Eberhard Neumann

Biophysical Chemistry
Max-Planck-Institut fur Biochemie
D-8033 Martinsried/Munchen F. R. Germany

1. Introduction

A living being may be viewed as a metastable entity: a transient individual appearance. It is less well known that this macroscopic metastability has microscopic molecular correlates, which are particularly apparent in the rapid biocommunication systems generating and transmitting electric signals in and between cells.

The bioelectric signals of nerve excitation, of nerve-muscle transmission and of neurostimulated release of hormones, are based on displacements and flows of ions (Na^+, K^+, Ca^{2+}, Cl^-) along and across the cell membranes. The ion movements are not purely passive charge diffusion; they are regulated by structural changes in channel proteins. It is remarkable that the functionally so important ion-conducting states of two widespread gating systems, the axonal Na^+-channel and the acetylcholine receptor-channel, are transient metastable states, spontaneously converting to inactivated (desensitized) non-conducting conformations.

Further, the generation or electric signals involves in both ion-flow gating systems at least one reaction step which is bimolecular.

These conclusions have largely been derived from an analysis of electrophysiological data. Recently it became possible to approach the peculiar shape of some electrophysiologically measured bioelectric signals, in particular by relaxation kinetic studies on isolated gating proteins. The application of a new analytical concept to the acetylcholine receptor-channel in sealed biomembrane fragments and reconstituted receptor-vesicle systems has shown that molecular origin of the 'ion-transport metastability' is the receptor protein itself.

In summary, we may state that the electrophysiological data and the results of molecular studies have revealed that both the rapid axonal and synaptic electric signals appear to reflect special cases of a more general, chemically dissipative control principle based on two molecular foundations: bimolecular activator-receptor reactions and shortlived structural metastability of the ion-conducting channel conformation.

2. Acknowledgement

The financial support of the Deutsche Forschungs-gemeinschaft, grant NE 227, is gratefully acknowledged.

References

1. P.L. Dorogi, E. Neumann (1980), Proc. Natl. Acad. Sci. U.S.A. 77, 6582–86, E. Neumann, J. Bernhardt, (1982), J. Physiologie (Paris).

Biological Implications of Liquid Membrane Phenomena

R.C. Srivastava, R.P.S. Jakhar, S.B. Bhise,
P.R. Marwadi and S.S. Mathur

Birla Institute of Technology and Science,
Pilani–333031 India

Kesting's liquid membrane hypothesis[1-3] which was originally propounded to explain enhanced salt rejection in reverse osmosis due to addition of surfactants like polyvinyl methyl ether in saline feed has been shown to be of significance in the systems of biophysical/biomedical interest.[4] The hypothesis has been exploited to generate bilayers of liquid membranes from lecithin, cholesterol and lecithin-cholesterol mixtures on a hydrophobic supporting membrane. The agreement between the passive transport data for the lecithin-cholesterol bilayer and that for biomembranes suggests that the liquid membrane bilayers, as generated in the present experiments could possibly act as a model system for the study of transport through biomembranes. In another set of experiments, role of liquid membrane phenomena in the mechanism of action of antipsychotic drugs has been investigated. It has been demonstrated that the resistance offered by the liquid membranes generated by the surface active drug plays significant role in its mechanism of action. This is a significant finding because the role of passive transport in the mechanism of action of antipsychotic drugs has so far been totally ignored.

A detailed summary of the work described here is given below.

Black lipid membrane[5] (BLM) is one of the widely investigated model system for studying transport characteristics of biomembranes. However its electrical resistance is of several order magnitude higher[6] and the ionic permeability is immeasurably low.[7] This is ascribed to the light molecular arrangement[8] in the BLM, while the fluid mosaic model[9] of membrane structure describes the lipid bilayer in biomembranes as a two dimensional oriented viscous solution. The most important characteristic of biomembranes is their fluid or dynamic nature. In view of this, in the present study, Kesting's liquid membrane hypothesis has been utilized to generate lipid bilayer membranes which have been found to possess passive transport characteristics, closer to biomembranes.

According to the liquid membrane hypothesis[3], a surfactant when added to water or aqueous solutions generates a liquid membrane which completely covers the interface at concentrations equal to or greater than its critical micelle concentration. The liquid membrane, thus generated influences mass transfer across the interface. In the present investigation, lecithin, cholesterol and their mixture are shown to form the liquid membranes at the interface. Experiments have been designed to demonstrate the formation of liquid membrane bilayers. Hydraulic conductivity, electro-osmotic velocity, streaming potential and streaming current are used as probes to prove the formation of liquid membranes and their bilayers. Data obtained on cationic permeabilities, resistance and cationic transport numbers of the liquid membrane bilayers generated by lecithin-cholesterol mixtures has been shown to match with the corresponding data on biomembranes.[10, 11]

A large number of antipsychotic drugs are known to be surface active[12–14] and hence capable of forming liquid membranes. In the present investigation Halopendol is selected as a representative antipsychotic drug. Using data on hydraulic conductivity coefficient, the phenomena of liquid membrane has been demonstrated. The study of permeability characteristics of various neurotransmitters like catecholamines, 547 hydroxytryptamine amino acids and few cations in presence of Haloperidol indicates that the passive resistance offered by haloperidol liquid membrane to the flow of these neurotransmitters notably contributes to the mechanism of its action. The orientation of the drug molecules in the liquid membrane is a determining factor for the permeability characteristics. The strengthening of hydrophobic core of the liquid membrane by gamma-amino-butyric acid has been demonstrated. This has been found to be relevant to the mechanism of the action of the drug.

References

1. R.E. Kesting, A. Vincent and J. Eberlin OSW R and D report No. 117 Aug. 1964.
2. R.E. Kesting, Reverse Osmosis Process Using Surfactant Feed Additives. OSW Patent Application SAL 830 Nov. 3, 1965.
3. R.E. Kesting, W.J. Subcasky and J.D. Paton J. Colloid Interface Sci. **28**; 156 (1968).
4. R.C. Srivastava and Saroj Yadav J. Non-Equilib. Thermodyn **4**; 219 (1979).

5. H.T. Tien, "Bilayer Lipid Membranes. Theory and Practice", Marcel Dekker Inc., New York, 1974; Chapter 11.
6. R. MacDonald Fed. Proc. **26**; 863 (1967).
7. H.J. Vreeman: Kon. Ned. Akad. Wetensch. Proc. Ser. B. **69**; 564 (1966).
8. N. Lakshminaryanaiah; "Transport Phenomena in Membranes", Academic Press, New York (1969). pp. 448.
9. S.J. Singer and G.L. Nicolson; Science: **175**, 720 (1972).
10. F.J. Brinley (Jr) and L.J. Mullins. J. Neurophysiol., **28**; 526 (1965).
11. A.S. Troshin, "Problems of Cell Permeability"; Pergamon 1966.
12. P. Seeman and H.S. Bioly Biochem. Pharmacol., **12**; 1181 (1963).
13. P. Seeman Pharmacol. Rev., **24**, 583 (1972).
14. A. Felmister J. Pharm. Sci., **61**; 151 (1972).

Ascent of Sap and Translocation by Means of Electrical Double Layers

M. Amin

School of Life Sciences
Jawaharlal Nehru University
New Delhi, India

Abstract

The vascular system of plants consists of vessels, tracheids, sieve tubes and plasmodesmata which have negatively charged surfaces. Electrical double layer phenomena becomes crucial to the mechanism of transport in such systems since the diameter of these organismal capillaries lies in the submillimeter rage. The throry of electrical double layers is applied to the problem of the ascent of sap and the translocation of metabolites in plants. It is shown that the electrostatic forces are sufficient to hold thin electrolyte films in the vessels against the gravitational pull, so that water can be transported upward without invoking negative pressures, which is the major draw back in the cohesion theory. Translocation of metabolites in the sieve tubes is also explained on the basis of electrical double layer theory. H^+ movement in the highly ordered structure of water at the double layers in the sieve tubes from sinks to sources causes a charge compensating movement of the mobile part of the double layer in the opposite direction. Strong thermodynamic coupling of the photosynthate fluxes to the fluxes of the constituents of the double layer provides a mechanism for the translocation of photosynthates in plants. Certain unifying features of the double layer theory for organismal capillaries are discussed which seem to provide an insight into the transport phenomena in the living systems in general.

Introduction

The theory of electrical double layers has found wide application in several branches of biology due to the fact that the cellular membranes usually bear a net negative charge at the normal pH. Whereas the influence of the electrokinetic potential on the ionic distributions close to the membranes and on the permeability properties of membranes has been the subject of fruitful investigations, the implications of charged surface in the vascular system of plants for the transport processes has not been fully appreciated. For the mechanism of translocation the involvement of localized electroosmosis at the sieve plates was proposed (Fensom, 1957; Spanner, 1958, 1975) but the idea was dropped in view of the missing driving forces. However, as will be shown here, it is possible to treat transport and translocation in plants on the basis of electrical double layer phenomena adopting an integrated network approach and utilizing the pH gradients, existing in the network owing to the different metabolic processes in the leaves and the roots and other sinks, as driving forces. In the first part of the present paper the phenomenon of the ascent of sap will be discussed. The second part will be devoted to the mechanism of translocation of photosynthates from sources to sinks. In the last section certain basic features of the double layer approach will be discussed which could be of relevance to all biological transport phenomena.

Ascent of Sap

The currently accepted cohesion theory is beset with many serious problems which justifies this attempt to find a new explanation for the phenomenon. While the involvement of transpiration as the driving force for the upward movement of water is doubtlessly correct, the concept of intact water columns, unbreakable at tensions of -20 bars, is totally untenable as we now proceed to show on the basis both of the experimental evidence and theoretical considerations. The evidenced gathered on tall trees (Haberlandt, 1914) shows, that apart from the newly formed xylem, the vessels and tracheids never have unbroken water columns but, instead, water columns alternating with usually much larger air columns, the so-called Jamin's chains. Even the few intact columns cavitate at a high rate under water-stress as found by Milburn in Ricinus (Milburn 1973). Transverse overlapping cuts, geometrically destroying the continuity of the water conducting tissue, do not impair the transport of water in the trees (Mackay and Weatherley, 1973) and the introduction of air-columns in the stems leads to only a slight increase in the tension while the normal water uptake continues unabated, (Scholander et. al., 1957). A major theoretical objection to the theory is the following: since the water taken up by the roots contains gases dissolved in it at 1 atm. these gases must come out of the water phase at negative pressures. The system of the water vapor in the bubble and the water around it is subject to the phase equilibrium conditions for water. A look at the phase equilibrium diagram of water (Barrow, 1973) shows that at temperatures higher than 0°C water will go over to vapor phase for all pressures less than 0 bar. Thus it must be the content of the dissolved gases which is crucial to the stability of water columns and not the cohesion of water molecules which is theoretically 18,000 bars, much beyond the requirements of the matter at hand. Furthermore it should not be possible for water to exist in the liquid phase in the Jamin's chains at negative pressures contrary to the experimental evidence. Indeed there is no unequivocal direct evidence for the occurence of negative pressures of less than even −2 bars in the stems; the method of the pressure bomb (Scholander, 1964) by which negative pressures have been recorded gives the total water retention capacity of twigs and not the actual hydrodynamic pressure existing in an intact operational water conducting channel.

It will be shown now that electrostatic forces operating in double layers at the internal surfaces of vessels and tracheids can hold thin films of electrolyte against gravity, that under conditions of low water stress these films can give rise to Jamin's chains and intact columns and that the difference in the water ₁potential between the leaves and the roots, generated by transpiration can cause the ascent of the xylem sap to the leaves. The available histochemical evidence (Northcote, 1972; Lâuchli, 1976) indicates that the nature has provided ideal conditions for the formation of double layers at the walls of vessels and tracheids. Pectin, which is a mator component of the cell walls, contains large number of carboxylic groups. These ionogenic groups are kept in a dissociated state by an assured supply of water from the neighboring cells through innumerable pits in the walls of the vessels and tracheids. The net negative charge density is known to give the cell walls the properties of cation exchangers and values of the surface potential $\phi°$ as high as −170 mV have been experimentally observed (Walker and Pitman, 1976).

In order to show that a thin film of electrolyte can be held in the lumen of the vessels and tracheids against gravity we may assume the following representative values: $\phi° = -150$ mV, $\epsilon = 20$, $T = 25°C$, Height $h = 100$ meters, and 1 mM KCl as sap. The Debye length is then about 100 Å and the surface charge density σ^- equal to 1.1×10^4 esu/cm². Adopting the Helmholtz-Perrin model we consider the counter-ionic layer to be localized as a positively charged surface with a density $\sigma^+ = -\sigma^-$, 100 Å away from the surface of the cell wall. The thickness of the film can be estimated as the distance from the surface for which $\phi(0) = \phi°/100$, which gives a value of about 400 Å. Consider a volume element dV having an area of 1 cm² and a thickness of 100 Å. The energy of the capacitor formed by σ^- and σ^+ for this volume is

$$(\pi\sigma^2/2\epsilon)\, dV = 0.96 \times 10^7\, dV \text{ ergs}$$

This has to be compared with the amount of work to be done against gravity in lifting the mass of $1 \times 1 \times 400$ Å $= 4\, dV$ to a height of 100 meters, which can be calculated to be $3.42 \times 10^7\, dV$ ergs. It is thus shown that the electrostatic energy of the double layer is of the same order of magnitude as the gravitational energy. This establishes the relevance of double layer phenomena at the walls of the vessels and tracheids to the mechanism of the ascent of sap.

Surface tension and the Zsigmondy effect (Zsigmondy, 1911), i.e. the condensation of water vapor in pores and fine capillaries at pressures less than normal for a given temperature, can lead the thin layers to coalesce and thus fill up the lumen of the cells. It should be noted that due to the dependence of the energy of the double layer on the square of the charge density the competitive capability of the double layer energy to the gravitation potential can be easily improved. For a conceivable value of 1 electronic charge per 100 Å^2 for instance it is seen to be $4.5 \times 10^7 \times dV$ ergs. Such values of σ^- would obtain in the freshly formed xylem and in the vessels at higher altitudes in which the carboxylic groups of the pectin molecules are neutralized to a lesser extent by Ca^{++}.

As regards the upward movement of the sap, the electrolyte in the thin layers at the walls of the vessels and the tracheids is subject to the electrochemical potential differences of its constituents in the whole vascular system of the plant. The major factor is the chemical potential of water which is lowered in the leaves by the transpiration process. Due to the continuity of the thin film in Jamin's chains the xylem sap would move upward from the roots to the shoot at rates as governed by the rate of transpiration. The relative thinness of the thin films is more than compensated by the much larger area of cross-section of the old xylem compared to that of the new annual ring with intact columns. The thin layer mode of ascent of sap can, therefore, serve as the durable mode for the supply of water and nutrients in plants. It should be noted that, in contrast to the cohesion theory, the double layer approach does not require the presence of negative pressures to explain the phenomenon of ascent of sap.

Translocation of Metabolites in Phloem

Sieve tubes, which constitute the channels for the translocation of photosynthates in plants, not only retain the cell membranes in their functional state but also have a parietal accumulation of endoplasmic reticulum (Parthasarathy, 1975). At the alkaline pH of the phloem sap the sieve tubes would, therefore, have an internal surface which is highly negatively charged. At the sieve plates and through the sieve plate pores again there is an accumulation of *ER* and of *P*-proteins so that while the cross-sectional area for volume flow is reduced the surface area for transport along surfaces is greatly enhanced. The stage is set in a way as to point towards a major role of electrical double layers in the process of translocation.

The major problem in phloem physiology is the apparent absence of any of the usual gradients of sufficient magnitude to account for the translocation of photosynthates from sources (photosynthetically active leaves) to sinks (roots, fruits, buds, etc.) (Wardlaw, 1974). Not only is the hydrostatic pressure difference originating from sucrose concentration gradient too low in magnitude to cause pressure flow (Milburn, 1975), there is no straight forward relationship between specific mass transfer and the velocity of the sap (Canny, 1975). The values of the sap velocity measured by different techniques differ widely from each other. While an average value of about 70 cm/hr is commonly assumed to obtain in sieve tubes, tracer experiments reveal a fast velocity component of 2000 to 5000 cm/hr and also large differences for the several constituents of the sap (Canny, 1973). Evidence of lateral leakage of water and potassium, bidirectional flow (Eschrich, 1975) in a single phloem bundle and the above mentioned facts clearly indicate that translocation in the sieve tubes cannot be treated as laminar flow in cylindrical pipes. It is a novel phenomena and the underlying mechanism can be expected to be correspondingly intricate.

In the present model we have adopted an integrated network approach to the phenomenon of translocation involving the whole vascular system including the medium distance symplastic transport in the rays. The electrochemical gradient required for translocation emerges from the consideration of basic metabolic processes in the leaves and in the sinks in terms pH differences. The theory of electrical double layers has been applied with a new input: the tangential charge compensating movement of the diffused part of the double layer in response to high frequency proton jumps from sinks to sources. Strong coupling of the fluxes of photosynthates to the K^+-flux in the double layer provides the basic mechanism for translocation. These points will now be elaborated further.

The metabolic processes of plant cells result in the intracellular production or consumption of H^+ (Smith and Raven, 1976). If the carban source for growth processes is CO_2 or hexose and the nitrogen source NH_4^+ there is an excess production of H^+. In the case of nitrate reduction there is a massive production of OH^-. The disposal of metabolically produced H^+ and OH^- in the bathing medium has been established for yeast cells and aquatic plants. In a land plant the possibility of disposal in the medium is restricted and the question of pH-stating in plants poses a fundamental problem which has not been solved so far (Ben Zioni, 1971). Since nitrate reduction takes place mainly in the leaves, the sources of photosynthates are alkaline with respect to the sinks, i.e., the roots, fruits and meristems and this is also reflected in a pH difference of about two units between the saps of phloem and xylem. One could resolve the dilemma of pH-stating both for the tissues of the sources and of the sinks by conceiving a migration of H^+ from the sinks to the sources and of OH^- in the opposite direction. The high frequency jump mechanism for H^+ in ice could also operate at charged interfaces in the vascular system of plants, since at such surfaces water exists in a highly ordered form (Bockris and Reddy, 1970). Due to the electrostatic field of the negative fixed charges there is a relatively high concentration of cations close to the surface as given by the Boltzmann distribution. As the radius of H^+ is negligible it would come to lie at the first layer of water in contact with the surface while monovalent cations form a diffused layer screening the charged surface. H^+ at the interface are subject to their electrochemical gradient along the plane of the surface and would move with their characteristic high mobility along the first layer of water molecules. The movement of protons down their concentration gradient will, however, stop due to the building up of an electrical potential in the opposite direction unless there is a simultaneous movement of either some anions in the same direction or of some cations in the opposite direction. As anions are repelled from the surface by the negative fixed charges these are not exposed to small field variations at the surface and cannot be thought to flow in pace with H^+. The diffused counterionic layer, on the other hand, is known to be the cause of high surface conductance at negatively charged surfaces (Matijević, 1974). This layer is also close to the site of the proton jumps and will instantaneously react to the electric field generated by such jumps by moving in the opposite direction thereby making a further movement of protons down their concentration gradient possible. The phloem sap is very rich in K^+ (Ziegler, 1975) and hence this cation is the most obvious candidate for the formation of the counterionic layer.

A precise localization of the mobile part of the double layer responsible for surface conductance is not possible but roughly it can be assumed to extend from the surface of shear to a distance corresponding to the Debye length χ^{-1}. For this layer, henceforth called the K-layer, we assume, for the sake of simplicity, an average concentration c_K and an average mobility u_K for K^+. Due to the negligible separation between the K-layer and the layer of water mediating H^+ transport we can further assume that the electric field component responsible for the tangential movement of H^+ and K^+ is the same. As will be justified later the concentration of K^+ in the sieve tubes is fairly constant. Though the K^+ concentration gradient is favorable for the movement of K-layer from the sources to the sinks, its contribution is omitted in the following calculation to bring out the role of pH difference as the driving force more clearly.

With the above mentioned assumptions the flux densities of H^+ and K^+ are

$$j_{H^+} = u_{H^+} c_{H^+} \left(RT \frac{d}{dx} \ln c_{H^+} + F \frac{dV(x)}{dx} \right) \tag{1}$$

and

$$j_{K^+} = u_{K^+} c_{K^+} F \frac{dV(x)}{dx} \tag{2}$$

The fluxes, J_H and J_K can be obtained from j_H and j_K by multiplication with the areas of cross-section S_{wl} and S_{Kl} of the first layer of water and the K-layer respectively. At steady state $J_H + J_K = 0$ and $dJ_i/dx = 0$ so that upon integrating from x_1 to x_2 we obtain

$$J_H = \sigma_H \left(2.3 \frac{RT}{F} \Delta pH - \Delta V\right) \tag{3}$$

$$J_K = -\sigma_K \Delta V \tag{4}$$

where

$$\sigma_H = u_H c_H S_{wl} F/(x_2 - x_1), \tag{5}$$

$$\sigma_K = u_K c_K S_{Kl} F/(x_2 - x_1), \tag{6}$$

$$\Delta pH = -(\ln c_H(x_2) - \ln c_H(x_1))/2.3 \text{ and}$$

$$\Delta V = V(x_2) - V(x_1)$$

By adding Eqs. (1) and (2) we get

$$V = \frac{\sigma_H}{\sigma_H + \sigma_K} \frac{2.3 RT}{F} \Delta pH \tag{7}$$

In terms of the pH gradient the fluxes are given by:

$$J_H = \frac{\sigma_H \sigma_K}{\sigma_H + \sigma_K} \frac{2.3 RT}{F} \Delta pH \tag{8}$$

$$J_K = \frac{\sigma_H \sigma_K}{\sigma_H + \sigma_K} \frac{2.3 RT}{F} \Delta pH \tag{9}$$

Eq. (9) gives the flux of K^+ from the sources to sinks in response to the movement of H^+ from the sinks to sources down its gradient. This movement of the K-layer can now be invoked as the driving force for translocation in two ways. Due to the viscosity dependent inter-layer friction the K-layer movement can cause the flow of the whole liquid phase in the lumen of the sieve tube with a velocity profile which is the inverse of that found in a laminar flow. The second possibility is by means of a coupling of the fluxes of the photosynthates with the flux of K^+ in the layer. Compared with the first mode of flow the second will be more selective and lead to appreciable differences in the flux rates of the phloem sap constituents. Since the sieve plates retard a volume flow at every junction the flux rates will be also affected by the number of sieve plate pores and their radii. It should be noted that the sieve plates will accentuate the K-layer coupled fluxes (larger S_{Kl}) while for the volume flow their presence has a negative effect.

The transport in the sieve tubes cannot be dealt with in isolation from the rest of the transport network of the plants. The K^+ flux, which in our model has to match the massive flux of H^+ corresponding to the total metabolic activity of the plant, would result in the accumulation of K^+ in the sink region contrary to the experimental evidence. K^+ is known, however, to move out of the sieve tubes laterally even in the source region. It is, therefore, imperative at this juncture to go into the anatomy of the vascular system in order to see what role could be played by the other components of the vascular system in the process of translocation.

Throughout the vascular system of plants there is a close association of the sieve tubes with the vessels and tracheids of the xylem. In the fine ramifications of the vascular system these two communicate over the transfer cells (Pate and Gunning, 1972) and modified companion cells. Where the distance between the two is greater, a chain of cells, starting with the Strasburger cell, in close contact with the sieve tubes via pits, followed by the procumbant ray cells, forming a low resistance symplastic pathway (Spanswick, 1976), and ending with the contact cells, again well connected to the vessels and tracheids by means of pits, establishes the link between them. Recalling that there is a high concentration of H^+ in the xylem compared to that in the phloem and knowing that the rays have a low resistance for H^+ and K^+ we can extend the H^+/K^+ exchange transport hypothesis to the rays. As H^+ move into the phloem from the xylem across the rays there has to be a charge compensating movement of K^+ from the phloem into the xylem. In the phloem H^+ will partially unite with OH^- while the rest move on towards the sources. K^+ would, on their part, substitute H^+ as the co-ions of NO_3^- and $H_2PO_4^-$ in the xylem and join in the upward flux of water caused by the transpiration process. The pool of K^+ in the leaves will thus be replenished and, at the same time, there will be no accumulation of K^+ in the sinks. The gradual removal of K^+ from the sieve tubes need not impair the translocation of the photosynthates coupled to K^+ so long the phloem sap is rich in K^+. K^+ deficiency would, however, have grave consequences. This, indeed is the case; translocation is extremely sensitive to K^+ deficiency (Amir and Reinhold, 1971). In the leaves K^+ reenters the sieve tubes thereby completing a full cycle in a loop of the network. In the sieve tubes the K-layer acts like a conveyor belt, driving the photosynthates towards the sinks. In the phloem and in the rays this conveyor belt is powered by the metabolically charged pH battery; in the xylem transpiration provides the driving force by means of the coupling of the K^+ to the flux of water. The K^+ concentration gradient between the phloem and the xylem as well as along the phloem itself thus established by transpiration helps in the conveyor belt movement. Its contribution can be easily incorporated in the flux equations, Eqs. (8) and (9).

The above presented model explains the experimental observations, mentioned at the beginning of this section, in a simple and obvious manner. The fast velocity component of the sap movement is that which is intimately coupled to the K-layer. The differences in the thermodynamic coupling coefficients of the fluxes of sap constituents with the K^+ flux explain their differential mobilities. The movement of K^+ in the loops of the network and, through electro-osmosis, also of water is the reason for the observed lateral movement of K^+ and water. Bidirectional flow can be explained by means of the cyclic flow of K^+, water and substances which can pass through the rays and return to the phloem via leaves or by the fact that the stylets of aphids cause the volume flow of the sap towards the low pressure zone while the K-layer mediated transport will continue from the leaves to the actual sinks. This model also provides a feedback mechanism between the shoots and the roots since the metabolic processes in the leaves will influence, by affecting the H^+ flux, the ionic uptake processes in the root cortex. Finally it resolves the dilemma of pH stating at the sinks and sources while at the same time providing the energy for translocation thus extrapolating, in essence, Mitchell's hypothesis (Mitchell, 1961) to the whole plant.

General features of the H^+/K-layer hypothesis

The hypothesis of long distance transport in organismal capillaries-vessels, tracheids, sieve tubes and plasmodesmata—by means of a mobile layer of counterions and high frequency H^+ jumps along ordered water molecules at the interface has certain interesting and significant implications for biological systems. If we consider the interaction of the common cations found in the living organisms with the charged groups at the surfaces of organismal capillaries we can classify them in four catagories: (1) the least electronegative cations occupying a position close to the surface at the inner Helmholtz layer, (2) those lying between the inner Helmholtz layer and the surface of shear, (3) those lying adjacent to the surface of shear towards the liquid phase and (4) the protons at the ordered layer of water in contact with the surface. Cations of the third category are free to move tangentially, while those in the second category have reduced mobility. The strongly bound cations of the first category are practically immobilized. For the commonly occuring carboxyl and phosphate groups in cell membranes and in organismal capillaries K^+ (also Rb^+) belongs to the third category, Na^+ to the second and Ca^{++} and Mg^{++} to the first. In terms of long distance transport the surface charges make the organismal capillaries cation selective. In this frame work cation selective pores in cell membranes are limiting cases (radius tending to ionic radii) of such organismal capillaries, with greatly enhanced selectivity. Since cell membranes invariably possess K^+ selective channels, K^+ has the distinction of being (next to H^+) the most mobile and abundant cation in all living organisms. Since the argument of compatibility in the selection of K^+ as the major cation cannot be applied in the direction from organismal to cellular level the transport along surfaces must be, in some way, of crucial importance even at the cellular and subcellular levels. In the context of our H^+/K^+-layer hypothesis it is natural to assume that this major role of K^+ is the same at the cellular level as in the long distance transport. The rapid movement of H^+ required for fast enzymatic reactions can take place from one location in the cell to another only if a charge balancing movement is provided at the same time along a surface such as endoplasmic reticulum. This contention is supported by the requirement of a high concentration of K^+ for protein synthesis to take place at a high rate (Lehninger, 1972). Protoplasmic streaming could also be understood as a side effect of K-layer movement caused by interlayer friction of the viscous plasmasol. The H^+/K-layer movement could thus be a fundamental feature of the living state of much wider and deeper implications than discussed here.

References

1. Amir, S., Reinhold, L.: Interaction between K-deficiency and light in ^{14}C-sucrose translocation in bean plants. Physiol. Plantarum 24:226–231 (1971).
2. Barrow, G.M.: Physical Chemistry. (Third Edition), McGraw-Hill Kogakusha Ltd. Tokyo (1973).
3. Ben Zioni, A., Vaadia, Y., Lips, W.: Nitrate uptake by roots as regulated by nitrate reduction products of the shoot. Physiol. Plantarum 24:288–290 (1971).
4. Bockris, J. O'M, Reddy, A.K.N.: Modern Electrochemistry. Vols. 1–2 Plenum Press, New York (1970).
5. Canny, M.J.P.: Phloem Translocation. London, Cambridge University Press (1973).
6. Canny, M.J.P.: Mass Transfer. Enc. Plant Physiol. 1:139–153 (1975).
7. Eschrich, W.: Bidirectional Transport. Enc. Plant Physiol. 1:245–255 (1975).
8. Fensom, D.S.: The bioelectric potentials of plants and their functional significance. Can. J. Bot. 35:573–582 1957).
9. Haberlandt, G.: Physiological Plant Anatomy. Macmillan and Co. Ltd. (1914).
10. Läuchli, A.: Apoplastic Transport in Tissues. Enc. Plant Physiol. 2B: 372–393 (1975).
11. Lehninger, A.L. Biochemistry. Worth Publishers Inc. New York (1972).
12. Mackay, J.P.G. Weatherley, P.E.: The effects of transverse cuts through the stem of transpiring woody plants on water transport and stress in leaves. J. Exp. Bot. 24: 15–28 (1973).
13. Matijevic, E. (Editor): Surface and Colloid Science. Vol. 7 John Wiley & Sons, New York (1974).
14. Milburn, J.A.: Cavitation studies on whole Ricinus plant by acoustic detection. Planta (Berlin) 112:333–342 (1973).

15. Milburn, J.A.: Pressure Flow. Enc. Plant Physiol. 1:328–353 (1975).
16. Mitchell, P.: Coupling of phosphorylation to electron and hydrogen transfer by a chemiosmotic type of mechanism. Nature (London) 191:144–148 (1961).
17. Northcote, D.H.: Chemistry of Plant Cell Wall. Ann. Rev. of Plant Physiol. 23:113–132 (1972).
18. Parthasarathy, M.V.: Sieve Element Structure. Enc. Plant Physiol 1:3–38 (1975).
19. Pate, J.S., Gunning, B.E.S.: Transfer Cells. Ann. Rev of Plant Physiol. 23:173–196 (1972).
20. Scholander, P.F., Ruud, B., Leivstad, H.: The rise of sap in tropical liana. Plant Physiol. 32:1–6 (1957).
21. Scholander, P.F., Hammel, H.T. Memmingsen, E.A., Bradstreet, E.D.: Hydrostatic pressure and osmotic potential in leaves of mangroves and some other plants. Proc. Natl. Acad. Sci. USA 52: 119–125 (1964).
22. Smith, F.A., Raven, J.A.: H^+ Transport and Regulation of Cell pH. Enc. of Plant Physiol. 2:317–346 (1976).
23. Spanner, D.C.: The translocation of sugar in sieve tubes. J. Exp. Bot. 9:332–342 (1958).
24. Spanner, D.C.: Electroosmotic flow. Enc. Plant Physiol. 1:301–327 (1975).
25. Spanswick, R.M.: Symplastic Transport in Tissues. Enc. Plant Physiol. 2B: 35–56 (1976).
26. Walker, N.A., Pitman, M.G.: Measurement of Fluxes Across Membranes. Enc. Plant Physiol. 2B: 93–129 (1976).
27. Wardlaw, I.F.: Phloem Transport: physical chemical or impossible. Ann. Rev. Plant Physiol. 25: 515–539 (1974).
28. Ziegler, H.: Nature of Transported Substances. Enc. Plant Physiol. 1:59–101 (1975).
29. Zsigmondy, R.: Z. anorg. Chem. 71: 356 (1911).

The Visuoperceptual Space

Balraj Bhatia

School of Environmental Sciences
Jawaharlal Nehru University
New Delhi—110067. India.

This presentation discusses the concept of what may be described as the visuoperceptual space in the visual system, and in which spatio-temporal excitation patterns generated by inputs from cortical neurons through a scaling mechanism, directly yield the perception of location, shape and size of objects and of their orientation to all the other objects in the visual field. The concept is based on the following lines of evidence.

1. Resolution of moving objects with the eyes fixed

In a recent paper (Bhatia, 1975) it was reported that the angular values of the minimum separable as a function of the angular velocity of a moving object differed with the variations in the distance between the observer and the moving objects. On the other hand, the linear values of the minimum separable as a function of the linear velocity was independent of the distance. In these experiments, 2 white rectangles on a black background moved in a horizontal direction. It was reported that beyond a certain critical velocity, the relation ship between the angular velocity, ω, and the minimum angular separation, α, at which the 2 rectangles appeared as separate was given by the equation, $\alpha = a + c\omega^2$ where a and c are constants characteristic of the individual. Plots of α against ω for 2 and 5 meter distances yielded dissimilar patterns due to differences in the values of the constants and the values of the critical velocities below which there was an abrupt fall in α also were different for the two distances. On the basis of the scaling mechanism proposed the 2 constants should change in a predictable manner. This can be illustrated by a simple model in which a pattern is projected at different distances.

Let us consider a curved line projected from slide A on to a screen at a distance of d meters. Another slide B projects a curved line at a distance of Fd meters, so that the images of both the lines projected on the screen are identical. The form of both the images can be described by, say, the following equation:

$$y = a + cx^2$$

the values of the constants a and c being the same for the two images. The lines on the slide A and B obviously do not have the same constants. Let the lines on the slide A (at shorter distance) be given by $y = a' + c'x^2$ and for the line on the slide B (at longer distance) by $y = a'' + c''x^2$.

Since the magnification of the line from slide B is F times that of the line on slide A, for any point on the line in the projected image, the corresponding x and y points in the line on slide A must be F times those on the slide B. The constants in the lines on the 2 slides must therefore be different and bear the following relationship to each other

$$Fa'' = a' \text{ and } Fc' = c''$$

Thus,

$$\frac{a'}{a''} = \frac{c''}{c'} = F$$

The existence of scaling mechanism to explain the results of the experiments on resolution of moving objects with the eyes fixed requires that the ratio of each of the constants in the equation relating the value of ω (or x) and α (or y) calculated for observations at distances of 5 m and 2 m should be 2.5 which is the ratio of the 2 distances.

The values of the ratios of the constants in respect of the 2 subjects are given in Table 1. The deviations from the predicted values were less than 10%, except in one case, where higher deviation can be explained as due to experimental error. Further, when the plots of the critical angular separation against the angular velocity were replotted for each of the 2

Table 1. Actual values and deviations from the predicted value of 2.5 in respect of a'/a'' and c''/c' in the equation $y = a' + c'x^2$ for 2m and $y = a'' + c''x^2$ for 5m distance, y and x being the critical angular separations in minutes of visual arc, and x, the angular velocity in deg./sec respectively with a', a'', c' and c'' as constants

Subject	Vision	a'/a''	dev. from 2.5	c''/c'	dev. from 2.5
1.	Central	2.5	2%	2.3	8%
	Peripheral	2.3	8%	2.5	0%
2.	Peripheral	1.9	24%	2.4	4%

Note: In case of central vision for subject 2, the relationship between y and x was rectilinear.

distances, using the respective constants, but with the x and y scales for the 5 meter distance expanded by a factor of 2.5 viz. the ratio of 2 distances, not only did the plots occupy the same position but also the points corresponding to the constant angular values below which there is an abrupt fall in the critical separation were also the same (Bhatia, 1981).

The results of the experiment have been examined in the light of simple examples illustrating the distortion in patterns by alterations in two-dimensional scale in order to emphasize that the scaling represents a form of projection from a reference point in a space at the terminal level of the visual system.

It is well known that the visibility of a moving object is reduced because of the distortions caused, by temporal delays in the retina, in the excitation pattern transmitted to the visual cortex. As the speed of the object is increased the distance between the rectangles has to be increased to maintain resolution of the figures. The experiments of the author have brought out the fact that whether or not the 2 rectangles appear as separate does not depend upon the degree of distortion and distance between the retinal images per se of the rectangles but upon the degree of distortion of the 'image' of the rectangles and of the distance between them at a level beyond the scaling mechanism. The latter may be conceived as a projection of excitation in a three-dimensional visuoperceptual space from a reference point or the oculoegocentre, the location of the 'image' in the space depending upon binocular and other cues.

Due to the operation of the scaling mechanism, the degree of distortion of the 'image' in the visuoperceptual space at a given linear speed and size of the object would be constant, in spite of variations in the size and velocity of the image on the retina with changes in distance. On the other hand, the degree of distortion of the image in the visuoperceptual space corresponding to a given size of retinal image and its velocity would vary with the changes in distance. The excitation in the visuoperceptual space may be termed beta activity to distinguish it from the excitation in the neurons which may be referred to as alpha activity.

2. Constancy of the minimum detectable size of a moving object with the eyes fixed

In a study on the threshold size of a moving object as a function of its speed, with the eyes fixed, the relationship between the two beyond a critical velocity was found to be according to the equation $0 = a + bV$ (Bhatia and Verghese, 1974). 0 is the threshold size in millimeters, V is the speed in centimeters per second and a and b are constants characteristic of the individual. According to this relationship the threshold size for detection at zero velocity is a. However, even under the conditions of the low brightness contrast (a reflectance of 76% for the object and 85% for the background) used in this experiment, the minimum size that could be detected under static conditions was much smaller than the value obtained from the equation. Further, the threshold size of the object at a given velocity beyond the critical velocity was independent of the distances when these were expressed in linear values, but it varied with the distance when the two were expressed in angular values (Bhatia and Verghese, 1963). These facts can be explained on the assumption that the criteria used for detection of the object are different for the objects moving beyond or below the critical velocity. This is substantiated by the observation that the perceptual experience of the object at threshold, when it was moving faster than the critical velocity was quite different from that when its velocity was below the critical velocity. The perceptual constancy beyond the critical velocity suggests that in the detection of the object under these conditions, the observer used criteria, whereby the limiting factors are dependent upon the phenomena in the visuoperceptual space. Below the critical velocity, the detection of the object is based on criteria where the limiting factors are determined by the phenomena at the retinal or pre-scaling level. As the angular velocity is increased, a larger angular size is required for the detection of the object. At a certain critical angular velocity which is dependent upon the distance and which corresponds to a linear velocity independent of the distance, the observer uses another clue for the detection of the object and the visibility of the object is now limited not by the factors in the retina, but by the phenomena at the visuoperceptual space. The relationship between the threshold size and the velocity (with the linear and angular values) below and above the critical velocity (linear and angular values) is diagramatically represented in Fig. 1a and b. For velocities above the

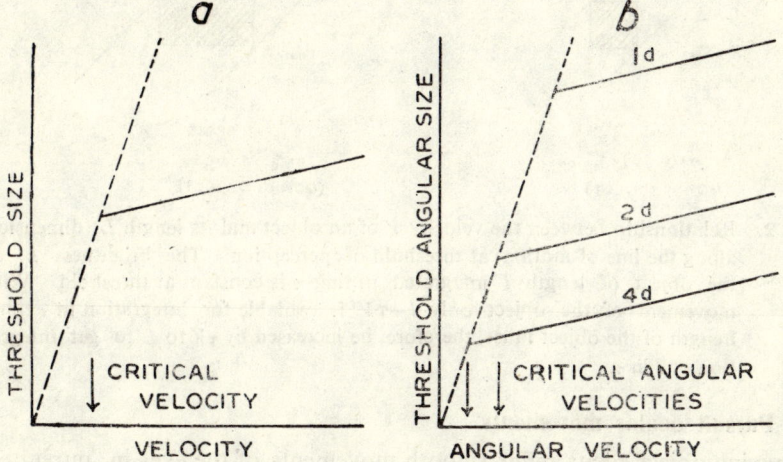

Fig. 1. Threshold size as a function (F) of velocity. (a) linear values, (b) angular values F (continuous line) for velocities above the critical level is determined by phenomena at post-scaling level. F (interrupted line) for velocities below the critical level is determined by phenomena at prescaling level.

critical value, it must be assumed that the detection of the object is determined by the value of the product of brightness and length of the object (the dimension along the direction of the motion of the object) integrated over a time τ and which must exceed a minimum value K. The equation representing the relation between the threshold size of the object and its speed can be explained with reference to Fig. 2; with the criteria of visibility used for detection of the object beyond the critical velocity, the object, when static, can be detected if the object with brightness level B and length l is exposed for a time period of τ. When the object is moving at a velocity V, the length of the object which can be integrated in time τ is reduced by τV and hence the product of the brightness, effective length and τ is less than K. The object cannot therefore be detected. In order to detect the object, its length must be increased by τV. Perceptual constancy means that the limiting factors are not in the retinal image which relates to the values of α and ω but in the post-scaling 'image' in the visuoperceptual space where $d'\alpha$ and $d'\omega$ relate to the linear size and velocity of the object, d' being the distance of the object from the observer as represented in the visuoperceptual space.

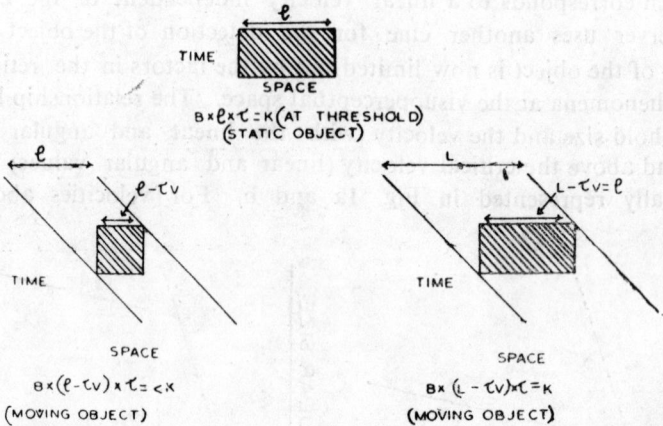

Fig. 2. Relationship between the velocity V of an object and its length L (dimension along the line of motion) at threshold of perception. The brightness B, of the object of length l integrated in time τ is constant at threshold. With movement of the object only $l-\tau V$ is available for integration in τ time. Length of the object must, therefore, be increased by τV to L to get integration of l in τ.

3. Pursuit ocular movements

A moving object elicits reflex smooth movements of the eyes in pursuit of the object. The pursuit ocular movements are replaced by jerky or saccadic eye movements with higher velocities of the object. Evidence was advanced by the author (Bhatia, 1957) that the pursuit ocular movements are not a manifestation of the fixation reflex as was popularly known for a long time, but are related to the specific excitation pattern in the visual system in response to a moving stimulus. The ability to detect a moving object with eyes free to move was found to be influenced by the length of the slit (dimension along the line of motion of the object), when the latter periodically appeared behind a slit, due to the limitation imposed by the slit on the maximum angular velocity of the eyes (Bhatia and Shastri, 1955). It was further observed (Bhatia, 1960) that the maximum angular velocity of the pursuit ocular movements is not related to the angular velocity of the object, which would have been the case if the pursuit movements were a manifestation of the fixation reflex. The maximum angular velocity of the eyes was given by

$$\frac{dx}{dt}\bigg|_{max} = \frac{57.3s}{d(bs + a)}$$

where 's' is the length of the slit (dimension in the direction of motion of the object), the object moving behind the slit, d is the distance between the observer and the object and a and b are constants characteristic of the individual. According to this equation, the maximum angular velocity of the eyes in response to a moving object is limited by the length (linear) of the slit and the distance between the observer and the object. Smaller the length of the slit and greater the distance, the less is the maximum angular velocity of the eyes. The fact that it is the physical length of the slit and not the length of the image on the retina which determines the maximum velocity of the ocular movements suggests that the 'image' at the post-scaling level of the visual system is involved. The inverse relationship with the distance appears to represent a mechanism to aid perceptual constancy of the speed of the objects.

4. Perception of the degree of curvature

In order to investigate the effect of changes in distance on the perceived degree of curvature, experiments were conducted to match cards carrying curves of different degrees presented at distance of 0.25, 0.5, 2 and 4 meters respectively against a standard curve at a distance of one meter. In order to ensure that the judgements by the observer were actually on the basis of the degree of curvature and not on the length of the line or the size of the card carrying the line, the curves of different degrees were drawn on cards of varying sizes. The judgements were made by binocular as well as monocular vision. Although for a given standard curve the degree of curvature in the retinal image varied markedly for different distances (Fig. 3) there was a high degree of perceptual constancy. The results are consistent with the hypothesis of a scaling mechanism in a visuoperceptual space.

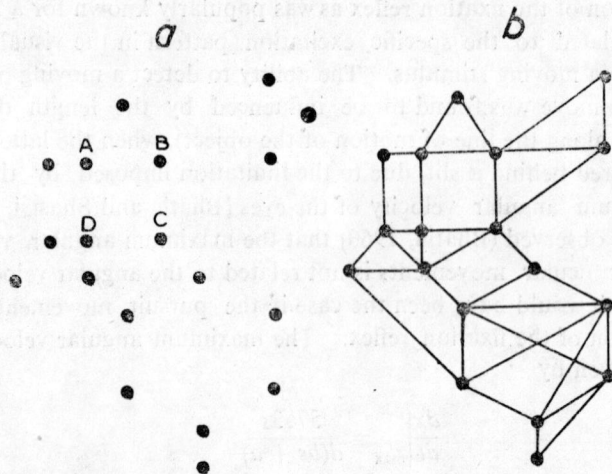

Fig. 3. Changes in the values of the constant b in the retinal image of a parabola with alternations in its distance from the observer. There is constancy of the perceived curvature of the line in spite of the changes in the retinal image.

5. Perception of retinal size and physical size

When we look at an object, say, a vertical wire A, at a distance of about 2 meters and another vertical wire B, slightly larger than and behind the wire A and located at a distance of 4 metres, then the wire B appears to be larger than the wire A in one sense and smaller than the wire A in another sense. This very common phenomenon can be readily explained on the basis of the scaling mechanism in the visuoperceptual space. The image of the wire A on the retina is slightly larger than that of the wire B. But due to the operation of the scaling mechanism in the visuoperceptual space, the 'image' A' (Fig. 4) of the wire A in the visuoperceptual space is smaller than that (B') of the wire B. When we see A smaller than B we are

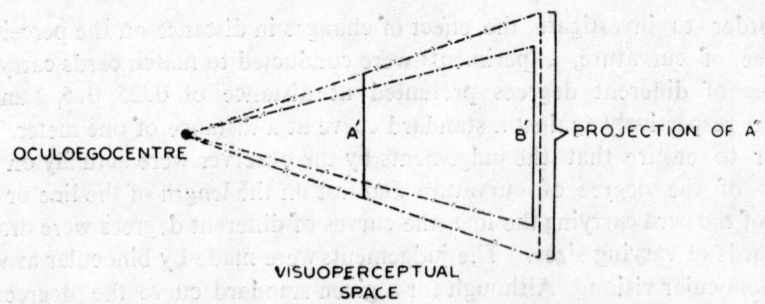

Fig. 4. Perception of object A at a distance of D and of object B (slightly larger than A) at a distance of $2D$, 'Image, B', of B in the visuoperceptual space is larger than the image, A', of A and at twice the distance from the oculoegocentre as compared to A', B' is smaller than the projection or shadow of A' at the same level as B'.

comparing the sizes of the sizes of the images A' and B', which correlate with the physical sizes of the object. When A appears larger than B, we are comparing B' with the shadow or the projection of A' on the background or the same distance as that of B'. The perception of the relative sizes therefore depends upon the criterion we use for comparing the different patterns in the visuoperceptual space.

6. Perception of geometrical figures in a set of dots

Consider a set of dots as shown in Fig. 5a. The exercise of little imagination reveals that many dots can be grouped in the form of geometrical figures such as equilateral, isosceles and right-angled triangles, squares, trapeziums, rectangles, parallelograms, quadrangles etc. Some of these are illustrated in Fig. 5b which has the same pattern of dots as in Fig. 5a. If the paper containing the dots in front of the eyes is tilted, the retinal image is distorted. For instance the paper may be so tilted that the dots A and B in the square $ABCD$ are further away from the observer than the dots C and D. The distance between A and B in the retinal image is less than that between C and D, and the distance between A and D or between B and C is less than that between C and D. We, however, still see the dots arranged in a square. In the perception of the geometrical figures in the set of dots, the movement of the eyes is not an essential condition. In a series of experiments conducted by the author, it was found that the judgement of relative position of the dots was possible with a high degree of accuracy even when the eyes were fixed on to one of the dots.

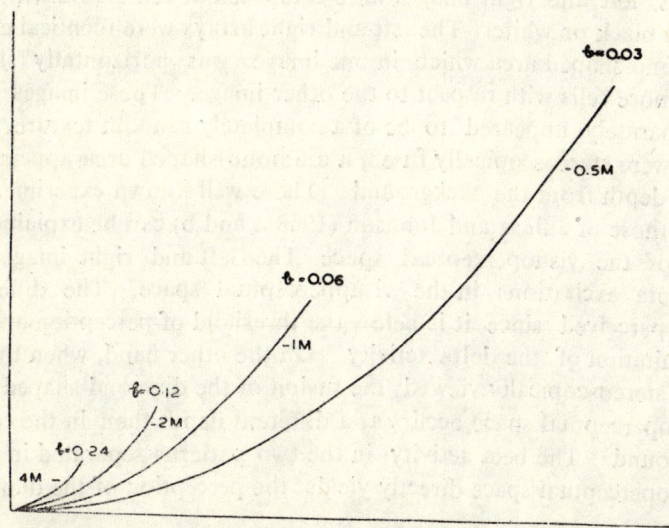

Fig. 5. Set of dots (a) and some of the many geometrical figures (b) that can be perceived in (a).

We have the capacity not only to group the dots into figures but also to examine many essential properties of the figures. For instance, we perceive four dots arranged in a square on the basis of the spatial relation between the dots. However, a square is known to have many other properties, some of which are: (a) its diagonals intersect each other at right angles, (b) the mid-point of one side of the square forms an equilateral triangle with its opposite side (c) the figure obtained by the lines joining the mid-points of all the sides is a square and (d) the line joining the mid-points of two adjacent sides is parallel to one of the diagonals and intersects the other at right angles. It is possible to perceive all this in a set of dots arranged as a square. It is true that the perception of these properties requires effort and, therefore, represents a particular type of activity. But unless the information is contained in the excitation pattern at some level of the visual system, it is not possible to perceive the various features in a set of dots. This leads to the inference that there must be geometrical isomorphism between the 'image' in three-dimensional perceptual space and the physical pattern. The activity corresponding to this image must be distinguished from the excitation activity in the neurons, on the one hand, and from that associated with the voluntary operations on the 'image' which yield the perception of geometrical images and of their various features. These may be referred to as beta, alpha and delta activities respectively.

In the experiments to demonstrate that stereoscopic depth perception can occur in the absence of any monocular form or other cues, Julesz (1960, 1961, 1964) has worked on random dot stereograms. If one of the experiments, left and right images were composed of cell arrays with each cell printed black or white. The left and right arrays were identical except for a diamond-shaped area which, in one image, was horizontally shifted by one or more cells with respect to the other image. These images when viewed separately appeared to be of a completely random texture. But when they were stereoscopically fused, a diamond-shaped area appeared at a different depth from the background. These well known experiments of Julesz and those of Julesz and Johnson (1968 a and b) can be explained on the basis of the visuoperceptual space. The left and right images give different beta excitations in the visuoperceptual space. The difference cannot be perceived since it is below the threshold of perception and due to the limitation of the delta activity. On the other hand, when the two images are stereoscopically viewed, the fusion of the diamond-shaped areas in the visuoperceptual space occurs at a different depth than in the case of the background. The beta activity in the two patterns separated in depth in the visuoperceptual space directly yields the perception of the diamond-shaped area.

7. Size-distance invariance hypothesis

The size-distance invariance hypothesis is consistent with the hypothesis of a scaling mechanism in a visuoperceptual space. According to the size-distance invariance hypothesis, the apparent size of an object is directly related to the retinal size and the apparent distance (Schlosberg, 1950; Gelinsky, 1951; Ittelson, 1951). The retinal projection or visual angle of a given size determines a unique ratio of apparent size to apparent distance. However, Kilpatric and Ittelson (1953) have presented evidence which appears contradictory to the invariance hypothesis. They have cited several conditions under which the relationship between apparent size, apparent distance and the visual angle is not consistently obtained. This does not invalidate the hypothesis of a scaling mechanism in a visuoperceptual space due to the following reasons. Firstly, the visuoperceptual space is not likely to be euclidean. It is well known that the size and shape constancy are valid only for short distances. Secondly, there is every likelihood that many situations tend to produce distortion or local perturbations in the visuoperceptual space which would result in deviations from the normal relationship between the apparent size and distance of an object. Thirdly, the evidence for and against the size-distance invariance hypothesis is based on subjective experiments involving judgements of size and distance, where the percentage of error can be very high. On the other hand, the author has inferred the existence of a scaling mechanism on experiments which do not involve a direct judgement of size and distance.

8. Current status of information on neurophysiology of vision

It is difficult to link the various facts about visual perceptions with the present knowledge on the neurophysiology of vision. The work of Hubel and Wiesel (1962, 1965, 1968) on the basis of extensive experiments on responses of single neurons of visual cortex to visual stimuli suggests that cortical units primarily function to extract different features present in a particular part of the retina. The neurons respond to stimuli such as a light bar with a specific orientation against a dark background or a horizontal edge bright above and dark below. In areas 17 as well as in areas 18 and 19 there are neurons specially sensitive to length and thickness of bright or dark lines and also to their orientation or to two lines meeting at an angle. Responses of what are called as 'complex' and 'hypercomplex' neurons are believed to be due to synthesis of neural circuits activated by single cells (Hubel, 1971). Further, Barlow et al (1967), Pettigrew et al (1968) and Hubel and Wiesel (1970) have demonstrated the existence of some binocular disparity sensitive units.

The findings of the author on the resolution of moving objects cannot be explained on the basis of recordings from single cortical units which indicate that information about the edges of the rectangles are already contained in the impulses from the retina. The temporal aspects of the excitations of the cortical neurons have not been investigated. However, if we assume that the temporal delays are responsible for the loss of information about the adjacent borders of two rectangles, the next question is how these are related to perceptual constancy of the resolution of the moving figure. The constancy together with a second order relationship between the velocity of the object and the minimum separable can only be explained on the basis of neural patterns with gradients of excitations which, after being processed through a scaling process are subjected to further modification in the visuoperceptual space, where either contours corresponding to the adjacent edges of the rectangles are 'constructed' or the gradients give place to uniform excitation corresponding to the fusion of the rectangles. It is a characteristic feature of the visual system that gradients of brightness in non-textured surfaces are not normally perceived. Such stimuli are generally perceived either as a uniform sheet of brightness or two different levels of brightness separated by a contour. Whether a pattern with gradients of excitation is perceived as a single sheet of uniform brightness or as two sheets of different levels of brightness separated by a countour is a function of the properties of the visuoperceptual space.

It is not likely that excitation gradients over a space excite a specific cortical neuron. Such a stimulus must lead to the stimulation of a number of neurons with different magnitudes of excitation.

The existence of an edge in the physical stimulus is not a prerequisite for the perception of an edge, although this is suggested by the neurophysiological findings. There are situations where an edge can be seen when there is none in the physical stimulus.

The present knowledge of neurophysiology also cannot explain the ability to perceive figures in groups of dots. No experiments have been reported on responses of single neurons to single dots or groups of dots. In addition to neurons responding to edges or lines there must be single neurons activated by dots or groups of dots at least in man or else we would not be able to perceive these. It is not possible that there is a cortical neuron for each of the specific patterns of dots, for then there is a need to have an infinite number of neurons even to cover a radius of 10' of the fovea centralis (Sutherland, 1968). However, since the dots can be localized in space with a high degree of accuracy, there must be a cortical neuron for each location in space for the foveal vision. A set of, say, five dots each with a diameter corresponding to the visual angle of the retinal unit must then excite five neurons in the cortex. Why then should there be a need for neurons which respond specifically to lines with different sizes and orientations. The answer to this may be that the neurons discovered by Hubel and Wiessel and others may be concerned not with the total information required for the immediate perceptual experience, but with the extraction of the minimum essential information required for storage as memory. The latter is required for recognition for which only certain features of the stimulus are sufficient, the rest of the information being redundant.

It needs to be emphasised that the failure to discriminate between two complex patterns does not necessarily imply that the cortical neurons do not 'see' any difference between the two patterns. The failure may be due to the limitation of the delta activity involved in the process of comparison of the two patterns. A brief visual presentation of a set of 12 dots would certainly produce a different excitation pattern in the visual system than that of a set of 10 dots, and yet the observer may fail to perceive this difference.

I am highly indebted to Prof. D.S. Kothari, Chancellor, Jawaharlal Nehru University for initiating me into the studies on visibility of moving objects when he was the Scientific Advisor to the Minister of Defence, his continued interest and encouragement in the work and helpful suggestions in the course of many discussions on the subject and allied fields during and after the development of the concept of visuoperceptual space.

References

1. Barlow, H.B., C. Blakemore, and J. Pettigrew, 1967. The neural mechanism of binocular depth discrimination. *J. Physiol.* '93: 327-42.
2. Bhatia, B. 1957. Eye movement patterns in response to moving objects. *J. Aviat. Med.* 28, 309-316.
3. Bhatia, B. 1960. Some factors determining the maximum angular velocity of pursuit ocular movements. *J. Opt. Soc. Am.* 50: 149-150.
4. Bhatia, B. 1975. Minimum separable as a function of speed of a moving object. *Vision Res.* 15, 23-33.
5. Bhatia, B. 1981. Some aspects of visual perception—the scaling mechanism. *J. Med. Life Sci. Eng.* (in print).
6. Bhatia, B. and M.L.N. Shastri. 1955. Some observations on the visibility of moving objects. *Bull. Nat. Inst. Sci.* (India) 10, 78-82.
7. Bhatia, B. and C.A. Verghese, 1963. Constancy of the visibility of a moving object viewed from different distances with the eyes fixed, *J. Opt. Soc. Am.* 54, 383-386.
8. Bhatia, B. and C.A. Verghese, 1964. Threshold size of a moving object as a function of its speed. *J. Opt. Soc. Am.* 54, 948-50.
9. Hubel, D.H. 1971. Specificity of responses of cells in the visual cortex. *J. Psychiatric. Res.* 8, 301-307.
10. Hubel, D.H. and T.N. Wiesel, 1962. Receptive fields, binocular interaction and functional architecture in the cat's visual cortex. *J. Physiol. London.* 160, 106-154.

11. Hubel, D.H and T.N. Wiesel 1965. Receptive fields and functiona architecture in two non-striate visual area (18 and 19) of the cat. *J. Neurophysiol.* 28, 229-289.
12. Hubel, D.H. and T.N. Wiesel. 1970. Stereoscopic vision in macaquel monkey. *Nature.* 225, 41-42.
13. Julesz, B. 1960. Binocular depth perception of computer-generated patterns. *Bell system Tech. J.* 39. 1125-1162.
14. Julesz, B. 1961. Binocular depth perception and pattern recognition. in E.C. Cherry, ed., *Information Theory* (French London Symposium, 1960) London: Butterworth.
15. Julesz, B. 1964. Binocular depth perception without familiarity cues. *Science.* 145, 356-362.
16. Julesz, B. and S.C. Johnson. 1968a. Stereograms portraying ambiguously perceived surfaces. *Prec. Natl. Acad. Sci.* 61, 437-441.
17. Julesz, B. and S.C. Johnson. 1968b. Mental holography: stereograms portraying perceivable surfaces. *Bell System Tech. J.* 49. 2075-2083.
18. Pettigrew, J.D., T. Nikara, and P.O. Bishop. 1968. Binocular interaction in single units in cat striate cortex; Simultaneous stimulation by single moving slit with receptive fields in correspondence. *Exp. Brain Res.* 6, 391-410.
19. Scholsberg, H. 1950. A note on depth perception, size constancy and related topics. (*Psychol. Rev.* 57, 314-317.
20. Sutherland, N.S. 1968. Outlines of theory of visual pattern recognition in animals and man. *Proc. Roy. Soc.* B. 171, 297-317.

Quantum-Mechanical Exploration of the Polymorphism and Microheterogeneity of DNA

Bernard Pullman, Alberte Pullman and Richard Lavery

Institut de Biologie Physico-Chimique
Laboratoire de Biochimie Théorique, associé au C.N.R.S.,
13, rue Pierre et Marie Curie, 75005 Paris, France

Introduction

Nucleic acids are the vehicles of life. In them is inscribed the genetic code, whose transmission maintains life and whose variations are responsible for the evolution of life. The structure of these substances, their biochemistry and biophysics, are therefore of fundamental importance for our understanding of the living state.

Primordial among the nucleic acids is DNA, which makes us what we are. Following the common belief, illustrated by numerous more or less artistic drawings, it is a beautiful *regular* structure whose main backbone is made of repeating identical units. In fact, it becomes more and more evident today that the situation is more complex and, because of this perhaps, more fascinating. Thus, it becomes well recognized that DNA may exist in a *number of allomorphic forms*, which differ among themselves in geometrical parameters and, what is particularly important, in conformation. This polymorphism may be due to external factors (humidity, salt concentration etc.) or, as discovered recently, induced by variations in the sequence of the base pairs. This last aspect of the problem is actively investigated presently, both experimentally and theoretically, in a number of laboratories. Its importance stems from the fact that it leads necessarily to the consideration of a possible *microheterogeneity* of native DNA as a function of the order of the base pairs along its helical axis. Such microheterogeneity could be of particular significance for the specificity of interaction or association of the nucleic acids with external reactants. Thus e.g., it is well known that when the potent carcinogen 3, 4-benzpyrene reacts with DNA, in vitro or in vivo, only one purine base in hundreds undergoes the reaction. Is this a purely statistical result or the outcome of locally increased reactivity, due to a particularly favourable specific microheterogeneity?

In view to assessing the possible role of the different known forms of DNA in such specificity we have undertaken recently in our laboratory a detailed examination of their structure with the double objective of evaluating quantitatively: (1) one of the main variable *geometrical* features associated with the conformational changes of DNA, namely the *accessibility* to reactive sites and (2) one of the main *electronic* features associated with these changes namely their *molecular electrostatic potential*, these two factors being expected to be among the most decisive ones in influencing the reactional possibilities of the nucleic acids. We have explored from this point of view A-DNA (1), B-DNA (2) and references given therein), "alternating" B-DNA (3), C-DNA (4), D-DNA (3) and Z-DNA (5, 6) and present here a rapid, synthetic view of some of the results obtained (see also 7).

Method

The techniques employed for calculating the accessibility (8) and the molecular electrostatic potential (2, 9) of nucleic acids have been described in detail in our previous publications and we shall therefore not restate them here.

We would like, however, to:

(1) indicate that the accessibilities (the accessible areas on the van der Waals surfaces of target atoms in the nucleic acids) have been evaluated with respect to a test sphere of radius 1.2 Å. It was demonstrated that these accessibilities may be considered as corresponding to an attack by a water molecule through one of its hydrogen atoms and as representing the upper limit of the atomic accessibilities, within the nucleic acids, towards molecular species (8).

(2) recall that five representations are available for describing the molecular potentials (2, 7). These are:

1. Point potentials.
2. Plane potentials (isopotential maps, whose minima represent the main site potentials at reactive centers, in particular at the purine and pyrimidine bases)
3. Radial potentials (computed generally in planes perpendicular to the helical axis)
4. Line potentials (an extension of the radial potentials across the whole width of DNA)
5. Surface envelope potentials (potentials on envelopes formed by the intersection of spheres centered on each atom of DNA with radii equal to the van der Waals radius of the atoms concerned, multiplied,

if desired, by a factor F; for practical reasons these are generally presented in the form of their projection on a two-dimensional "window" (8)).

We shall be concerned here essentially with the site, line and surface envelope representations of the potentials.

The model helices of the different DNAs studied represent each one complete turn of the appropriate conformation. The input data for the computations have been taken from the following original sources: A and B-DNA (10), alternating B-DNA (11), C-DNA (12), D-DNA (13), Z_I and Z_{II}-DNAs (14).

Results and Discussion

A) In the grooves of DNA

One of the essential distinctive features which differentiate the various allomorphs of DNA and which has a decisive influence on the distribution of their molecular electrostatic potential and the accessibility to their reactive sites is the position of the helical axis with respect to the complementary base pairs, adenine-thymine and guanine-cytosine. The situation concerning this position is summarized schematically in figure 1 from which it is evident that the DNAs may be divided into two groups: (1) one involving the B and C-DNAs (and also the alternating B-DNA), in which the helical axis passes through the pairs of hydrogen-bonded complementary bases and (2) the other, involving A, D and Z-DNAs, in which the helical axis is situated outside these pairs. In this last situation we may distinguish two possibilities, one, characteristic of A-DNA, in which the helical axis is located in what in the conventional language is called

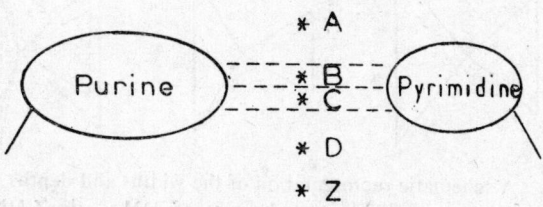

Fig. 1. The position of the helical axis with respect to the base pairs in different conformers of DNA.

the major groove and the other, characteristic of D and Z-DNAs, in which this axis is located in the minor groove.

This displacement of the helical axis with respect to the base pairs has two major consequences for the problem that we are investigating here:

(1) One concerns its influence on the *shape of the grooves* of the nucleic acids. The attention centered until recently essentially on B-DNA, considered and probably justly so as biologically the most significant form of DNA, has accustomed us to believe that the nucleic acid double helix is associated with two grooves which although referred to as "major" and "minor" are nevertheless of rather comparable width and depth. In fact, this particular situation is the consequence of the approximately central positioning of the helical axis with respect to the base-pairs in this form of DNA. The displacement of the axis in the other forms has a drastic influence on the nature of the grooves. Thus, the displacement of the axis towards the major groove (or of the base pairs towards the minor groove), as in A-DNA, renders the minor groove very shallow and very wide, while the major groove becomes very deep but narrow. On the contrary, the displacement of the helical axis into the minor groove (or of the base pairs into the major groove) renders the major groove shallow and wide and the minor groove deep (and narrow in D-DNA, but narrow or wide in Z-DNA following the type of phosphate considered). The situation is summarized schematically in Fig. 2.

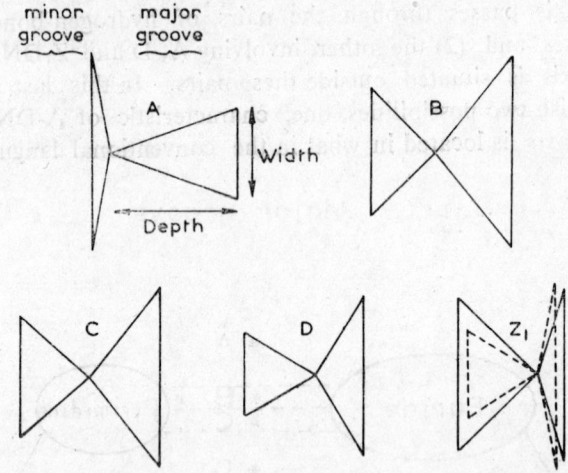

Fig. 2. A schematic representation of the widths and depths of the grooves in different conformers of DNA. In Z-DNA the full lines refer to the GpC phosphates and the dotted lines to the CpG phosphates.

(2) The second consequence, determined in fact by the previous one, concerns the distribution of the electrostatic molecular potential produced

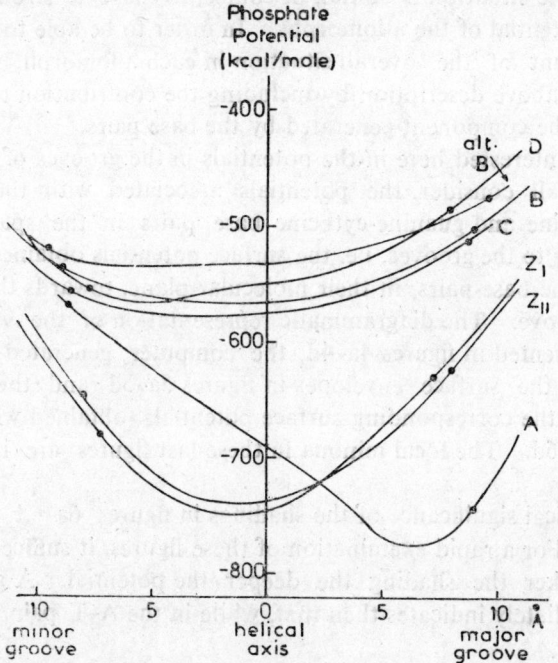

Fig. 3. The potential due to the phosphate backbone along the mM line in different conformers of DNA. The dots represent the radii of the helices.

by the phosphate backbone of the different allomorphs of DNA. This consequence is important because this distribution represents numerically the principal component of the global potential. Fig. 3 represents the potential due to the phosphate backbone alone produced along a line (line potential) perpendicular to the helical axis and passing through the center of it in the models studied (one helical turn of the allomorphs) and bisecting the major (M) and minor (m) groove of each allomorph (7). It is seen that for the related conformations of B-, alternating B- and C-DNA, in which the base pairs are located more or less symmetrically about the helical axis, the mM line potential exhibits a shallow variation with only a slight minimum on the side of the minor groove. A very different situation prevails for A-DNA. In this conformer, unique in having the base pairs largely displaced into the minor groove, the mM line potential shows a strong asymmetry, with the minimum inside the major groove, close to the backbone. (A recent study has demonstrated a similar although slightly less pronounced situation in A-RNA (15)). The Z conformers, exhibit an inverse effect. The displacement of their base pairs into the major groove produces again a noticeable asymmetry of the phosphate potential, but this time to the advantage of the minor groove.

This variable situation is bound, of course, to have a strong effect on the global potential of the allomorphs. In order to be able to give a more precise account of the overall situation in each allomorph, we need to complete the above description by including the contribution to the global potential of the component generated by the base pairs.

As we are interested here in the potentials in the grooves of the nucleic acids, we shall consider the potentials associated with the individual adenine-thymine and guanine-cytocine base pairs in the space elements corresponding to the grooves, i.e. the surface potentials obtained by looking sideways at the base-pairs, in their molecular plane, towards the major or the minor groove. The diagrammatic representation of the views of the pairs is represented in figures 4a–4d, the computer generated perspective drawings of the surface envelopes in figures 5a–5d and the "window" projection of the corresponding surface potentials (obtained with $F = 1.7$) in figures 6a–6d. The local minima in these last figures are indicated by the letters M.

The numerical significance of the shadings in figures 6a–6d is indicated in Table 1. For a rapid examination of these figures, it suffices to remark that the darker the shading the deeper the potential. A glance at the figures immediately indicates then that, while in the A–T pair the deepest

Table 1. Shadings used for surface envelope potentials of Fig. 6.

SHADING	POTENTIAL (kcal/mole)
☐	30 ↓
◩	10 ↓
▨	0 ↓
▩	−10 ↓
⌗	−20 ↓
⌗⌗	−30 ↓
	−41

Quantum-Mechanical Exploration of the Polymorphism   375

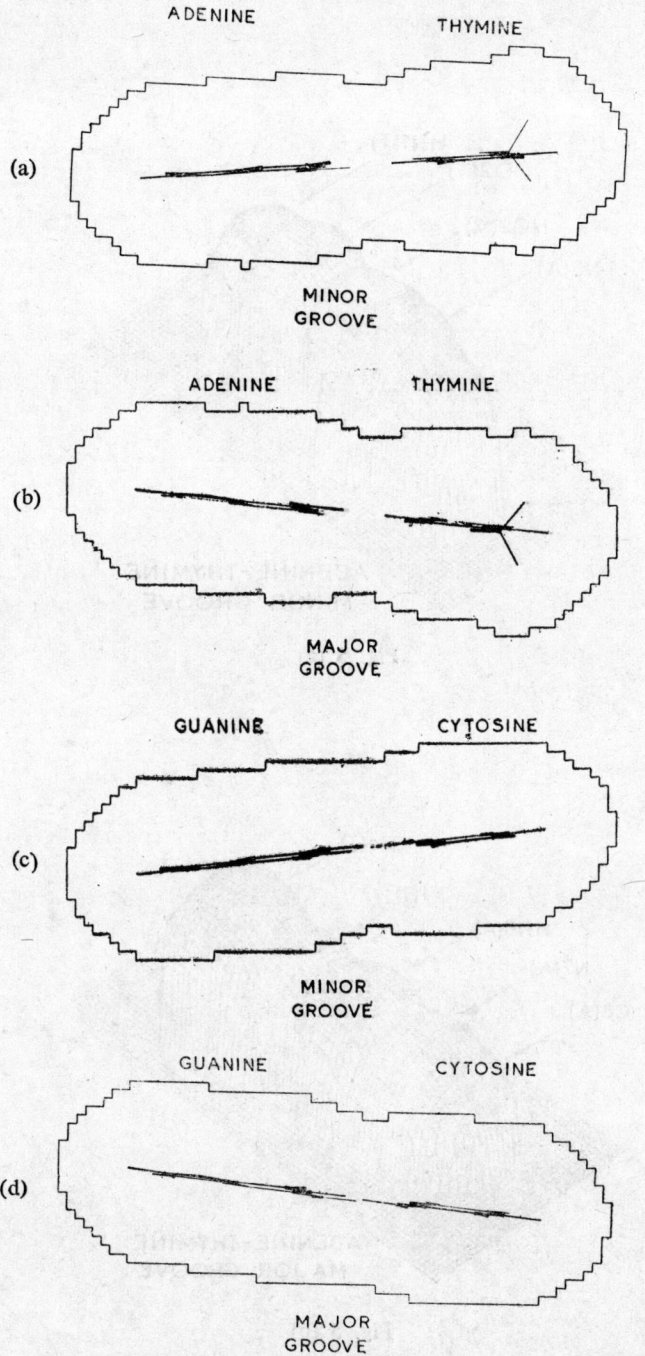

Fig. 4. Diagrammatic representation of the side views on the complementary base pairs.

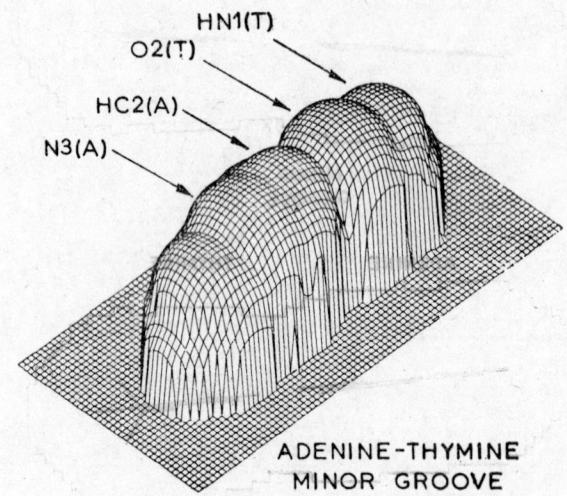

ADENINE-THYMINE
MINOR GROOVE

Fig. 5 (a)

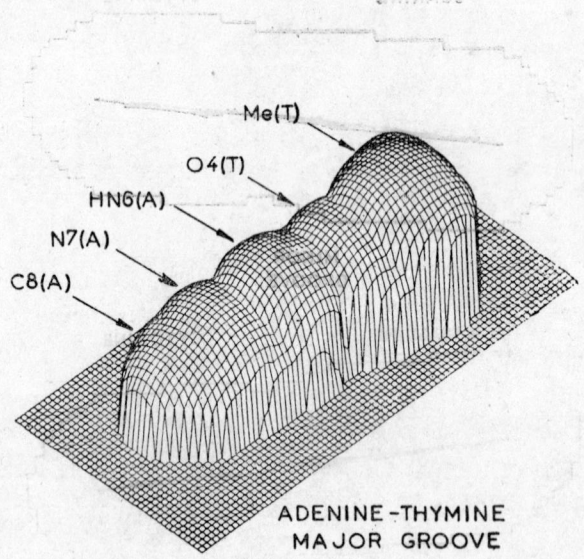

ADENINE-THYMINE
MAJOR GROOVE

Fig. 5 (b)

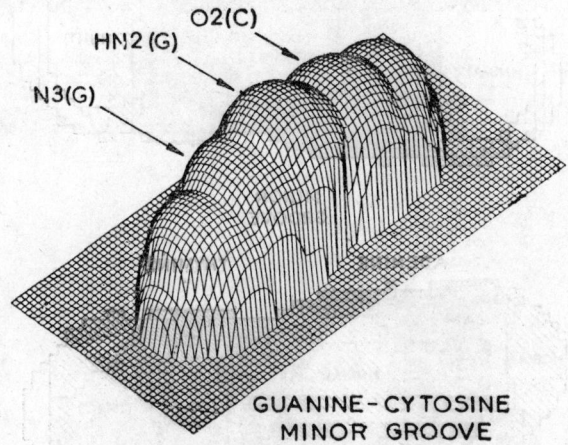

Fig. 5 (c)

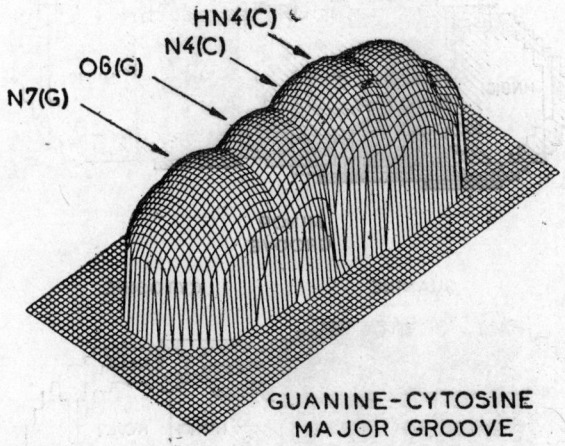

Fig. 5 (d)

Fig. 5. Computer generated perspective drawings of the suface envelopes corresponding to the side views of Fig. 4.

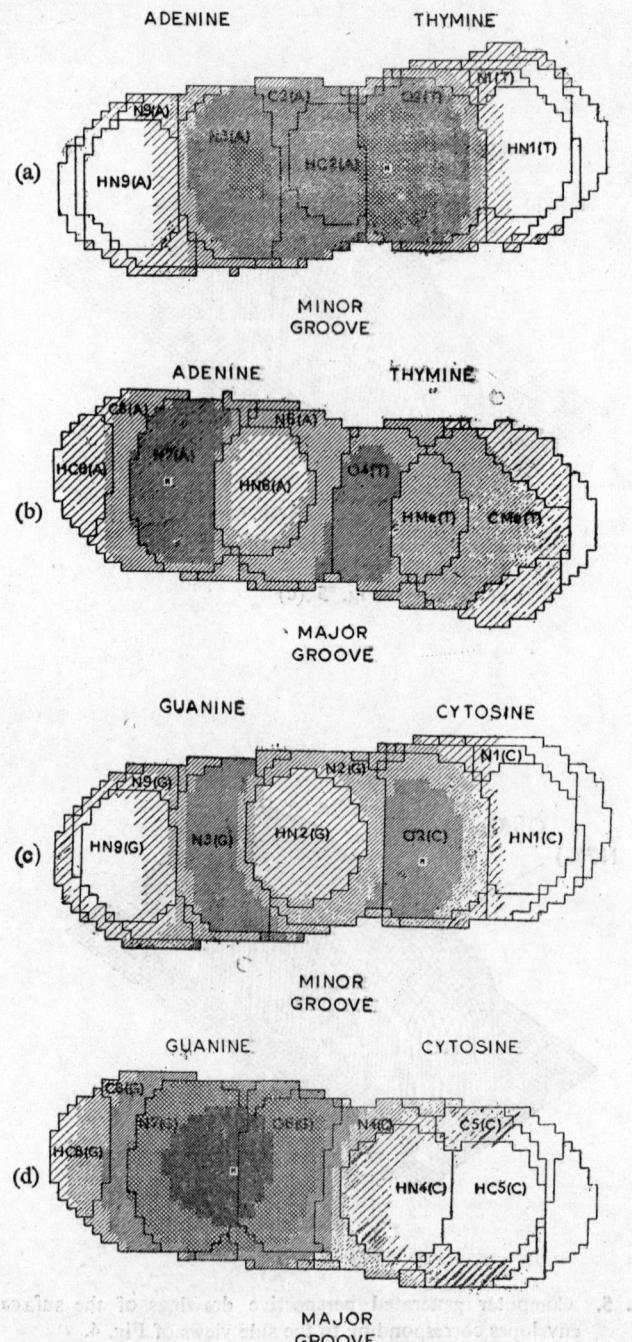

Fig. 6. Surface envelope potential: side view in the plane of the base pairs (For the significance of the shadings see Table 1).

Notations: N7(G)=nitrogen 7 of guanine, etc; HC8(G)=hydrogen atom bound to C8 of guanine, etc.; CMe(T)=carbon atom of the methyl group of thymine; HMe(T)=hydrogen atom of the methyl group of thymine.

minimum is associated with its minor groove (at the surface of 02 of cytosine), it is associated with the major groove (at the surface of N7 of guanine) in the G C pair. The numerical values of these minima (and of the maxima) are summed up in Table 2 from which it becomes also evident that the altogether deepest minimum is localized in the major groove of the G–C pair.

Table 2. Surface potential minima and maxima in isolated complementary base pairs (kcal/mole)

Base pair	Minor groove		Major groove	
	Minimum	Maximum	Minimum	Maximum
AT	−25.0	24.6	−16.9	16.9
GC	−14.7	27.0	−39.9	19.4

The difference (major groove minimum-minor groove minimum) is positive (8 kcal/mole) for the AT base pair and negative (-26 kcal/mole) for the GC base pair. The asymmetry in the distribution of the potential on the two sides of the base pairs is thus small but not negligeable. Whether it will have an effect on the overall distribution of the potential in the grooves will depend on whether it will be able to perturb the asymmetry due to the component of the potential produced by the phosphates. A detailed study (7) shows that in B, alternating B and C-DNAs, in which the contribution of this last component in the region of the base pairs presents only a very small variation, the effect of the base-pair is sufficient to produce the predominance, from the point of view of the potential depth, of the minor groove in A–T sequences and of the major groove in GC sequences. On the contrary, this effect is unsufficient to influence the distribution of the potential in A-D- and Z-DNAs in which the overall groove potential is thus independent of the base sequence and contains the deepest minimum in the major groove in A-DNA and in the minor groove in Z-DNA. Explicit computations on the appropriate models of DNA (summed up in Table 3) confirm this situation. (The case of D-DNA is particular: in principle its deepest minimum should always reside, as for Z-DNA, in its minor groove, whatever the base sequence. Explicit computations for the $F=1.7$ surface envelope locate it, however, in the major groove (Table 3). This exceptional situation is due to the extreme narrowness of the minor groove in D-DNA, which is practically occluded from the surface envelope).

This state of affairs is bound to have significant consequences for the interaction of the different forms of DNA, whether separated or inters-

Table 3. Surface envelope potentials in the various DNA conformations

Form of DNA	Base sequence	Surface potential minima (kcal/mole)		Absolute values M = major m = minor
		Minor groove	Major groove	
A	(dG.dG)	−700	−790	M>m
	(dA.dT)	−696	−784	M>m
Alternating-B	(dA–dT).(dA–dT)	−605	−572	m>M
B	(dG.dC)	−603	−632	M>m
	(dA.dT)	−625	−598	m>M
	(dI.dC)	−623	−622	m≈M
	(dG–dC).(dG–dC)	−594	−613	M>m
	(dA–dT).(dA–dT)	−631	−599	m>M
C	(dG.dC)	−632	−645	M>m
	(dA.dT)	−656	−614	m>M
D	(dG.dC)	−569	−596	M>m
	(dA.dT)	−561	−586	M>m
Z_I	(dG–dC).(dG–dC)	−735	−684	m>M
Z_{II}	(dG–dC).(dG–dC)	−680	−658	m>M

persed within a continuous chain, with external reactants, in particular with non-intercalative, medium or large ligands. In fact, we have demonstrated recently the decisive role of the location of the surface potential minimum for the interaction of different natural and synthetic nuclei acids with the oligopeptide antibiotics netropsin and distamycin A (16).

These antibiotics bind to B-DNA, preferentially in the minor groove of AT rich sequences, and to poly (dA.dT). They do not bind to A-DNA or to poly (dG.dC) (17, 18, 19). Their binding is thought to involve interactions between their charged terminal groups and the phosphates of DNA and also hydrogen bonds between their amide groups and the nucleophilic base atoms of DNA, N3(A) and 02(T). This binding scheme is insufficient, however, to explain by itself the very variable affinities of the antibiotics for nucleic acids with different base-pair sequences or conformations.

We have been able to show that the binding of the antibiotics correlates with the presence of the most negative surface potential minimum in the minor groove of the nucleic acids. Such is the case for poly (dA.dT) with which the antibiotics bind most strongly. It is also the case for poly (dA-dT). poly (dA-dT), which exists most probably in the alternating-B conformation. The binding with this sequence is somewhat weaker than with poly (dA.dT) and it is interesting to note (Table 3) that the minimum of potential in the minor groove is also somewhat weaker in this alternating copolymer than in poly (dA.dT). Inversely, the minor groove potentials are disfavoured in poly (dG.dC) and in A-DNA of any sequence and the antibiotics do not bind to these species.

It is possible that similar considerations may be important quite generally for nucleic acid-protein recognition processes.

(B) At the base active sites

We would like now to consider briefly the results concerning the state of the principal *reactive sites* at the bases, in relation to their potential and accessibility. We shall center our attention in particular on the nucleophilic atoms for which the significance of the potential is the most straightforward. Because of the multitude of such atoms in the nucleic acids we cannot discuss all of them here and shall only present an example. More complete information including the numerical values of the potentials and accessibilities for all the major reactive sites in the different types of DNA may be found in the references (2-7).

The most striking representation of the results is probably the one given in Figure 7 for the illustrative case of the C8 atoms of the nucleic acid purines. It has the advantage of containing simultaneously the information relevant to both the potential and the accessibility of these sites in the different types of DNA and it offers thus ther instantaneous classification with respect to the two investigated properties. The predictive information contained in Figure 7 may be used for the comparison of the different types of DNA whether in separate forms or interspersed within, say, a predominantly B-DNA conformation.

As an example of the utility of such a quantitative, comparative and simultaneous study of the potentials and accessibilities of the nucleophilic sites of the bases we may quote a problem raised recently in conjunction with the discovery of Z-DNA. The analysis of the crystallographic data and the construction of the corresponding models have suggested to their authors (5, 6) that a number of reactive sites on the bases should be more accessible in Z DNA than in its B-DNA counterpart. The observation concerns in particular atoms such as N7(G), 06(G) and C8(G) known to be the receptor sites for covalent bond formation with a series of carcinogenic compounds (see e.g. 20, 21). This situation led these authors to hypothesize that Z fragments interspersed within B-DNA could form particularly sensitive targets for the action of such carcinogens. They paid particular attention to C8(G) known to be the preferred site of action of the carcinogenic N-2-acetylaminofluorene.

Our calculations confirm and quantify the increased accessibility of a

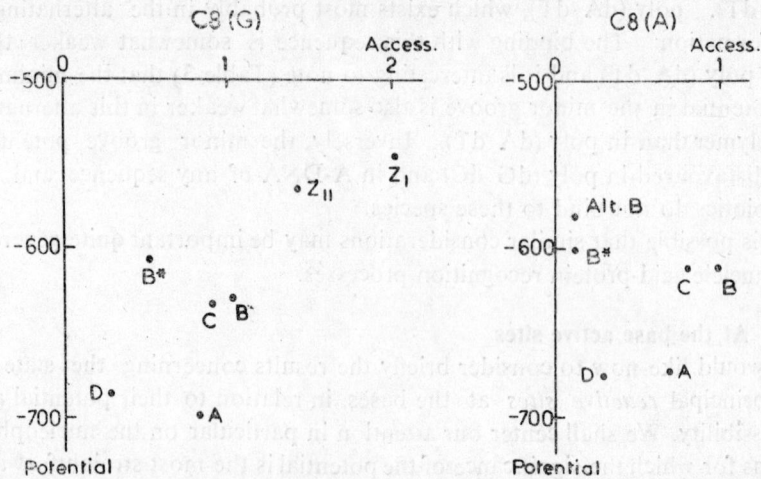

Fig. 7. Potentials (kcal/mole) and accessibilities ($Å^2$) of C8 of the purines in the different conformers of DNA. In this figure B refers to the complementary homopolymers poly (dC.dC) and poly (dA.dT) and B* to the alternating copolymers poly (dG–dC). poly (dG–dC) and poly (dA–dT). poly (dA–dT), all in the B-DNA conformation.

number of important acceptor sites of Z-DNA, in particular of C8, with respect to the same site in the related B-DNA (see Fig. 7). The comparison of the molecular electrostatic potentials of the same sites in the same species leads, however, to the conclusion that this potential decreases for the C8(G) receptor site in Z-DNA's with respect to B-DNA. We are therefore faced, as concerns this position, with the dilemma of which of the two factors, accessibility or molecular electrostatic potential, will dominate for its behaviour toward an attacking electrophilic species.

A very recent publication (22) demonstrates that the reactivity of poly (dG–DC). poly (dG–dC) for binding the carcinogenic N-2-acetylaminofluorene at the C8 position of guanine residues is substantially smaller in conditions in which the polymer exists in the Z-form then when it exists in the B-form. This result may be considered as an indication of the preponderance in this case of the effect of the decrease of the electrostatic potential at C8 of guanines in Z-DNA with respect to B-DNA, which thus seems to dominate over the greater accessibility of that position in Z-DNA.

This example, complicated because of the divergence of the indications based on potential and accessibilities, illustrates the way in which figure 7 may be used. Predictions are, of course, straightforward when the indications based on the potential and accessibilities are parallel.

It is thus evident that the present theoretical analysis of two essential characteristics of the polymorphic forms of DNA, its molecular electrostatic potential and the accessibility to its reactive sites, offers a tool for the understanding of the differences in the biochemical reactivity of these forms and of the variations of this reactivity as a result of the microheterogeneity of native DNA.

Acknowledgement

The work was supported by the National Foundation for Cancer Research (USA) to which the authors wish to express their deep thanks.

References

1. Lavery, R. and Pullman, B. (1981) Nucl. Acids Res. **9**, 4677–4688.
2. Pullman, A. and Pullman, B. (1981) Quart. Rev. Biophys. **14**, 289–380.
3. Lavery, R., Pullman, B. and Corbin, S., Nucl. Acids Res., in press.
4. Lavery, R., Corbin, S. and Pullman, B., Theoret. Chim. Acta, in press.
5. Zakrzewska, K., Lavery, R., Pullman, A. and Pullman, B. (1980) Nucl. Acids Res. **8**, 3917–3932.
6. Zakrzewska, K., Lavery, R. and Pullman, B. (1981) in Biomolecular Stereodynamics, ed. R.H. Sarma, Adenine Press, N.Y., 163–184.
7. Pullman, B., Lavery, R. and Pullman A. (1981) Europ. J. Biochem., in press.
8. Lavery, R., Pullman. A. and Pullman B. (1981) Int. J. Quant. Chem. **20**, 49–62.
9. Pullman, A. and Pullman, B. in Chemical Applications of Atomic and Molecular Electrostatic Potentials (1981) P. Politzer and D.G. Truhlar eds., Plenum Press, N.Y., 381–405.
10. Arnott, S. and Hukins, D.W.L. (1972) Biochem. Biophys. Res. Comm. **47**, 1504–1509.
11. Klug, A., Jack. A., Viswamitra, M.A., Kennard, O., Shakked, Z. and Steitz, T.A. (1979) J. Mol. Biol. **131**, 669–680.
12. Arnott, S. and Selsing, E. (1975) J. Mol. Biol. **98**, 265–269.
13. Arnott, S., Chandrasekaran, R., Hukins, D.W.L., Smith, P.J.C. and Watts, L. (1974) J. Mol. Biol. **88**, 523–533.
14. Wang, A. H-J., Quigley, G.J., Kolpak, F.J., van der Marel, G., van Boom, J.H. and Rich, A. (1980) Science **211**, 171–176.
15. Corbin, S., Lavery, R. and Pullman, B., Biochim. Biophys. Acta, in press.
16. Pullman, B. and Pullman, A., Studia Biophys., in press.
17. Zimmer, Ch. (1975) Progr. Nucleic Acid Res. Mol. Biol. **15**, 285–317.
18. Gursky, G.V., Tumanyan, V.G., Zasedatelev, A.S., Shuze, A.L.,

Grohovsky, S.L. and Gottikh, B.P. (1977) in Nucleic Acid-Proteins Interactions (H.J. Vogel Ed.) Academic Press, N.Y., 189–217.
19. Reinert, K.E., Geller, D. and Stutter, E. (1981) Nucl. Acids Res. 9, 2335–2349.
20. Pullman, B. and Pullman, A. (1980) in Carcinogenesis; Fundamental Mechanisms and Environmental Effects, eds. B. Pullman, P.O.P. Ts'o and H. Gelboin, Reidel, Holland, 55–66.
21. Pullman, A. and Pullman, B. (1980) Int. J. Quant. Chem. Quant. Biol. Symp. 7, 245–259.
22. Santella, R.M., Grunberger, D., Weinstein, J.B. and Rich, A. (1981) Proc. Natl. Acad. Sci. USA 78, 1451–1455.

PCILO Investigation on the Structure of Intermediate Species of the Photoreaction Cycle of Bacteriorhodopsin

Anil Saran and M.M. Dhingra

Chemical Physics Groups
Tata Institute of Fundamental Research
Homi Bhabha Road, Bombay 400 005

Introduction

Bacteriorhodopsin (bR) is a chromoprotein having retinal (aldehyde of Vitamin A) as chromophore and it occurs in the cellular purple membrane of halobacteria[1-3]. It is structurally similar to rhodopsin[4,5], but its biological role is quite different. Besides the wellknown photosynthetic system—the chlorophylls, bacteriorhodopsin is the only other such system known to exist in nature. A series of very complex oxidation-reduction processes in chlorophylls convert the absorbed light energy into electrochemical proton gradient across the membrane which contains the chlorophylls. The situation in bacteriorhodopsin is, however, simpler: the absorbed light energy is directly converted into an electrochemical proton gradient across the membrane. The energy, thus, stored is then used by the cells to synthesize ATP and to drive other transport processes. In this way, bacteriorhodopsin functions as a light driven proton pump.

Bacteriorhodopsin is a relatively small protein having 247 amino-acids and a molecular weight of around 27000 daltons. Khorana et al.[6] and Ovchinnikov et al.[7] have independently determined the primary sequence of this protein. Their results indicate that the chromophore retinal is attached to the ε-amino group of lysine which is 41 residue distant from the N-terminal amino-acid of the protein molecule. The hydrophobic aminoacids constitute about 62% of the polypeptide chain which is largely buried in the membrane. The electron micrographic and electron diffraction studies [8-10] have shown that bacteriorhodopsin contains seven helical segments of about 40 Å long and 10 Å apart extending nearly across the membrane which is about 45 Å in width and almost perpendicular to the membrane plane.

The photochemical studies on bacteriorhodopsin indicate that bR has a strong absorbance band around 568 nm which is about 200 nm red shifted from the band (~370 nm) which is normally observed for a retinal Schiff base. This large red shift has been demonstrated to be partially due to the protonation of the Schiff base[11-13] and partially due to environmental factors arising from the non-covalent interactions with the polypeptide chain of the protein molecule. The actual mechanism to explain this significant red shift (~200 nm) however, still remains to be established. When bacteriorhodopsin is kept in the dark for about 2 hours the absorbance maximum shifts from 768 nm to 558 nm and there is a concomitant decrease in the intensity by about 15%. This form of bacteriorhodopsin is known as dark-adapted bacteriorhodopsin (bR_{558}^{DA}). The light-adapted bacteriorhodopsin (bR_{568}^{LA}) is restored within a few seconds when the dark-adapted form is exposed to moderate light intensities. On chemical extraction with organic solvents, bR_{568}^{LA} yields a nearly exclusively all-*trans* retinal while bR_{558}^{DA} yields equal amounts of all-*trans* and 13-*cis* retinals[14].

Bacteriorhodopsin undergoes a cyclic photoreaction and the kinetics of this photoreaction have been studied by low temperature absorption spectroscopic and flash photolytic methods[15-18]. A tentative scheme of the photoreaction cycle of bacteriorhodopsin[14] is shown in Fig. 1. It is believed that a *trans-cis* isomerization is the primary photochemical event followed by the deprotonation of the Schiff base and reprotonation at some later stages of the photoreaction cycle.

The complete photoreaction takes about 10 m sec. Becher and Ebrey[19]

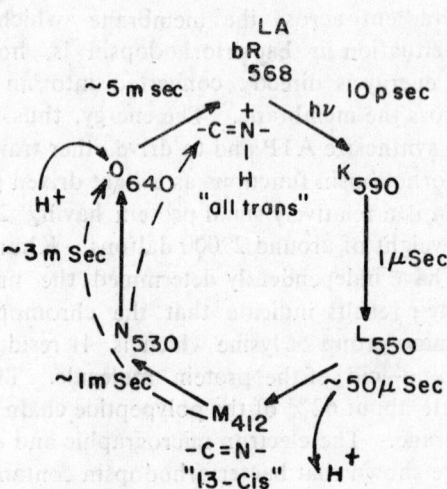

Fig. 1. Scheme of the photoreaction cycle of bacteriorhodopsin involving the different intermediate species[14].

have suggested that two protons are translocated during each cycle but it is not yet established. In addition to flash photolysis and low temperature absorption studies mentioned above, resonance Raman studies[20-22] have also been utilized to work out the details of photoreaction cycle. In spite of these studies the interferences drawn are far from satisfactory. Many questions still remain to be answered:

(a) What is the number of the intermediate species (see Fig. 1) in the photoreaction cycle?
(b) What is the structure of these intermediate species?
(c) Is there a *trans-cis* isomerization accompanying the photochemical event?
(d) If it is so, then what is the source of energy to effect a *cis-trans* isomerization at some other stage of photoreaction cycle to explain the existence of all-*trans* retinal in bR_{568}^{LA}? and
(e) What is the number of protons (one or two) translocated during each photoreaction cycle?

To answer some of these questions we have recently carried out PCILO computations on the conformations of retinal[23] as well as protonated and deprotonated Schiff bases of all-*trans* and 13-*cis* retinals with *n*-pentylamine[24] as model Schiff base of chromophore in bacteriorhodopsin.

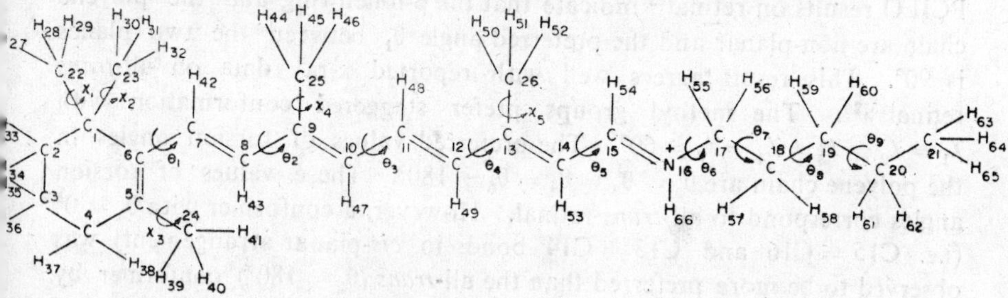

PROTONATED SCHIFF BASE

Fig. 2. Schematic diagram of the protonated Schiff base of all-*trans* retinal with *n*-pentylamine[24]. The various torsion angles which determine the conformation of the molecule are indicated with arrows.

Fig. 2 shows the schematic diagram of the protonated Schiff base of all-*trans* retinal with *n*-pentylamine and the various torsion angles have been defined as[23,24]:

$$X_1 = C2 - C1 - C22 - H27$$
$$X_2 = C2 - C1 - C23 - H30$$
$$X_3 = C4 - C5 - C24 - H39$$
$$X_4 = C8 - C9 - C25 - H44$$

$$\chi_5 = C12 - C13 - C26 - H50$$
$$\theta_1 = C5 - C6 - C7 - C8$$
$$\theta_2 = C7 - C8 - C9 - C10$$
$$\theta_3 = C9 - C10 - C11 - C12$$
$$\theta_4 = C11 - C12 - C13 - C14$$
$$\theta_5 = C13 - C14 - C15 - N16$$
$$\theta_6 = C15 - N16 - C17 - C18$$
$$\theta_7 = N16 - C17 - C18 - C19$$
$$\theta_8 = C17 - C18 - C19 - C20$$

and

$$\theta_9 = C18 - C19 - C20 - C21$$

with the *cis*-planar arrangement of the terminal bonds being taken as torsion angle equal to 0°. The clockwise rotation of the distant bond relative to the near bond is taken as the positive value of the torsion angle.

Results

Conformation of retinal

PCILO results on retinal[23] indicate that the β-ionen ring and the polyene chain are non-planar and the preferred angle θ_1 between the two planes is 90°. This result agrees well with reported x-ray data on all-*trans* retinal[25,26]. The methyl groups prefer staggered conformation with $\chi_1 = \chi_2 = \chi_3 = \chi_4 = \chi_5 = 60°$. The preferred values of torsion angles in the polyene chain are $\theta_2 = \theta_3 = \theta_4 = \theta_5 = 180°$. These values of torsion angles correspond to all-*trans* retinal. However, a conformer with $\theta_5 = 0°$ (i.e. C15 = C16 and C13 = C14 bonds in *cis*-planar arrangement) was observed to be more preferred than the all-*trans* ($\theta_5 = 180°$) conformer by about 0.8 kcal/mole and the reason for this is the close proximity of carbonyl oxygen atom O16 to the methyl group attached C13[23]. The above mentioned results on retinal have been then utilized for the studies on protonated and deprotonated Schiff bases of all-*trans* and 13-*cis* retinal.

Conformation of Schiff base of all-*trans* retinal

Deprotonated Schiff base

First a two-dimensional ($\theta_7 - \theta_6$) conformational energy was constructed with $\chi_1 = \chi_2 = \chi_3 = \chi_4 = \chi_5 = 60°$, $\theta_1 = 90°$, $\theta_2 = \theta_3 = \theta_4 = \theta_5 = \theta_8 = \theta_9 = 180°$ and the preferred values of θ_6 and θ_7 were obtained[24]. Next the θ_6 value obtained from the above step was fixed and ($\theta_8 - \theta_7$) conformational energy map was constructed and the preferred values of θ_7 and θ_8 were obtained. By using the θ_6 and θ_7 values, thus, obtained ($\theta_9 - \theta_8$) conformational energy map was constructed and the preferred values of θ_8

and θ_9 were obtained. At the end of these steps the preferred values of $\theta_7 = \theta_8 = \theta_9 = 300°$ were obtained and these values were considered while constructing each $(\theta_6 - \theta_5)$ conformational energy map for deprotonated and protonated Schiff bases of all-*trans* retinal[24].

Fig. 3 shows the $(\theta_6 - \theta_5)$ conformational energy map constructed for deprotonated Schiff base and one can see from this map that there are six global minima having exactly the same energy at $(\theta_6, \theta_5) = (120°, 30°)$, $(120°, 180°)$, $(120°, 330°)$, $(240°, 30°)$, $(240°, 180°)$ and $(240°, 330°)$. This map is very flexible and all the six minima are enclosed within 2 kcal/mole isoenergy curve. The barrier height between one to another minima is of the order of only 1 kcal/mole.

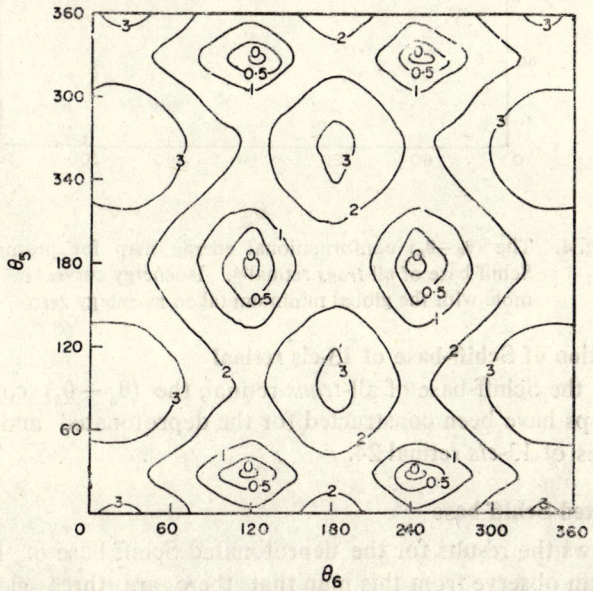

Fig. 3. The $(\theta_6-\theta_5)$ conformational energy map for deprotonated Schiff base of all-*trans* retinal[24]. Isoenergy curves in kcal/mole with the global minimum taken as energy zero.

Protonated Schiff base

The $(\theta_6-\theta_5)$ conformational energy map for protonated Schiff base of all-*trans* retinal is shown in Fig. 4. This map reveals that the conformational flexibility in this case is very restricted as that compared to the deprotonated Schiff base shown in Fig. 3. There is only one global minimum in contrast to six minima for the deprotonated Schiff base at $\theta_6=120°$ and $\theta_5=180°$. There are two energy regions about 2 kcal/mole higher in energy at $\theta_5=180°$ and $\theta_6=240°$ and $300°$. Thus, a comparison of the results presented in Figs. 3 and 4 clearly indicates that deprotonation greatly enhances the conformational flexibility in the chromophore of bacteriorhodopsin.

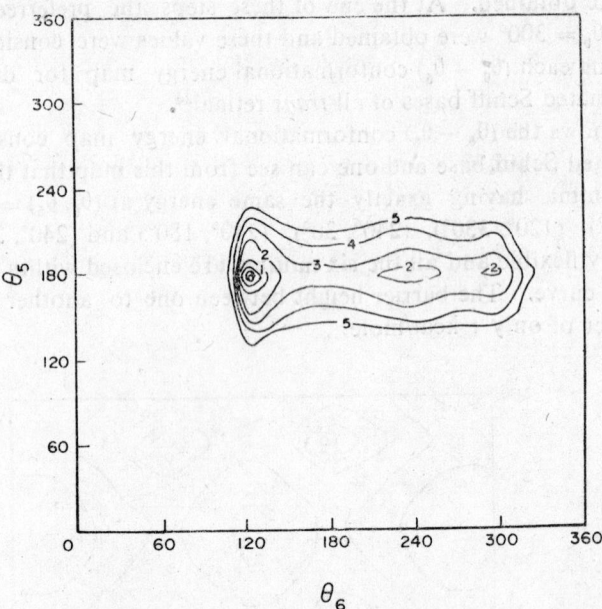

Fig. 4. The ($\theta_6-\theta_5$) conformational energy map for protonated Schiff base of all-*trans* retinal[24]. Isoenergy curves in kcal/mole with the global minimum taken as energy zero.

Conformation of Schiff-base of 13-cis retinal

Similar to the Schiff-base of all-*trans* retinal, the ($\theta_6 - \theta_5$) conformationall energy maps have been constructed for the deprotonated and protonated Schiff bases of 13-*cis* retinal 24.

Deprotonated Schiff base

Fig. 5 shows the results for the deprotonated Schiff base of 13-*cis* retina and one can observe from this map that there are three global minima in this case as compared to six in all-*trans* retinal case (Fig. 3). These minima occur at $\theta_6 = 120°$ and $\theta_5 = 150°$, 210° and 330° and they are encircled by a single 0.5 kcal/mole isoenergy curve. In addition, there are three local minima within 0.5 kcal/mole isoenergy curves at $\theta_6 = 240°$ and $\theta_5 = 30°$, 150° and 210°. The energy of these local minima is exactly 0.4 kcal/mole above the global minimum. Thus, in this case also there are six regions of low energy regions (within 0.4 kcal/mole).

Protonated Schiff base

Fig. 6 shows the map constructed for protonated Schiff base of 13-*cis* retinal and it can be seen that this map is very similar to the map for all-*trans* retinal shown in Fig. 4. There is only one global minimum at $\theta_6 = 120°$ and $\theta_5 = 180°$ with two low energy regions about 2 kcal/mole higher in energy at $\theta_5 = 180°$ and $\theta_6 = 240°$ and 300°. In this case again,

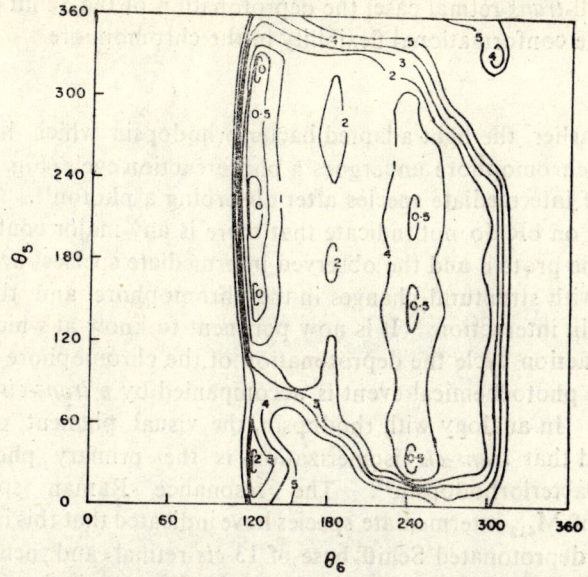

Fig. 5. The $(\theta_6-\theta_5)$ conformational energy map for deprotonated Schiff base of 13-*cis* retinal[24]. Isoenergy curves in kcal/mole with the global minimum taken as energy zero.

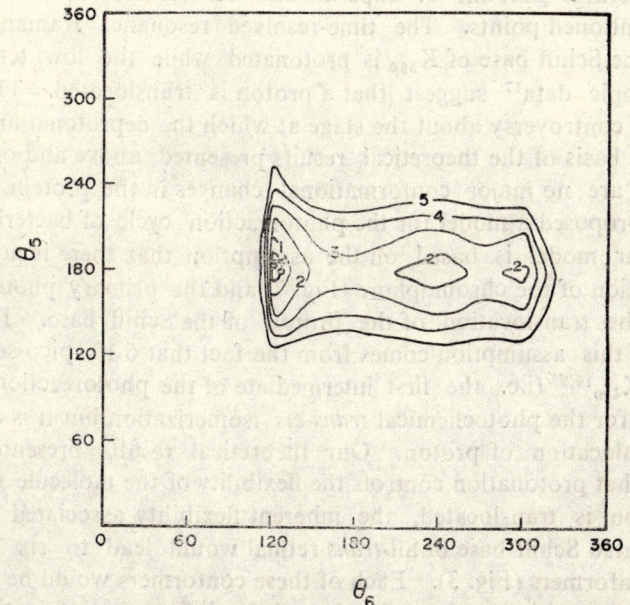

Fig. 6. The $(\theta_6-\theta_5)$ conformational energy map for protonated Schiff base of 13-*cis* retinal[24]. Isoenergy curves in kcal/mole with the global minimum taken as energy zero.

similar to all-*trans* retinal case, the deprotonation of the Schiff base greatly enhances the conformational flexibility in the chromophore.

Discussion

As stated earlier, the light adapted bacteriorhodopsin which has all-*trans* retinal as a chromophore undergoes a photoreaction cycle (Fig. 1) involving a number of intermediate species after absorbing a photon[14]. The experimental data on bR do not indicate that there is any major conformational change in the protein and the observed intermediate species are probably associated with structural changes in the chromophore and the chromophore-protein interaction. It is now pertinent to know at which stage of the photoreaction cycle the deprotonation of the chromophore occurs and whether the photochemical event is accompanied by a *trans-cis* isomerization or not. In analogy with rhodopsin, the visual pigment of the eye, it is believed that *trans-cis* isomerization is the primary photochemical event in bacteriorhodopsin[27]. The resonance Raman spectroscopic studies[22,28] of M_{412} intermediate species have indicated that this intermediate is probably deprotonated Schiff base of 13-*cis* retinal and hence support the *trans-cis* isomerization hypothesis. If this hypothesis is correct then the deprotonation must occur before conversion of M_{412} intermediate, i.e. $L_{550} \rightarrow M_{412}$ or $K_{590} \rightarrow L_{550}$ or the photochemical isomerization is followed by deprotonation. Hence, the structure of K_{590} species has been the target of considerable amount of experimental studies in order to resolve the above mentioned points. The time-resolved resonance Raman data[20,21] indicate the Schiff base of K_{590} is protonated while the low temperature spectroscopic data[17] suggest that a proton is translocated. Thus, there is a lot of controversy about the stage at which the deprotonation occurs.

On the basis of the theoretical results presented above and on the fact that there are no major conformational changes in the protein molecule, we have proposed a model for the photoreaction cycle of bacteriorhodopsin[24]. Our model is based on the assumption that there is no *trans-cis* isomerization of the chromophore (Fig. 7) and the primary photochemical event is the translocation of the proton of the Schiff base. The justification of this assumption comes from the fact that 6-10 pico-second rise time of K_{590}[16,29] (i.e. the first intermediate of the photoreaction cycle) is too short for the photochemical *trans-cis* isomerization but it is consistent with translocation of proton. Our theoretical results presented above indicate that protonation controls the flexibility of the molecule and once the proton is translocated, the inherent flexibility associated with the deprotonated Schiff base of all-*trans* retinal would lead to six minimum energy conformers (Fig. 3). Each of these conformers would be stabilized by interaction with protein molecule and the differential interaction would regulate the λ_{max} of the chromophore.

The most attractive feature of our model (Fig. 7) is that one does not

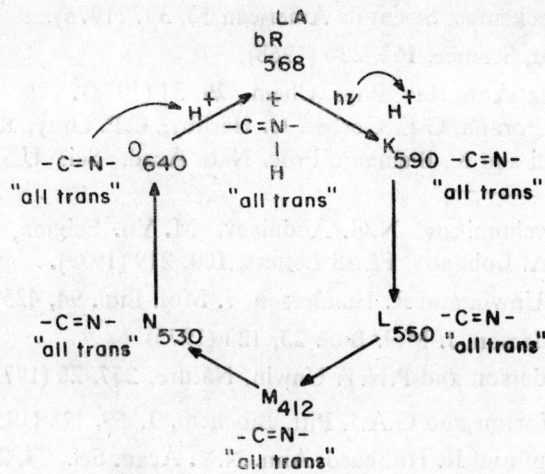

Fig. 7. Proposed model for the photoreaction cycle of light-adapted bacteriorhodopsin (bR_{568}^{LA}).

have to invoke the *cis-trans* isomerization at some stage of the photoreaction cycle. This is one of the unanswered questions in the existing model (Fig. 1) in which *trans-cis* isomerization is believed to be the primary photochemical event. Even if the *cis-trans* isomerization takes place, the question remains to be answered: what is the source of energy which is of the order of about 25 kcal/mole. As stated earlier the short rise time (6—10 picosecond) of K_{590} intermediate is consistent with translocation of proton rather than to the *trans-cis* isomerization. These two arguments augment the validity of our proposed model. An additional support for our model comes from the evidence of chromophore mobility in bacteriorhodopsin. Sherman and Caplan[30] have observed significant rotational rate constant which was found to be of the order of 20 sec^{-1} at room temperature. This is too high to be attributed to the rotation of the protein molecules and thus, suggest an internal conformational change of the chromophore.

Acknowledgement

The authors wish to thank Prof. R. K. Mishra for inviting them to present this paper at this interesting seminar on "Living State".

References

1. W. Stoeckenius and R.J. Rowen, J. Cell. Biol. **34**, 365 (1967).
2. D. Oesterhelt and W. Stoeckenius, Nature N.B. **233**, 149 (1971).

3. W. Stoeckenius, Scientific American **13**, 337 (1976).
4. G. Wald, Science, **162**, 230 (1968).
5. B. Honig, Ann. Rev. Phys. Chem., **29**, 31 (1978).
6. H.G. Khorana, G.E. Gerber, C. Herliby, C.P. Gray, R.J. Anderegg, K. Nihei and K. Biemann, Proc. Natl. Acad. Sci. U.S.A. **76**, 5046 (1979).
7. Y.A. Ovchinnikov, N.G. Abdulaev, M. Yu. Feigina, A.V. Kiselev and N.A. Lobanov, FEBS Letters, **100**, 219 (1979).
8. P.N.T. Unwin and R. Henderson, 7. Mol. Biol. **94**, 425 (1975).
9. R. Henderson, J. Mol. Biol. **23**, 123 (1975).
10. R. Henderson and P.N.T. Unwin, Nature, **257**, 28 (1975).
11. R.A. Morton and G.A.J. Pitt, Biochem. J. **59**, 128 (1958).
12. A. Kropf and R. Hubbard, Ann. N.Y. Acad. Sci., **74**, 266 (1958).
13. H. Suzuki and Y. Kito, Photochem. & Photobiol. **15**, 275 (1972).
14. W. Stoeckenius, Acct. Chem. Res. **13**, 337 (1980).
15. M.C. Kung, D. Devault, B. Hess and D. Oesterhelt, Biophys. J. **15**, 907 (1975).
16. S.L. Shapiro, A.J. Campilo, A. Lewis, G.J. Perreault, J.P. Spoonhower, R.K. Clayton and W. Stoeckenius, Biophys. J. **23**, 383 (1978).
17. M.L. Applebury, K.S. Peters and P.M. Rentzepis, Biophys. J. **23**, 215 (1978).
18. J.B. Hurley, T.G. Ebrey, B. Honig and M. Ottolenghi, Nature, **270**, 540 (1978).
19. B. Becker and T.G. Ebrey, Biophys. J. **17**, 185 (1977).
20. J. Terner, A. Campion and M.A. El-Sayed, Proc. Natl. Acad. Sci. U.S.A. **74**, 5212 (1977).
21. J. Terner, C.L. Hseih, A.R. Burns and M.A. El-Sayed, Proc. Natl. Acad. Sci. USA, **76**, 3046 (1979).
22. B. Aton, A.G. Doukas, R.H. Callender, B. Becher and T.G. Ebrey, Biochemistry, **16**, 2495 (1977).
23. M.M. Dhingra and A. Saran, Proc. Indian Acad. Sci. (Chem. Sci.) in press.
24. M.M. Dhingra and A. Saran, Proc. Indian Acad. Sci. (Chem. Sci.) in press.
25. R. Gilardi, I.L. Karle and J. Karle, Nature, **232**, 187 (1971).
26. T. Hamanaka, T. Mitsui, T. Ashida and M. Kakudo, Acta Crystallogr. **B28**, 214 (1972).
27. R. Hubard, D. Bownds, and T. Yoshizawa, Cold Spring Harbor Symp. Quant. Biol. **30**, 301 (1965).

28. M.A. Marcus and A. Lewis, Biochemistry, **17**, 4722 (1978).
29. K. Peters, M.L. Applebury and P.M. Rentzepis, Proc. Natl. Acad. Sci. USA **74**, 3119 (1977).
30. W.V. Sherman and S.R. Caplan, Nature, **265**, 273 (1977).

Role of Secondary Interactions in the Anomalous pK Values of Histidine in Proteins

Ratna S. Phadke, R.V. Hosur and Girjesh Govil

Tata Institute of Fundamental Research
Bombay 400 005, India

Introduction

Most enzyme functions involve extrarction of an electron, proton or water from the substrate molecules. These reactions are catalysed by the acid/base functional groups present in the amino acid side chains of polypeptides. The pK values of these side chains are considerably different, some times of the order of 2–3 units from those in the amino acid itself. Therefore, the nature of acid base catalysis of enzymes at a particular pH can be quite different from that in the original amino acid residue. For instance the titration curves of His(146) in oxyhemoglobin has pK at 6.8 which is not very different from that in the amino acid itself ($pK_{int} = 6.38$). However, in de-oxyhemoglobin which has a relative shift of iron atom by about 0.75 Å with respect to oxyhemoglobin, the same residue titrates at 7.9 [1]. This occurs because of the fact that the relative movement of the polypeptide chains brings a charged residue ASP(94) in the immediate neighbourhood of the titrable group. In short: (1) The dissociation constants of acid-base amino acid side chains in protein differ from their intrinsic values. (2) pK of a acid/base group is influenced by the specific environment due to intramolecular peptide-peptide interaction. (3) The spatial folding of the protein gives rise to changes in the environment and hence in the pk values. (4) Intermolecular electrostatic interactions are likely to be dominant factor in variations in the pK values and can be calculated using semiempirical theories [2].

Methodology

We have undertaken a study involving semiempirical approach to calculate the influence of electrostatic interactions on pK values of acid-base functional groups in proteins and polypeptides whose crystal structures are known. The methodology is described below.

The pK value of an acid-base group in a protein environment can be written as

$$pK = pK_{int} - \Delta E/2.303\, RT \tag{1}$$

pK_{int} = the pK value of a acid/base functional group, let us say in the amino acid itself. Thus, for His, $pK_{int} = 6.38$

$$\Delta E = E_b - E_a \tag{2}$$

is the difference between the interaction energies of the acid form E_a and the base form E_b of the amino acid with its environment in the polypeptide or protein. It may be pointed out that the acid and the basic forms of the amino acid under question differ in the presence or absence of a proton. Thus, there is a change of one atomic unit of charge as the proton is removed from the acid. Consequently, there will be a large change in the electrostatic interactions with the environment, while interactions such as van der Waals, stacking, etc. will be quite similar in the two cases. Hydrogen bonding interaction can contribute in some cases since on protonation of the amino acid residue, an additional hydrogen becomes available for forming a hydrogen bond in cases where a proton acceptor is already present in the neighbourhood (within $2.5 - 3$ Å distance). The electrostatic interaction, which is a dominant factor can be subdivided into monopole-monopole, dipole-monopole and dipole-dipole [3]. Thus,

$$\left.\begin{array}{l} \text{Monopole} - \text{Monopole} \\[4pt] W_{00} = \dfrac{q_i \cdot q_j}{|R_{ij}|} \\[6pt] \text{Monopole} - \text{Dipole} \\[4pt] W_{01} = \dfrac{q_j\,(\mu_i \cdot R_{ij})}{|R_{ij}|^3} \\[6pt] \text{Dipole} - \text{Monopole} \\[4pt] W_{10} = - \dfrac{q_i\,(\mu_j \cdot R_{ij})}{|R_{ij}|^3} \\[6pt] \text{Dipole} - \text{Dipole} \\[4pt] W_{11} = \dfrac{\mu_i \cdot \mu_j}{|R_{ij}|^3} - \dfrac{3(\mu_i \cdot R_{ij})(\mu_j \cdot R_{ij})}{|R_{ij}|^5} \end{array}\right\} \tag{3}$$

where q_i, q_j are monopoles and μ_i, μ_j are dipoles which can be obtained from molecular orbital calculations (4, 5). R_{ij} are distances between two charges.

$$\begin{aligned} E &= E_{es} + E_{hb} \\ &= W_{00} + W_{10} + W_{10} + W_{11} + E_{hb} \end{aligned}$$

where E_{hb} is contribution due to hydrogen bonding. The procedure is summarised in Fig. 2. A peptide fragment such as shown in Fig. 1 has

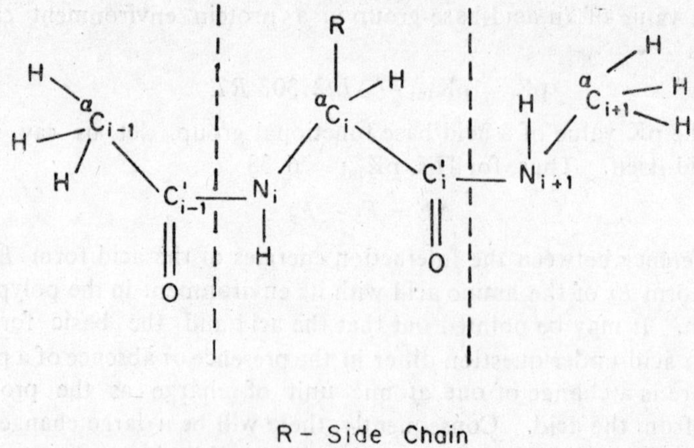

Fig. 1. The polypeptide fragment used for CNDO/2 calculations.

been generated using standard bond lengths, bond angles and dihydral angles. Coordinates of all the atoms in this fragment has been calculated using standard coordinates calculation programme. Coordinates so generated have been used as an input to CNDO/2 calculations which yields atomic monopoles (q_i) and dipoles (μ_i). We have carried out calculations on fragments for all the common amino acid residues which occur in nature. In an actual protein, some of these residues would be present around the particular titrable group in question. The environment of the titrable group is generated using the x-ray programme developed by R. Feldmann. The monopoles and dipoles of the titrable group and its environment are oriented by suitable transformations using the crystal structure data on the protein under study. Amino acids within 10° A radius of the particular titrable group are taken into account. The method is quite general in nature in the sense that it can be applied to any protein of interest whose three dimensional structure is known to estimate pK values of any titrable group.

As a particular case we have carried out calculations in the case of Ribonuclease S whose x-ray crystal structure data is available [6]. Let us recall at this juncture, the facts that (1) the imidazole ring pK value falls into the physiological pH range and that (2) protonation-deprotonation reaction of histidine is considered to be essential for numerous enzyme reactions, (3) histidine ring hydrogens absorb at very low field in NMR because of the deshielding effect of the nitrogen atoms in the imidazole ring and thus NMR data on the pK values of this ring in several proteins are available [2, 7—8].

In Ribonuclease S, there are four histidines at residue numbers 12, 48, 105 and 119. Out of these residues His- 12 and 119 recide near the active site. Residue 48 is deeply buried and its environment is known to undergo a conformational transformation on protonation of the said residue, hence

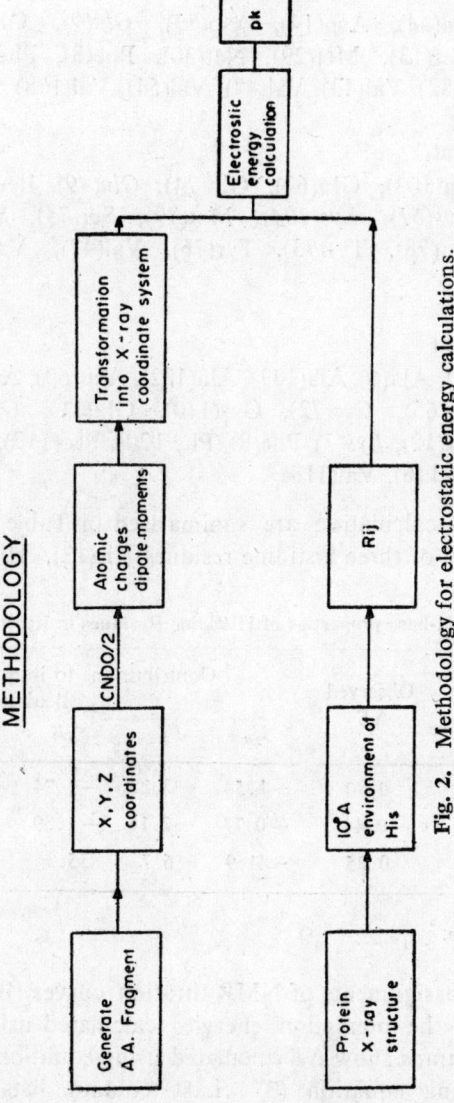

Fig. 2. Methodology for electrostatic energy calculations.

this residue is not considered in calculation [9]. The 10°A environment of residues 12, 105, 119 are listed below. The numericals in paranthesis indicate residue number and the charged groups are underlined.

His(12) environment

Arg(33), Asn(44), Asp(14), Asp(83), *Glu(9)*, Gln(11), His(48), His(119), Met(13), Met(29), Net(30), Phe(8), Phe(46), Phe(120), Thr(45), Thr(82), Val(43), Val(47), Val(54), Val(108)

His(105) environment

Ala(122), Asn(103), Gln(60), Gln(74), *Glu(49)*, Ileu(81), Ileu(106), Ileu(107), *Lys(61)*, *Lys(104)*, Met(79), Ser(75), Ser(77), Ser(80), Ser(123), Thr(78), Tyr(73), Tyr(76), Val(47), Val(57), Val(108), Val(124)

His(119) environment

Ala(4), Ala(5), Ala(6), Ala(109), Ala(122), Asn(67), Asn(71), *Asp(121)*, Cys(58), Cys(65), Cys(72), Cys(110), Gln(11), *Glu(9)*, *Glu(111)*, Ileu(107), His(12), *Lys(7)*, Phe(8), Phe(120), Pro(117), Thr(3), Tyr(73), Val(108), Val(116), Val(118)

The results and calculation are summarised in Table I. Table I lists observed pK values of three histidine residues-His(12), His(105), His(119)

Table I. Acid-base properties of Histidine Residues in Ribonuclease S

Res. No.	Observed pK	Observed ΔE	Contributions to interaction energies (calculated)				
			E_a	E_b	ΔE	ΔE_{hb}	$\Delta E + \Delta E_{hb}$
12	5.8	0.80	−1.54	−3.28	−1.74	2.5	0.76
105	6.7	−0.44	−0.77	−2.16	−1.39	2.5	1.11
119	6.2	0.25	−41.9	−6.7	35.3	0	−41.9

Energy in kcal/Mole.

as per the recent assignments of NMR titration curves [10]. In the third column are listed the interaction energies calculated using equation (1). The following columns show ΔE calculated using equation (2). E_b and E_a are calculated using equation (3). Last column lists the calculated energies. One observes that the agreement for the residue His(12) is very good whereas for other residues it is not so. This may be due to possible contormational changes in the environments of these residues on protonation. The actual utility and merit of this methodology can be judged only when calculations are carried out on several proteins whose pK values are well established. In general this method may be helpful in predicting pK values of acid/base titrable groups in proteins, and also in assigning NMR peaks to particular residues from their titration curves. Other applications involve assessment of the nature of peptide-peptide interactions in proteins and polypeptides and detecting conformational changes which may accompany titration of acid/base group.

References
1. Kilmartin, J.V., Breen, J.J., Roberts, G.C.K. and Ho, C. (1973). Proc. Nat. Acad. Sci. U.S.A. **70**, 1246.
2. Tanford, C. (1960), Adv. Protein Chem. **24**, 447–545.
3. Rein, R. (1973), Adv. Quantum Chem. **7**, 335.
4. Pople, J.A., Beveridge, D.L. (1970) Approximate Molecular Orbital Theory, McGraw-Hill, New York.
5. Pople, J.A., Santry, D.P., Segal, G.A. (1965) J. Chem. Phys. **13**, S129.
6. Wyckoff, H.W., Tsernoglou, D., Hanson, A.W., Knox, J.R. Lee, B., Richards, F M. (1970), J. Biol. Chem. **245**, 305.
7. Roberts, G.C.K., Jardetzky, O. (1970), Adv. Protein Chem. **24**, 447.
8. Dwek, R.A. Nuclear Magnetic Resonance in Biochemistry: Applications to Enzyme Systems, Clarendon Press, Oxford (1973).
9. Ruterjans, H., Wintzel, H. (1969), Eur. J. Biochem. **9**, 118.
10. Shindo, H., Hayes, M.B., Cohen, J.S. (1976), J. Biol. Chem. **251**, 2644.

Life and Mind in the Universe

George Wald

Professor of Biology Emeritus
Harvard University
Cambridge, Mass

In my life as a scientist I have come upon two major problems which, though rooted in science, though they would occur in this form only to a scientist, project beyond science, and are I think ultimately insoluble as science. That is hardly to be wondered at, since one involves consciousness, the other cosmology.

The consciousness problem was hardly avoidable by one who has spent most of his life studying mechanisms of vision. We have learned a lot, we hope to learn much more; but none of it touches or even points however tentatively in the direction of what it means *to see*. Our observations in human eyes and nervous systems and in those of frogs are basically much alike. I know that I see; but does a frog see? It reacts to light; so do cameras, garage doors, any numbers of photoelectric devices. But does it *see*? Is it aware that it is reacting? There is nothing I can do as a scientist to answer that question—no way that I can identify either the presence or absence of consciousness.

I believe that to be a permanent condition that involves all sensation and perception. Consciousness seems to me to be wholly impervious to science. It does not lie as an indigestible element within science, but just the opposite: science is the highly digestible element within consciousness, which includes science as a limited but beautifully definable territory with the much wider reality of whose existence we are conscious.

The second problem involves the special properties of our universe. Life seems increasingly to be part of the order of nature. We have good reason to believe that we find ourselves in a universe permeated with life, in which life arises inevitably—given enough time—wherever the conditions exist that make it possible. Yet were any one of a number of the physical properties of our universe to change—some of them basic, others seeming trivial, almost accidental—that life, which seems now to be so prevalent, would become impossible, here or anywhere. It takes no great imagination to conceive of other possible universes, each stable and workable in itself, yet lifeless.

How is it that, with so many other apparent options, we are in a universe that possesses just that peculiar nexus of properties that breeds life? It has occurred to me lately—I must confess with some shock at first to my scientific sensibilities—that both questions might be brought into some degree of congruence. This is with the assumption that mind, rather than emerging as a late outgrowth in the evolution of life, has existed always, as the matrix, the source and condition of physical reality—that the stuff of which physical reality is composed is mind-stuff. It is mind that has composed a physical universe that breeds life, and so eventually evolves creatures that know and create: science-, art- and technology-making animals. In them the universe begins to know itself. Also such creatures develop societies and cultures—institutions that present all the essential conditions for evolution by natural selection variation, inheritance (mainly Lamarckian), competition for survival so introducing an evolution of consciousness parallel with, though independent of, anatomical and physiological evolution.

Superpositions, Coherence and Choice: Towards a Physics of Life*

E.C.G. Sudarshan†

*Presented at the International Conference on Physics of the
Living State, All India Institute of Medical Sciences
New Delhi, December 1981

Introduction

Physics is the ordering of our perceptions and the consequence of such ordering on the faculty of perception and the constellation of ideas that constitute our communicable universe. In as much as it deals with *all* our perceptions no domain of human experience is outside the domain of physics whether it be lauded as scientific or denigrated as mystical, paranormal or supernatural. As the body of our knowledge grows the depth and subtlety of physical models increase and more phenomena become subject to physics. Physics is structured by refined perceptions and the conceptual framework associated with them.

1. Time and Temporal Evolution

The fundamental concept which dominates our physics today is that of dynamics. While Galileo and Newton were concerned with motion of ponderable isolated bodies we see dynamics in its general form as the *stringing together* of sequences of configurations (kinematical states) labelled by a monotonically increasing parameter, identifying the configurations as being the successive states of the *same* system and expressing this succession in terms of definitive rules. This "stopping of the world" is what gives the notion of the abstract dynamical system; and the law of succession expressed in mathematical form are the equation of motion. For a standard dynamical system the equation of motion is stated in a manner that applies to all configurations: there is a clear separation between initial conditions and equations of motion.

†Chalmers' Jubileum Professor 1982, on leave of absence from Centre for Particle Theory, University of Texas, Austin, TX 78712, USA.

Seen thus the very notion of a dynamical system implies a set of restrictions. *The world is an orderly place because the configurations* at different times *are not freely chosen*. Given an initial configuration the later configuration is not arbitrary; one specific configuration would be chosen by virtue of the equations of motion. The future holds nothing unknowable: *dynamical evolution is the reexpression of the initial state unfolding in an orderly manner*.

In Newtonian physics the initial configurations required were conveniently taken to be the positions and velocities of all the particles, the equations of motion expressing the accelerations as functions of the positions and velocities. Since the evolution is step-by-step in time it is possible to point to the accelerations multiplied by the masses as the forces which *cause* the motion. But this view of *cause* (force) *and effect* (acceleration) arise from a *fragmentation of perception* in seeing the system as consisting of parts: for a closed system is the cause of its own evolution.

The use of global dynamical laws like Hamilton's Principle which derives the equation of motion as a variational Euler-Lagrange equation extremising the action integral is the first step towards the recognition that dynamics can be formulated in a manner distinct from the Newtonian format. It had already been seen that for generalizing the scope of dynamics so that rigid and deformable bodies could be brought under its purview one had to deal with constraints between coordinates of particles. But the simpler of these constraints could be handled by introducing generalized coordinates which were unconstrained but in terms of which the configurations could be fully described. But not all constraints were so "integrable". The existence of such nonintegrable constraints was a serious incompleteness to the development of canonical dynamics. It was this gap which was solved by Dirac in his theory of constrained dynamical systems. This epoch making discovery has been the major development of dynamics since the work of Lagrange and Hamilton.

2. Dynamical Evolution and Separability in Relativistic Physics

As an example of the need to revise the "obvious" preconceptions that we have about physical systems when the scope of physics is extended to new domains and new principles are incorporated let us digress to consider a technical problem: When dealing with general dynamical structures including interaction between particles in the context of the special theory of relativity new problems are encountered. Traditionally the time indicated on a clock can be used as the monotonic labelling parameter. The configurations include the specification of positions and velocities at the *same time*. But distant simultaniety is not relativistically invariant.

We could continue to use the clock time as the label but insist on the canonical dynamical transformations themselves to deal with this problem. Dirac had given a formulation of the essential principles of such a relativistic canonical system. This involves the use of a Lie algebra of canonical transformations associated with the relativistic transformations constituting the ten-parameter Poincaré (inhomogeneous Lorentz) group to reflect the change of reference frames. The Hamiltonian is one of the generators of this Lie algebra corresponding to time translations. While retaining its functional form in terms of the primary variables the Hamiltonian describes different "sections" of the set of unfolding states of the many particle system. Such an implementation of the principle of relativity together with canonical description is straight-forward.

There are two additional requirements that are desirable. The first one is that of separability. When a $(n+1)$-particle system is considered we restrict attention to those states in which the $(n+1)$-particle is arbitrarily far away we could expect the system to behave as an autonomous n-particle interacting system without any reference to the $(n+1)$th particle. This requirement is not as simple to implement as one would have expected but it can be precisely stated and its consequences worked out.

An investigation of systems allowable under such conditions has been carried out by a team of scientists working in Bangalore. The results are startling: For a system of three or more particles they cannot all interact with each other if manifest covariance and separability both apply. In other words "out of sight" is not "out of mind" in such systems.

3. Coherence

When physics is to be applied to sentient beings there can be no doubt that the metabolic biochemistry, nerve electrochemistry and muscular biophysics must apply standard physicochemical principles. There are however, same novel or unusual features that must be given special consideration. Coherence, noncommutativity and superposition; and spontaniety: all these are to be as much part of our theoretical framework as chemical thermodynamics or molecular physics.

Coherence is the correlation of parts; as such it is there in classical dynamics and is often evidenced in folk dances and chorus lines. But coherence comes into its own in wave phenomena where amplitudes are relevant but physically measured quantities are quadratic in the amplitude. Thus, for example, in the propagation of light the quantity subject to the usual photometric technique is the local intensity of light, or the flux of light which is quadratic in the amplitude. But the propagation law is for the amplitude itself. For example, if we interpose a screen in the path of the light beam in which two nearby narrow parallel slits are cut, the resulting intensity pattern on a screen further downstream shows a series of bright and dark fringes: these interference fringes are equally spaced and represent a redistribution of the intensity that was to be expected in terms of the level of illumination on the slits. The width as well as the existence of fringes can be quantitatively accounted for in terms of the wave properties of light: More about this later.

But the interesting point is that the intensity and the striped fringe pattern on the screen are not fully determined by the intensity on the slits. If the illumination is from two different sources, even though of the same exact intensities, the interference pattern would wash out and we could have a more or less uniform distribution on the screen. We account for this by recognizing the concept of "coherence". We say that the light vibrations from different sources are not coherent but from different parts of the same wavefront are coherent.

Coherence is a property of the system which pertains to the relation between parts of a whole which is not necessarily explicit at each time but which can be revealed by a suitable physical evolution. Classical photometry has *no concept of coherence* since in its austere simplicity *it does not have a place for that which cannot be directly measured*. Classical wave optics has a concept of coherence since underlying the photometric intensity it has the notion of the complex wave amplitude. Quantum optics continues to have this conceptual under spinning. It is not only in religion that faith in things unseen provides a more complete description of reality!

Coherence, so introduced while it manifests itself in the influence on the unfolding of events, is not a "cause" in the sense that a force causes acceleration. Rather, it is an inherent element in the description that reveals itself by the dynamical unfolding, in the continuing articulation and paraphrase of the state of the system according to its own natural law. Invariably, coherence is important precisely because there are aspects of the specification of the state of a system unknown at one stage but by no means unknowable at a different stage. The best mechanism we know of mathematical description is to introduce amplitudes and their superpositions. To these we now turn.

4. Superpositions and Noncommutativity

We are used to systems that can be specified in a precise and unambigous fashion. And the ideal of classical kinematic description is to be able to specify the system by all its attributes: to be able to "print all the news". There is *nothing unknown and unknowable* at that time but *which becomes revealed later*. Dynamical development in this picture is the movement from one configuration to another.

When one is dealing with a wave we do have the possibility of superposition. Two waves can coexist and the result is yet another wave. But the resultant is qualitatively not different from the original wave: we could "subtract" by adding an opposite amplitude. In this the notion of *superposition* is distinctively *different from* the arithmetical *addition* of objects. But the really important aspect of superposition comes when there is an unobservable "phase": after all, velocities add but we have no need to invoke coherence there.

A careful examination of classical physics shows that the treatment of a physical system involves two entirely distinct categories: the dynamical *variables* (like position, velocity, mass etc.) *and* the dynamical *processes* (like the action of boosting or accelerating a system). The former alone are accorded substantiality: they alone are given values in specifying the state of a system. The processes cannot have values along with the quantities since one is the change of the other: in mathematical parlance, they do not "commute". One acts decisively on the other. Classical physics finds this no problem since it deals entirely with quantities. As we have already mentioned in an earlier section, we may associate dynamical variables with processes: energy with time evolution or angular momentum with rotation. But these are *associations, not identifications*.

5. Identification of Substance and Process: Quantum Theory

It is the remarkable genius of this century that this distinction between process and substance has been erased in a theoretical framework which promotes the above associations to identification. The processes and quantities not commuting means that they cannot be both specified quantitatively at the same time. Classical physics took care of this by choosing always to give values to quantities and ignoring processes: it was consistent in this denial. But the new quantum physics decided to accord equal role to them and identify both with the familiar dynamical variables. For example the process of displacing the position is identified with the momentum. It follows that position and momentum cannot commute: the celebrated Heisenberg uncertainty relation is one expression of the unusual world picture that emerges.

It follows that a quantum system must always contain information that cannot be directly apprehended at any instant but which becomes available at a later stage: a true case of revelation!

It is often said that quantum physics is difficult to comprehend since it is so far from our usual classical experience and must forever remain in the language of mathematics. I find it difficult to accept this view. Our own experience of the perceived universe is that we not only perceive an outer universe but act on that perception to get newer world views. This complex of perceptions and the perception of the transmutation of perceptions is much more akin to quantum physics than classical physics. It is rarely that we are sure of the content of our awareness in a manner that we can articulate it. Anyone who can express himself fully about his total perception in ordinary language is either decieving us or decieving himself. Quantum world order is much more intuitively appealing to us than classical world order.

Very often people attempt to make use of classical world pictures and models for depicting a quantum system. In such cases inevitably distortions must emerge. This is the origin of the notorious quantum uncertainties. If we attempt to depict the spherical surface of the earth globe on flat sheets of paper in making maps, however commendable and commonsense the effort, it must result in distortions, separating regions which are geographically contiguous or distorting shapes and areas. We know that there is no fissure in the Pacific ocean separating the Mongolian Peninsula from Alaska if the map presents them to be so far apart. In the same sense the quantum uncertainties are artifacts of description devoid of fundamental significance. It is noteworthy that the most relevant examples of quantum probabilistic behaviour almost always have a classical stochastic detectors like a counter in the experimental arrangement.

6. Beyond Cause: Creativity and Choice

The final topic that I wish to discuss is the *transcedence of cause*: the question of spontaniety. We value such spontaniety as a mark of liveliness and celebrate it as creativity. By the nature of creativity there can be no "cause" for it: we can only *facilitate* creativity and spontaniety, *not cause* it. Along with spontaniety is the volitional component of "option". It is a remarkable fact that neither classical nor quantum physics (and hence all the physical sciences) has a niche for options. We have two very poor substitutes: statistical probability and randomness. But I wish to emphasize that *uncertainty or randomness is not choice.* As modern biology intends to consider life it may not involve options or choices; but *I am aware of choices*; and I guess others are. If this perception is not an illusion then some living systems contain options: and present day physics is incomplete in not *providing* a niche *for choices* and to properly *distinguish* it from chance.

I am not suggesting for a moment that we are not to consider the possibility that causes not found may not be understandable in certain cases as the dynamical unfolding of unappreciated coherence or tendencies. But I am suggesting that ignoring of the existence of genuine (uncaused) choices is a mutilation, a veritable bed of Procrustus.

7. Towards a Complete Cosmology

A satisfactory scientific cosmology would involve *all* our perceptions, be it of the very large or very small. But it would also involve the direct perceptions of the experimenter. The arbitrary exclusion of a significant segment of our experience and perceptions is unscientific. It is true that adequate care must be exercised in refining perceptions and controlling variability. But all reproducible and communicable experiences must be included within the scope of scientific investigations. State of the art instrumentation and modern mathematical techniques should warrant careful reexamination of coherence phenomena and instances of choice: of tendencies as well as emergent qualities. It is my experience and that of others that as perception is refined more extended coherence manifests and yet more creativity manifests too.

It is probably not too farfetched to suggest that "option" and "coherence" are as much essentials of life as metabolism, genetic imprinting and reproductive invariance. Rather than fall into the absurdity of ignoring phenomena until a theory can accomodate it, it may be a worth while attempt to recognize these characteristics of life in general.

Acknowledgements

I thank Chalmers University of Technology and Professor Jan Nilsson for their hospitality where this paper was prepared for publication. I am grateful for discussions with R.L. Kapur, R.K. Mishra, N. Mukunda, I. Prigogine, B.R. Seshachar, K.P. Sinha and Maharshi Mahesh Yogi who have enlightened me on the unknown but knowable.

The Universe and the Living State: An Information Theoretic Relationship

K. P. Sinha

Indian Institute of Science
Bangalore 560 012, India

1. Introduction

The three most important fundamental problems of modern science are (i) the nature and evolution of the universe (ii) the nature of the elementary particles along with their interactions and (iii) the nature and evolution of the living state. Perhaps, it is not well appreciated that answers to all these questions are interconnected. The ultimate reality may embrace the microcosmos and macrocosmos of both the living and the non-living world in one unifying description.

A serious study of the origin of the universe (cosmology) has now become a respectable scientific pursuit typified by the recent book of Weinberg[1]. It is to be hoped that the future of the universe along with the origin and fate of the living world and consciousness will hold the attention of serious scientists.

There is now growing realisation that the origin of life is not geocentric in the sense that only the planet earth provides the narrow range of conditions under which life as we know would have evolved. Such conditions might be existing in a large number of objects in various galaxies around their stars. Some believe that even the 1000 billion comets provide better conditions for the evolution of the living state than the earth. Furthermore, the discovery of several organic molecules in interstellar space in recent years have emboldened some authors to speculate that interstellar dust clouds may be infested with living cells in a frozen dormant state[2]. Thus the occurrence of life (coherence, information) bearing systems seems to be fairly widespread in the universe than hither to supposed. We are, in fact, faced with ever expanding vista of life, consciousness and memory. And naturally, the notion of value and purpose arises, which is forbidden in Monod's world of chance and necessity[3]. But recently there has been a strong defiance of Monod's anathema by

Dyson[4] and he has encouraged others to do so. In what follows, we shall keep this in mind.

2. A Model for the Information Content of the Early Universe and the Living State

The fact that there is a very large information content in the universe at present is of great significance. It has been reemphasized by Hoyle[5] recently in the context of cosmology. The information content is made manifest in the living systems and some non-living systems which are in ordered and coherent state. A human being alone accounts for 10^{23} information units or bits for each moment of awareness lasting about a second (Dyson[4]). Coming down to cellular and molecular level, we have about 20 distinct amino acids and 2000 different enzymes each containing a specific distribution of these amino acids in the polypeptide structure. These specific enzymes are crucial for the evolution of life in the universe from cell to man. Then there are other specific molecules such as RNA and DNA and the genes and the cell. The chance of getting the enzymes alone from a completely random process would be less than $10^{-40,000}$. As Hoyle[5] has remarked this infinitesimal probability cannot be accounted for even if the entire universe consisted of organic soup. Thus the random chance process cannot produce living systems. We need an initial input of enormous information content.

In this context let us look at the models of the origin of the universe. First we have the steady state theory of Hoyle and coworkers in which the universe has been there all the time no beginning and no end, timeless and eternal. While this eternity may cope with the infinitesimal probability ($10^{-40,000}$) along with its appealing philosophy there are serious observational difficulties such as lack of agreement with redshift-magnitude relation, count of radio sources, microwave background and its thermalization. In addition, the steady state concept involves a creation field. In short the model has been ruled out except for a few of its protoganists.

Next we have the widely accepted and reigning standard model[1] in which the universe started with a big bang at time $t = 0$, from an initial singularity when the matter-energy density ρ and the temperature T are both infinite. There was total chaos and complete lack of knowledge at $t = 0$, and hence the information content was zero. Nevertheless, current observational data, red shift, microwave background, expansion rate of the universe etc. strongly support this model apart from some serious problems of the initial state. Some of these arise from the extrapolation of the observational data backward in time to $t = 0$ and the assumption that Einsteins classical equations of eneral Relativity remain valid in this extreme high energy and short distance regime. The problems that need sorting out are[6]: (i) the singularity itself (ii) horizon problem, namely, partitioning the universe in causally disconnected regions (assuming, of course, that no signal faster than light can be transmitted). The horizon goes as αt. Going backward in time to the singularity ($t \to 0$), the horizon shrinks to zero and all particles get causally disconnected. This also leads to total loss of information. Yet how do we find so much information now and independent regions of the universe have identical dynamics and matter content (to 1 part in 10^3). (iii) entropy problem. The entropy per baryon S_B is estimated to be

$$S_B = \frac{\text{Number of microwave photons }(n_\gamma)}{\text{Number of baryons }(n_B)} = 10^8.$$

If local irreversible processes are ignored the above value is independent of the cosmological time. The present uniformly expanding adiabatic situation offers no clue either for the existence or the moderate value $S_B \sim 10^8$. This has led to the suggestion of chaotic cosmology (Misner[7] and Rees[8]) involving random cosmological initial states along with dissipative mechanisms for matter and radiation. However, instead of solving the problem, this model reveals another serious difficulty. The dissipative effects will build up in the past in denser matter and radiation and close to the singularity there will be entropy catastrophe.

All these factors suggest that the initial conditions of hot big-bang model are not valid. The initial state of the universe must have been near zero entropy (quiescent) ordered state which is singularity free. The observed entropy ($S_B \sim 10^8$) and the dominance of matter over antimatter developed subsequently by processes involving elementary particles.

3. A New Model of the Initial State

In discussing a model which can answer these questions, we shall start with the singularity problem. It is now realised that Einsteins classical equations are not valid in the high energy, the high density and very short distance regime. Further, we live in a quantum world and quantum effects along with elementary particle physics must be taken into account during the early stages of the universe. The assumption that the coupling constants of elementary particle interactions remained unchanged even under the extreme conditions of the early universe (when the possibility of various kinds of phase transition exists) is not justified.

Our work in the last several years[9-14], has shown that at density of the order of 10^{17} g cm^{-3} (occurring within hadrons) there is a phase transition in which gravitons (or spin-2 gluons) become massive and the gravitational constant is renormalized such that the gravitational fine structure constant

$$\gamma = \frac{GM_p^2}{\hbar c} \sim 1$$

where M_p is the proton mass. This gives $G \to G_f \sim 10^{38} G_N$; G_N is the

Newtonian constant. The same value is obtained if we make the proton compton length (its quantum fuzziness) and 'strong' Schwarzschild radius compatible (equal). This has been called super (strong) gravity and is consistent with quantum mechanics and operates within short range $\sim 10^{-14}$ cm One can formulate Einstein-type equations but with a very large cosmological constant (which is related to the mass of the spin-2 boson).

$$\Lambda_f \sim (m_f c/\hbar)^2 \sim 10^{28} \text{ cm}^{-2}$$

One very important consequence of this is that the collapse of the universe is arrested at the density

$$\rho_{max} = \Lambda_f c^2/4\pi |G_f| \sim 10^{17} \text{ g cm}^{-3},$$

and minimum radius.

$R_{min} \sim 10^{13}$ cm (size of the solar system) owing to the repulsive scalar component of the super strong gravity. Thus the universe is free from singularity. Let us look at the characteristic quantum (strong) gravitational quantities at this stage:

$$L_{pl}(s) = (\hbar G_f/c^3)^{1/2} \sim 10^{-14} \text{ cm}$$

$$M_{pl}(s) = (\hbar c/G_f)^{1/2} \sim 2 \times 10^{-24} \text{ g}$$

$$t_{pl}(s) = (\hbar G_f/c^5)^{1/2} \sim 10^{-24} \text{ sec}$$

$$T_{pl}(s) = \frac{1}{k_B}(\hbar c^5/2G_f)^{1/2} \sim 10^{13} \text{ K}$$

Thus the characteristic length, mass and time are scaled upto the elementary particle level unlike those obtained for the weak (G_N) case. As it is meaningless to talk of time sharper than this quantum limit of 10^{-24} sec, we cannot go to earlier times. Also the collapse has been arrested and there is no $t \to 0$ limit. At this epoch, the horizon is of the order of 10^{-14} cm and elementary particles are in causal contact and there is flow of information (interaction) between them.

What is the significance of the strong Planck temperature $T_{pl} \sim 10^{13}$ K? Now at the prevailing density $\rho_c \sim 10^{17}$ g cm^{-3}, hadrons would overlap largely and dissociate. The entities occurring in this state are leptons, quarks and gluons, and any other fundamental objects. Thus at this epoch we have a degenerate Fermi sea of quark like objects (lepto quarks) along with gluons. There is a possibility of a phase transition to a superfluid state owing to one gluon exchange between quark and antiquarks. It is instructive to estimate the Fermi temperature of this sea. With total number[15] of lepto quarks $N \sim 10^{90}$ and volume determined by $R_{min} \sim 10^{13}$ cm, the Fermi wave number $k_F \sim 2 \times 10^{17}$ cm^{-1}. This gives a Fermi temperature $T_F \sim \left(\frac{\hbar c}{K_B}\right) k_F \sim 10^{17}$ K. Now the ratio $\frac{T_F}{T_{pl}(s)} \approx 10^3$ to 10^4. This is the approximate ratio in all the known superfluid Fermi liquids.

A $T_c \sim 10^{13}$ K for a hadronic superfluid has been obtained earlier by us[12]. This suggests that at this era (10^{-24} sec) the universe is most likely in an ordered superfluid state with great deal of coherence and information content. This is the initial superfluid fireball (or big bag). This is the true ground state (true vacuum). The curvature of space-time and the presence of a large cosmological term oppose the symmetry restoration action of temperature. However, density and temperature fluctuation may lead to certain reactions. In this context the most important reaction is the Yoshimura[16] processes involving violation of both baryon conservation and CP or T invariance. This, as has been shown, will give the observed matter—antimatter asymmetry and $S_B = \dfrac{n_r}{N_B} \sim 10^8$. The reactions for these processes heat up the fireball leading to the big bang (or bursting of the bag) and adiabatic expansion starts. The above mentioned moderate entropy produced then remains constant to the present epoch. The universe still retains some remnants of the initial order inside the hadrons, and various living and non-living structures which are in coherent ordered state. Thus the information content that we see to day is related to the information content of the early universe.

4. Concluding Remarks

We have discussed a model of a relationship of the information content of the living state and that of the early universe. It is based on the assumption that the distribution of the living state is fairly widespread in the universe.

Let us come to the question of the information content of highly evolved systems (human mind) namely consciousness. Scientists such as Wigner[17] and others[18] have put forward the idea that mind (consciousness) must involve new degrees of freedom in conrtadistinction to the physical brain. Dyson[4] has posed the question whether the basis of consciousness is matter or structure. Can we have conscious structure or sentient computer, which are not associated with biological molecules inside the brain of the human being.[22] My personal reaction to these questions, which I have given on earlier occasions also[19], is the following. If consciousness is associated with the coherent state of new degrees of freedom, these must arise from elementary particle level say, basic fermions, or bosons or in their supersymmetric state. Now physicists have been extremely inventive in the past in giving new degrees of freedom to particles e.g. spin, flavour, and colours. So why not give another degree of freedom like choice or discrimination to elementary particles, for example to an electron. However, unlike other attributes, we do not have a quantitative measure for this new degree of freedom and the law of interaction and quantum restrictions like the exclusion principle. Once we have this, it may be possible to describe consciousness as the ordered state of this new degree of freedom of constituent entities of the structure. Then the possibility of sentient structures in interstellar space cannot be ruled out.

5. Acknowledgement

I should like to thank Professors E.C.G. Sudarshan and N. Kumar for constructive comments.

References

1. S. Weinberg. The First Three Minutes (Basic Books, New York 1977).
2. F. Hoyle and C. Wickramsinghe. The Origin of Life (University of Cardiff Press, Cardiff, U.K. 1980).
3. J. Monod. Chance and Necessity (Alfred A. Knopf, New York, 1971).
4. F.J. Dyson. James Arthur Lectures on Time and its Mysteries (New York University, 1978).
5. F. Hoyle. Steady-State Cosmology Re-visited (University of Cardiff Press, Cardiff, U.K. 1980).
6. J.D. Barrow. Sci. Prog. **65**, 129 (1978).
7. C W. Misner. Astrophys. J., **151**, 431 (1968).
8. M.J. Rees. Phys. Rev. Lett., **28**, 1669, (1972).
9. C. Sivaram and K.P. Sinha. Lett. Nuovo Cim. **8**, 324 (1973).
10. E.A Lord, K.P. Sinha and C. Sivaram. Prog. Theor. Phys. **52**, 161 (1974).
11. K.P. Sinha and C. Sivaram. Proc. Int. Conf. Frontier of Theor. Phys. (McMillan, New Delhi 1977).
12. K.P. Sinha, C. Sivaram and E.C.G. Sudarshan. Found. Phys. **6**, 717 (1976).
13. C. Sivaram and K.P. Sinha, Physics Reports **51**, 111 (1979).
14. U. Raut and K.P Sinha. Int. J. Theoret. Phys. **20**, 69 (1981).
15. R. Brout, F. Englert and P. Spindel. Phys. Rev. Lett. **43**, 417 (1979).
16. M. Yoshimura. Phys. Rev. Lett. **41**, 281 (1978).
17. E.P. Wigner. Symmetries and Reflections (MIT Press 1970).
18. D.S. Kothari, J.N. Tata Lecture I.I.Sc. (1979).
19. K.P. Sinha. Int. Conf. on Evolution and Consciousness, MIU, Fairfield (1975).

Knowledge and the Evolution of the Living State

Jerzy A. Wojciechowski

Department of Philosophy
University of Ottawa
Ottawa
Canada

Introduction

The purpose of this paper is to discuss the relationship between knowledge, in particular, conceptual knowledge, and the phenomenon of life. The relationship has, among others, a temporal dimension. It extends into the past and into the future. It is the future of this relation which is of particular interest and invites reflection. In essence, by this paper we draw attention to a body, the corpus of knowledge, which affects the phenoenology exhibited by the living state. They constitute a part of a larger system. Their relation in this context is explored and similarities and dissimilarities between organisations point out to help draw up parameters for this self-organising ensemble for further study.

Man identifies with life. Life is of fundamental importance to him. He may look at life from the Western perspective and perceive it as an essential value underlying all other values, or he may view it in the Eastern tradition as bondage and suffering. In either case, man wants to understand the mystery of life and to do something about it. Either he wants to master it so as to perpetuate it, or he wants to break the chain of suffering and destroy the self so as to reach the state of Nirvana.

Beyond feeling concern about life man can increasingly do something about it. His influence on life: his personal existence, that of his species, and life in general, grows daily and acquires previously undreamt of proportions. The greater man's impact on life, the greater is his responsibility for it and to it. Man's impact on life is proportional to his ability to act which, in turn, is proportional to his knowledge. Consequently, the relationship between responsibility for life, on the one hand, and knowledge, on the other, may be expressed in the form of the following law:

Law I: Man's responsibility to life is proportional to existing knowledge.

The more man interferes in life in general, the better he must understand it. This is why, a discussion of the relationship between knowledge and life becomes increasingly more important.

Let it be stressed that it is impossible to study the living state in its totality without paying special attention to the rational activity of man. It is the most perfect of all the activities of living beings and has the greatest potential for evolution. This activity is influenced not only by the organism of man, i.e. by his biological nature, but also, and increasingly, by past rational activity and its products. Consequently, a study of man as a living being has to involve the discussion of his rational activity and its consequences.

The Knowledge-Life Relationship: General Remarks

The present paper is an illustration of this fact. It proposes to discuss the relationship between knowledge and life from the perspective of the *problematique* of the man-knowledge relationship. In order to do justice to the complex totalities of the phenomena studied here, the systemic approach will be used.

Life and knowledge being organised systems share some essential characteristics and differ in some other.

Life can be described as a concrete, systemic, dynamic, hierarchic, purposive, self-reproducing, evolving phenomenon. In its totality life is a system of systems.

The body of knowledge which we propose to call the knowledge construct, KC for short, is another system of systems. Besides similarities, there are of course, important differences between the system of life and the system of knowledge. Conceptual knowledge may be described as an abstract, systemic, hierarchic, man-produced, purposive, life-assisting phenomenon. Both elements of the knowledge-life relationship, KLR for short, are systemic and evolving. So is the relationship itself.

One cannot discuss the consequences of the KLR without discussing the nature of the relationship itself. The central question here concerns the role of knowledge in relation to life. In this respect knowledge does essentially two things: it represents life and it influences life. It increases the capacities of rational organism to evolve and speeds up their evolution.

The present predicament of mankind with its multiple and growing ecological, social, political, and economic problems, indicates that besides positive effects of knowledge, there are many negative consequences as well. Knowledge facilitates life, is a life enhancing device, but it also threatens life in various ways. No other living being harms life nearly as much as man. The relationship between this ability and knowledge may be expressed in the form of a law which is implicitly contained in Law I:

>   Law II: Man's potential to influence life positively or negatively is proportional to the KC.

Through conceptual knowledge life becomes conscious, therefore reflective, self-conscious, and self-directive. It becomes also more open to change, and acquires the capacity to destroy itself. In other words, life gains mastery over itself.

The living state is made up of two very unequal parts, namely man and all the other living species. One species against a million. But the one species contains all the levels of perfection of the other million species.

The Existential System of Man

In order to explore the relationship between knowledge and life it is necessary to analyse the relationships existing between man, knowledge and the ambient world. The following propositions express the main aspects of the situation under study and should serve as premises for the discussion: (a) man in himself is a binary system, i.e. composed of organism and intellect: (b) the system "man" is included in a hierarchy of natural system: inanimate and animate; (c) man forms with other elements new systems not provided by nature; (d) the existential system of man, ESM for short, is composed of man, ambient nature, the knowledge construct, and products of man's rational activity-both concrete, ex. tools, and abstract, ex. laws.

The ESM has certain noteworthy characteristics. Let us mention first the characteristic due to its being a system, the general laws of systems apply to it. A systemic approach to the study of the ESM is therefore justified. The nature of the system may be described as follows. The system is dynamic, self-energizing, growing in size, complexity and causality. It is negentropic, and evolving at an ever faster pace.

The growth of the system is mainly the result of human activity. Man's actions transform all the relationships existing between the parts of the system. Because the relationships within the ESM change continually, they make past insights and explanations insufficient and demand continual intellectual effort to be understood. The present ecological crisis is the best illustration of this need and a timely warning.

The situation seems to be paradoxical. The more man thinks and acts, the more the system changes. The more it changes, the greater the challenge it poses to man's intellect—the more man has to think to understand it. The more he understands it, the greater becomes his impact on the system and the more the system changes forcing him to think even more. It may be objected that this reasoning is not generally applicable; that there exist modes of intellectual knowledge which do not appreciably change the ESM. Contemplation as is traditionally practiced in India for instance, does not seem to produce external effects and to have an impact on the ESM but to produce wisdom instead; it therefore does not act as an organiser.

The ESM changes and it changes ever faster. It is an open system. Its openness grows with the growth of the KC, i.e. with human progress. Future states of the system become, therefore, increasingly different from any present state and more difficult to predict adequately. The greater the knowledge, the more unpredictable becomes the human conditions. Apparently, the demiurgic powers of man produce a heraclitean situation in which everything changes and nothing remains the same.

The human factor becomes an increasingly important and determining part of the system. The system becomes more and more man-made. Through his rational activity man produces knowledge, and culture and thus transforms the natural environment. It is important to realize that the system has not been planned, at least not by man, nor was it forseen by him. The system is a by-product of human activity. In fact, until recently man was not aware of its existence.

The system replaces nature as man's everyday habitat. Man is not in a simple, direct relationship with nature anymore. The direct relationship between organisms and nature characterizes the infra-human world of plants and animals. They co-exist harmoniously with their environment, i. e. with nature of which they are an integral part. Through the continual exercise of his rational powers man has excluded himself from nature's self-regulating mechanism which ensured ecological balance. Having elevated himself above the order-producing interplay of the forces of nature, man cannot return to it. Because of his rationality, he has lost nature's paradise forever.

It is a particular achievement of man that he has replaced the direct relationship of organism to nature, which is characteristic of the stage of natural, i.e., non-rational evolution, by a more dynamic, more rapidly evolving, and increasingly more complex relationship: man, culture, nature. This relationship is in fact three relationships in one: (a) man-nature; (b) man-culture, i.e. man—the sum total of intellectual and material products; (c) culture-nature. The complex relationship between man and culture in general, and the KC in particular, is responsible for much of what man does and of what happens to him.

Although the ESM makes possible the human mode of life and its evolution, the understanding of the workings of this influence is difficult. The causal impact which the system exercises on all its component parts is not only complex but changing as well. It is a function of the totality of the system and of its parts. The change and the growing complexity of the ESM are, as earlier noted, the mainly unintended result of man's rational activity. This fact may be expressed in the form of the following laws:
Law III: The size and complexity of ESM are proportional to the level of

Knowledge and the Evolution of the Living State   421

Table I   The Existential System of Man (ESM)

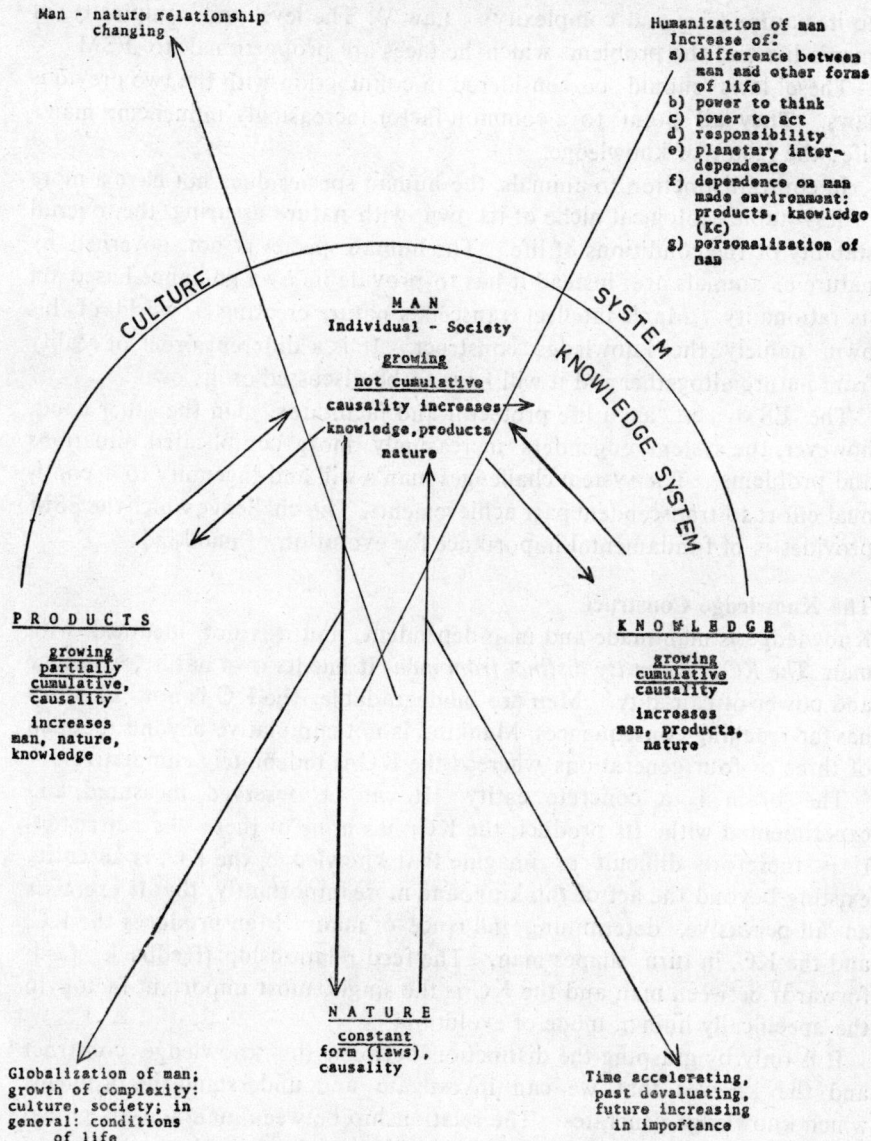

All arrows indicate feed relationships

Knowledge — the word is used here in the sense of the sum total of intellectual products: subjective and objective, scientific or not, past and present.

System: the knowledge system (KS) — man — knowledge, the culture system — man — knowledge — products, the existential system of man (ESM) — man — knowledge — products — nature.

rational activity. Law IV: The causality of ESM on man is proportional to its nature (size and complexity). Law V: The level and complexity of man's life and the problems which he faces are proportional to ESM.

These laws should be considered in conjunction with the two previous laws. They all point to a common factor increasingly influencing man's life: the factor of knowledge.

In contradistinction to animals, the human species does not have a more or less stable ecological niche of its own, with nature assuring the overall stability of the conditions of life. The human species is not governed by nature as animals are; instead it has to provide its own guidance based on its rationality. Man's intellect transcends nature creating a world of his own, namely the knowledge construct. It is a different order of reality from nature altogether and it will have to be discussed on its own.

The ESM acts as a life protector and facilitator. On the other hand, however, the system engenders increasingly more complicated situations and problems. The system challenges man's will and ingenuity to a continual effort to transcendent past achievements. The challenge which the ESM provides is of fundamental importance for evolution of mankind.

The Knowledge Construct

Knowledge is man-made and man-dependent, but it is not identical with man. *The KC is an entity distinct from man.* It has its own nature, existence and power of causality. Men are biodegradable, the KC is not. This fact has far-reaching consequences. Mankind is not cumulative beyond the span of three or four generations whereas the KC is indefinitely cumulative.

The brain is a concrete entity. It can be observed, measured, and experimented with. Its product, the KC, has none of these characteristics. It is therefore difficult to imagine that knowledge, the KC, is an entity existing beyond the act of thinking, and more importantly, that is exercises an all-pervasive, determining influence of man. Man produces the KC, and the KC, in turn, shapes man. The feed relationship (feedback, feedforward) between man and the KC is the single, most important factor in the specifically human mode of evolution.

It is only by grasping the distinction between the knowledge construct and the knower that we can investigate and understand the problems which knowledge generates. The relationship between man and his global intellectual product is not simple; neither is it univocally determined, nor necessarily always advantageous to the knower. From the biological point of view the appearance of the KC is a radical novelty. There is nothing like it in nature.

In order to understand the man-knowledge relationship, it is necessary to view man and knowledge as elements of a system composed of knowers and the body of knowledge. Let us call it the knowledge system. The fact that it is a system allows us to deduce a crucial consequence, namely, knowers and knowledge are parts of a whole which is bigger than either of these two parts. The nature of the whole which results from the constituent elements is not identical with any one of them.

Because Man and KC are as the whole a system, the relation possesses the general characteristics of systems. The elements of the knowledge system are interrelated by complex feed-back and feed-forward relationships. Man produces knowledge, and knowledge in turn shapes man; knowers and knowledge are Parvara whole.

The more coherent and evolved is the system of knowledge the more the next step in the development of knowledge is facilitated and influenced by the existing system.

In other words, the greater the KC, the greater is the tendency to guide the development of knowledge, to make it predetermined and selective. At best this is a partial blessing.

Fortunately for man and for the living state in general, the KC is not the only factor influencing the development of knowledge. The external world forces man to adjust many of his ideas and thus keeps open the possibility of the evolution of knowledge. At the basis of the KC's impact lies the relationship between ideas and mind. The mind produces ideas and in turn is influenced by them. The mind becomes more and more its own maker.

It is through the agency of the mind that the KC acts on man and influences individual and group behavior using relations involving four elements:

$$\text{human condition} \begin{smallmatrix} \rightarrow \text{mind} \rightarrow \\ \searrow \text{behavior} \nearrow \end{smallmatrix} \text{ideas (KC)}$$

The KC is impersonal; intellection is always personal. Individuals as individuals are incapable of building a KC of significant scope. Only society can be the necessary and adequate basis for the KC, because only society can assure the accumulation and the preservation, from generation to generation, of the results of all the acts of knowledge. In order that society fulfill its role as the base for the development of the KC, something else is necessary, namely that *knowledge be externalized, communicated and stored.*

The act of knowing is always environmentally conditioned. The more knowledge the organism has, the better are its chances of survival. The storing of knowledge is, therefore, a biological necessity and an evolutionary device of prime importance.

The storing of knowledge has many important consequences. One of them concerns man's control of knowledge. Although a KC is produced by man, once formed it acquires a life of its own. It becomes independent of its makers, and those who formed it cannot control it completely. Nor can they control its impact on man and nature, let alone foresee the whole range of its potential effects.

The interdependence of man and the KC is not static but dynamic. It forms a self-energizing, negentropic, evolutionary system.

In order to make the nature of these systems more intelligible, let us compare briefly organisms, KCs and cultural systems (Table II).

Table II

Organism (in general)	Knowledge Construct	Cultural System
1. Concrete	1. Abstract	1. Combination of concrete and abstract elements.
2. Subsystems organic	2. Subsystems-conceptual	2. Subsystems: organic, artifactual, conceptual.
3. Type of system: acting (dynamic)	3. Type of system: pattern (passive)	3. Type of system: dynamic and passive.
4. Type of causality: efficient	4. Type of causality: formal and final	4. Type of causality: efficient formal and final.
5. Product of biological evolution	5. Product of "rational" evolution	5. Product of biological and "rational" evolution.
6. Energy throughput information throughput	6. Information throughput energy throughput	6. Varying ratio between the throughput of energy and of information.
7. Biodegradable	7. Non biodegradable	7. Some elements biodegradable, others not.
8. Definite lifespan	8. Indefinite lifespan	8. Indefinite lifespan.
9. Non cumulative (specific limits)	9. Cumulative (large, widely varying limits)	9. Cumulative in its totality, but not in all the parts.
10. Genetically determined	10. No genetic code	10. No genetic code.
11. Finite growth capacity	11. Indefinite growth capacity	11. Either finite or indefinite growth capacity.
12. Does not change its specific determinations	12. Evolves	12. Evolves.
13. Forms ecosystems	13. Forms ecosystems	13. Does not form ecosystems.

The Storing of Knowledge and its Consequences

Generally speaking, there are two ways of preserving knowledge: biological and cultural. The first consists in encoding it in the genetic system and transmitting it through heredity. It is not externalized and does not produce a knowledge construct distinct from the knower. The second way of preserving knowledge consists in externalizing knowledge and preserving it by artificial devices. The difference between these two ways of storing knowledge accounts for the difference in the nature and the

TABLE III

EXTERNALIZATION OF KNOWLEDGE AND ITS CONSEQUENCES

Subjective world		Objective World			
Sender	Expression of knowledge	Receptor	Storage of Knowledge	Knowledge Consequences	Culture
Knower	Non-verbal manifestation (body language) — Direct Communication Speech →	One or few knowers only synchronic communication (simultaneous), no communication through time	Storage in knowers Consequences: 1. No storage outside of knowers 2. Storage limited by the capacity of memorization 3. Limited retrieval 4. Subjective acts: memorization, retrieval 5. No verification of acts or contents possible	Knowledge Construct I → Consequences: 1. Not cumulative beyond limits of memory 2. Stable: capacity for growth limited	Stable state culture (non literate cultures)
Personal acts of knowledge 1. Sensation 2. Intellection	Verbal externalization (language) — Indirect Communication writing, print, etc. →	Indefinitely many receptors Knowers or devices Synchronic & diachronic communication	Storage outside of knowers Consequences: 1. Unlimited storage capacity 2. Unlimited retrieval 3. Objective acts: storage, retrieval 4. Verification possible	Knowledge Construct II → Consequences: 1. Cumulative 2. Capacity for growth unlimited	Progress (cultural evolution)

pace between biological and cultural evolution.

Through the externalization of knowledge the KC is built up. This, in turn, makes the knowledge system, the culture system, and the ESM constantly: (a) more complex, (b) more dynamic, (c) more rapidly evolving. The systems are: (a) nonsymetric, (b) open, (c) nondeterministic. For these reasons the systems' behavior is largely unpredictable. The outcome of the development of knowledge.

Table IV

Sense knowledge	Conceptual knowledge
Cannot be externalized, therefore:	Can be externalized, therefore:
1. cannot be communicated directly through language	1. can be communicated directly through language
2. cannot be stored outside of the knower	2. can be stored outside of the knower
3. cannot be communicated dia-chronically	3. can be communicated dia-chronically
4. does not become an element of human environment	4. becomes an element of human environment
5. is not additive nor cumulative beyond personal capacities of retention	5. is additive and cumulative beyond personal capacities of retention
6. does not form a KC	6. forms a KC
7. does not evolve appreciably through generations	7. evolves at an ever increasing pace
8. does not form an evolutionary system with man	8. forms an evolutionary system with man
9. is always personal, cannot become impersonal	9. can become impersonal
10. does not survive the individual, i.e. is bio-degradable by nature	10. can survive the individual, is not bio-degradable, is not absorbable by nature
11. does not alter nature, does not disrupt ecological balance	11. alters nature, disrupts ecological balance
12. is mainly a product of natural not of cultural, evolution	12. is a product of cultural evolution
13. does not produce culture	13. produces culture
14. is not an exclusively human mode of knowledge	14. is an exclusively human mode of knowledge

Table V

System of Man	System of Knowledge
I. Individual	I. Concept
1. product of nature	1. most natural product of intellect
2. potential element of:	2. potential element of
(a) human relationships	(a) logical relationships
(b) systems: family, society	(b) systems: judgment, reasoning
II. Family	II. Judgment
A. Composed of:	A. Composed of:
(1) individuals differing by:	(1) logical elements differing by:
(a) sex	(a) meaning
(b) age	(b) role
(2) relationships of:	(2) relationships of:
(a) interdependence: physical, emotive	(a) interdependence: logical
(b) order	(b) order
(c) hierarchy	(c) hierarchy-logical
(d) compatibility and incompatibility:	(d) compatibility and incompatibility:
(i) within the family	(i) within the judgment
(ii) of the family with other individuals, families and societies.	(ii) with other concepts, judgments and reasonings
B. Having the quality of being good or bad. Their quality affects society.	B. Having the quality of being true or false. Their quality affects reasoning
III. Society	III. Reasoning
1. Composed of:	1. Composed of:
(a) individuals	(a) concepts
(b) families	(b) judgments
(c) relationships	(c) relationships
2. In comparison to family, greater:	2. In comparison to judgments, greater:
(a) complexity	(a) complexity
(b) variety of arrangements of elements	(b) variety of arrangements of elements
(c) capacity of cumulation of knowledge and experience	(c) capacity of cumulation of knowledge
(d) power to act	(d) power to explain
(e) possibility of inadequate organisation	(e) possibility of inadequate organization
(f) more complex relationship between individual and society, both ways	(f) more complex relationship between concept and reasoning, both ways.

An individual can retreat to a desert and spend his life in contemplation, limiting, if he so wishes, his physical needs to a minimum. But humanity cannot do the same. A very important consequence for the state of knowledge follows from this fact. It may be expressed in the form of a law:

Law VII: The need for efficacious knowledge and activity is proportionate to the size of mankind.

The more men there are, the more they must know and the more efficient and organized must be their behavior. More and more varied knowledge is needed, not less.

Knowledge is an ambivalent factor. In itself it is a perfection, but the body of knowledge is also a force both a positive for those who have knowledge and negative for those who do not, a tremendous power. We may express it in the form of a law.

Law VI: Knowledge stratifies society and mankind. This has always been true, but it becomes increasingly more obvious.

Contrary to what is often explicitly or implicitly believed, *knowledge is the most powerful factor creating inequalities among men.*

Also we may state another conclusion.

Law VII: The time available for solving man-produced problems is inversely proportional to man's demiurgic capacities.

The rather obvious consequence of this relation is the existence of a limit to the process of acceleration of progress as we have known it until now. We may well be approaching this limit rapidly.

Variety and development of knowledge constructs

The striking fact about knowledge is the variety of the KCs. What actually exists is not one universal and uniform knowledge construct but a multiplicity of different systems of knowledge, each one peculiar to a given culture. Let us call them cultural knowledge constructs, CKCs for short.

Through his most specifically human activity, man engenders a system of which he is a part, but the nature and causality of which does not coincide with his own.

The problem of man engendering a system whose nature and causality is not identical with his own is so important and surprising that it demands further elucidation. The whole: man-effect is a totality of a higher order of complexity than man himself. The causality of a whole is a function of its nature. Consequently, the causality of a totality of a higher order is not identical with that of a lower order. The situation above described may be generalized further and expressed in the form of the following law:

Law VIII: Whenever agent A produces and effect E, the resulting system S is more complex than A or E, both in its nature and its causality.

The law may provide an insight into the process of evolution in general, and an elucidation of the increase of the rate of evolution of mankind in particular. Moreover, it may explain why the effects of man's rational activity pose so many problems.

One consequence is that there is a growing disproportion between the capacity of individual knowers to know and the knowledge construct. The individual knower becomes relatively more and more ignorant of existing knowledge. Thus the progress of knowledge generates ignorance—a rather unforseen consequence of knowledge. Moreover, it renders the individual knower increasingly more dependent on the knowledge construct and on other knowers who possess knowledge which he does not master. The more knowledge there is, the less we can be independent of each other.

The central problem of our times

So far as adverse effect or otherwise of KC are concerned, let us stress at this juncture that it would be rather futile to single out as the culprit, one particular kind of knowledge, for instance, physical sciences or technology. As we are rapidly learning, the so called "soft" sciences are potentially more dangerous than the "hard" ones. The more value based and value oriented knowledge is, and the closer related its subject matter to the human subject, his nature or his behavior, the greater its potential to influence man, to help him or to harm him.

In the light of what has been just said, and in keeping in mind that knowledge engenders the need for more knowledge, it is plausible to conclude that *the central issue of our times is the relationship between man and his knowledge*. The future of mankind will largely depend on its ability to solve this problem satisfactorily.

Probably the most far-reaching consequence of the creation of culture is the exclusion of man from nature's self-regulating system and for then "How did this happen?"

The idea of the future is logically linked with the notion of progress. Progress demands effort. More progress means more individual and group effort and more subordination to the requirements which expanding knowledge and the production at progress imposes upon man.

The increasing influence of knowledge on man is a manifestation of a general and fundamental phenomenon which can be expressed in the form of a law:

Law XII: human behavior shapes human behavior.
The behavior of civilized man is more conditioned by the consequences of rational behavior than was that of primitive man. Humanity is increasingly of its own making.

The evolution of man

The intellect is a change-engendering faculty. No such faculty seems to exist in the realm of life outside of man. This is why man's relation to and role in the process of evolution is different from all other organisms.

Man has always been an open system. He has always coexisted with some level of knowledge; originally knowledge was transmitted genetically and expressed in instinctive behavior. But man was not destined to remain indefinitely on this level of knowledge. The feedback relationship between man and knowledge increased the openness of the human system. Man is a complex, multilevel structure and his openness is complex and multilevel, being both biological and psychological. He receives and transforms various kinds of energy and of information.

In animals the intake of information is mainly directed to the intake of energy and both result from and serve their biological needs. As indicated earlier, for them, energy input is greater than information input. Man however is in a different situation. For him, the information input is greater than the energy input. Moreover, while the latter is more or less constant, the former grows continuously.

The greater the KC, the more open becomes the system of man. Man's potential indeterminateness increases, thereby increasing the potential for evolution. The relationship between the openness of a system and its potential for evolving is fundamental enough to be expressed in the form of a law.

Law XX: The potential of a system for evolving is proportional to its openness.

Human evolution is not only different from biological evolution, it also proceeds at a much faster, and increasingly accelerating pace. The situation may be represented by the following relationships:

A. $\dfrac{\text{intellectual evolution}}{\text{biological evolution}} > 1 \to$ indefinitely large number ∞

B. $\dfrac{\text{biological evolution}}{\text{geological evolution}} > 1 \qquad \qquad \infty$

C. $\dfrac{\text{geological evolution}}{\text{cosmic evolution}} > 1 \qquad \qquad \infty$

D. $\dfrac{A}{B} > 1;$ E. $\dfrac{B}{C} > 1;$ F. $\dfrac{D}{E} > 1$

An interesting consequence of these relations concerns time. Biological time is the rate of biological change. It is fairly constant. This is not the case for the time proper to the KC. Ideas are generated at an increasingly rapid pace and in ever greater numbers. And they exist indefinitely long. Consequently, the ESM exists in a time different from that of the rest of organisms.

We have indicated earlier, that human progress, as we know it, may be approaching a limit. Progress is the result of the development of knowledge. It is therefore fair to say that as far as the development of knowledge is concerned mankind has reached a crucial stage. The situation bears some resemblance to that which occurred when man the gatherer and the hunter began to put the soil to systematic use and became a farmer. Then as now, the factor underlying the change was the need for greater efficiency resulting from a growing demand. Whether soil or human brains, a natural resource had to be put to a better use. The great difference between the two situations is due to the nature of the resource. While arable land exists in a finite quantity which cannot be extended beyond a certain limit, it is difficult to envisage analogical limits to the activity of the brain.

Economists speak about two types of resources: renewable and non-renewable. In fact there are three types: non-renewable, renewable and growing. The third type, of particular interest to our analysis, being the grey matter of the brain and its product—the KC. Not only the amount of grey matter in the world and the KC are growing, but more and more their growth is the result of the conscious and deliberate efforts of man. Through rational activity mankind moves to a situation of ever greater knowledge construct, more and better trained brains and more efficient, implosive intellectual cooperation. The consequences of this crucial fact are many. One of them concerns the science of economics. If the founders of this science, especially Ricardo, understood this fact and the role of the intellect in the production of wealth, liberal and Marxist theories of economy would not have developed as they did.

Conclusion

The existential system of man is gaining momentum. The system is implosive, selfenergizing and negentropic: It generates constraints, tensions, human energy and capacity to act. It grows in size, complexity and casual impact on all elements of the system, forcing man to evolve towards higher levels of rationality, consciousness, responsibility, ability to cope with complex problems, creativity, and harmonious coexistence with fellow men resulting in fruitful cooperation on an ever larger scale, and the planetary interdependence of mankind. The system has already drastically changed the conditions of man's life, and has had a profound impact on the living state in general. It will continue increasingly to exercise its evolution—engendering causality. Until now the great problem for man, at least for the most active among men, was to subordinate nature. In the future, the central problem will be to subordinate man's demiurgic powers to some higher ideal, so as to make the system compatible with nature and with man's higher vocation.

Having achieved in and through man self-reflexive and self-directive capacities, the living state will have to engage more and more in self-controlled evolution. Being the conscious element of nature, man will have to assume increasingly his responsibility as keeper of nature and helmsman of the evolutionary process.

Bibliography

Calhoun, J.B. (1971). Space and the Strategy of life. In A.H. Essr (Ed.), *Behavior and Environment: The Use of Space by Animals and Men*, Plenum Press, N.Y., pp. 329–387.

Calhoun, J.B. (1973). Revolution, tribalism and the Cheshire cat: three paths from now. *Technological Forecasting and social change.* 263–282.

Dubos, R.J. (1972). *A God within*, Scribner, N.Y.

Dubos, R.J. (1971). *Man Adapting*, Yale University Press, New Haven.

Esser, A.H. (1978). Designed communality: a synergic context for community and Privacy. In A.H. Esser and B.B. Greenbie (Eds.), Design for *Communality and Privacy*, Plenum Press, N.Y. pp. 9–49.

Gray, W. (1979). Understanding creative thought processes: on early formulation of the Emotional-Cognitive Structure Theory. *Man-Environment Relations*, Vol. 9, 1, pp. 3–14.

Gray, W. (1979). Emotional-cognitive structuralism and system-precursor theory in improving man-environment relation. In R.F. Ericson (Ed.), *Improving the Human Condition, Quality, and Stability in Social Systems*, Society for General Systems Research, Louisville, K.Y. pp. 952–960.

Koestler, A. (1967). *The Ghost in the Machine*, Pan Books, London, p. 421.

Laszlo, E. (1972). *Introduction to Systems Philosophy*, Harper & Row, New York, p. 72.

Lukasiewicz, J. (1972). The ignorance explosion: a confrontation of man with the complexity of science-based society and environment. *Transaction of the New York Academy of Sciences*, Series II, Vol. 34, 5, pp. 373–391.

Maruyama, M. (1963). The second cybernetics; deviation-amplifying mutual causal processes. *American Scientist*, 51, pp. 164–1979; 250–256.

Maruyama, M. (1972). Endogenous research and poly-ocular anthropology. In Regina E. Holloman and S. Arutiunov (Eds.), *Perspectives on Ethnicity*, Mouton, Hague.

McLean, Paul D., (1977). In the evolution of three mentalities. In S. Arieti and G. Chrzanowski (Eds.) *New Dimensions in Psychiatry: A World View*, 2, John Wiley & Sons, N.Y. pp. 306–328.

McLean, P.D. (1978). A mind of three minds: educating the triune brain. In *Education and the Brain*, Seventy-seventh Yearbook of the National Society for the Study of Education, the University of Chicago Press, Chicago, pp. 308–342.

Meyer, F. (1974). *La surchauffe de la croissance*, Artheme Fayard, Paris, 140 p.

Pattee, H.H. (Ed.), (1973). *Hierarchy Theory*, George Braziller, New York, 156 p.

Prigogine, I. (1976). Order through fluctuation: self-organization and social system. In E. Jantsch and C.H. Waddington, *Evolution and Consciousness*, Addison-Wesley Publ. Co., Reading, Mass.

Wojciechowski, J.A. (1975). The ecology of knowledge. In N H. Steneck (Ed.), *Science and Society, Past, Present and Future*, The University of Michigan Press, Ann. Arbor, MI, pp. 258–302.

Wojciechowski, J.A. (1978). Knowledge as a source of problems, can man survive the development of knowledge? *Man-Environment Systems*, Vol. 8, 6, pp. 317–324.

Wojciechowski, J.A. (1978). La crise de la culture on la crise de la rationalite? *Bulletin de la classe des sciences, Academic Royale de Belgique*, 5e serie, Tome LXIV, pp. 478–488.

Wojciechowski, J.A. (1980). Man and knowledge: one or two systems? In Bela Banathy (Ed.), *Systems Science and Science*, Proceedings of the 24th Annual Meeting of the Society for General Systems Research, Louisville, KY, pp. 427–438.

Wojciechowski, J.A. (1981). Ethical implications of medical genetics. In *Monograph*, Institute for Theological Encounter with Science and Technology, St. Louis, Mo., pp. 51–64.

Wojciechowski, J.A. (1981). The Knowledge Environment, Evolution, and Trenscendence. Nature and System, Vol. 3, 6.

Author Index

Abdulaev N. G., 394
Abdulnur S., 61
Abe K., 318, 319
Adams K. J., 217
Ahrens P. B., 334, 335
Aleksander I., 307
Alfray T., 140
Alfsen A., 87
Allen R. E., 20
Allen R. H., 140
Amin M., 344
Amir S., 352, 354
Ananiew E. V., 261
Anderegg R. J., 394
Anderson D. B., 140
Anderson W. F., 253
Anisouica A., 260
Anosov D., 111
Applebury M. L., 394, 395
Arnold V. I., 111
Arnott S., 241, 253, 383
Ashby W. R., 307
Ashida T., 394
Aspinall D., 61
Aton B., 394
Avez A., 111
Axilrod B. M., 47, 61

Babcock A. K., 307
Babloyantz A., 13, 112, 116, 119, 121, 141, 150, 316, 319
Bakowska J., 310, 319
Balasubramanian D., 62
Balinsky B. I., 308, 309, 319, 325, 333
Balis M. E., 174
Baltimore D., 260
Balischeffsky H., 227, 231
Barlow H. B., 365, 367
Barrow G. M., 346, 354
Barrow J. D., 416
Beaty B. R., 174

Beck J. S., 183
Becker B., 386, 394
Beeman K. L., 260, 261
Bender W., 260
Benedetti E. L., 171, 174
Beng H., 261
Benzioni A., 349, 354
Berge C., 306
Bernard C., 59
Bernhardt J., 341
Berns A. J. M., 261
Bernstein V. C., 277
Berridge M. J., 218
Berrill N. J., 311, 319
Bersinger T. J., 34
Berteaud A. J., 87
Bessem C., 334
Beveridge D. L., 402
Bhatia B., 356, 358, 359, 360, 361, 367
Bhaumik D., 61, 77, 184, 192, 193
Bhaumik K., 53, 57, 61, 70, 77, 184, 192
Bhise S. B., 342
Biemann K., 394
Billetter M. A., 260
Bilz H., 183
Bioly H. S., 344
Birdsall D. L., 253
Birnbaum J., 335
Bishop J. M., 260, 261
Bishop P., 368
Blair D. G., 261
Blakemore C., 367
Bockris J., 349, 354
Bodiss J., 141
Bogomolni R. A., 234
Bohr G. R., 30, 34
Boltzmann L., 14, 15, 91, 110
Bonner J. T., 310, 319
Boom J. H. Van., 253, 383
Borckmans P., 121
Borgens R. B., 287, 289

Author Index

Born M., 111
Boulton A. P., 88
Bownds D., 394
Boyd D. B., 62
Boyer P. D., 234
Braden T., 194
Breen J. J., 402
Bremermann H. J., 47
Brice M. D., 277
Bridson P. K., 253
Brinley F. J. Jr., 344
Britten R. J., 306
Brorien W. J., 261
Brout R., 69, 417
Brown D. D., 306
Brown G. H., 61
Brownell A. G., 334
Brumberger M. N., 62
Burns A. R., 394
Burny A., 260
Bush G. L., 306
Buttner H., 183
Bykov G. V., 54, 62

Calhoun J. B., 433
Callahan R., 261
Callender, 394
Campbell R. D., 316, 319
Campilo A. J., 394
Campion A., 394
Cannon W. B., 57, 59, 62
Canny M. J. P., 349, 354
Caplan A. I., 333, 334
Caplan S. R., 393, 395
Cardillo T. S., 306
Careri G., 183
Carlson S. S., 306
Casten R. G., 314, 315, 319
Cavender J. A., 307
Chadam J., 10
Chaki T. K., 192
Chan S. C.., 334
Chance B., 234
Chandrasekaran R., 253, 383
Chandraekhar, S., 69
Chaube S., 333
Chaudhry S. S., 61
Chen H. H., 68
Chen J., 261
Cherry L. M., 306
Chevallier A., 333
Chikaraishi D. M., 306
Christ B., 333

Christensen H. N., 31, 32, 33, 34
Clarke J. J., 183
Clayton R. K., 394
Clegg J. S., 78, 87, 88
Coehn A., 217
Coffin J. M., 260
Cohen J. S., 402
Cohen M. H., 312, 315, 319, 320, 321
Colbow K., 192
Cole R. K., 260
Conti R. G., 334
Cooke R., 83, 87, 88
Cooke, 309
Coon H., 333
Cooper L. N., 336, 337, 338, 339
Cooper M. S., 286, 289
Cope F. W., 33
Corbin S., 383
Cormack D. H., 334
Cotterill R. M. J., 192
Courbage, M., 111
Coustomitros C., 111
Cowan J., 148, 150
Cox R. P., 171, 174
Crawford J. L., 253
Crick F. H. C., 167, 174, 313, 319, 320
Crowe J. H., 78, 87
Curran P. J., 141

Dancis J., 174
Daniel E. E., 34
Darwin C., 14, 146, 322, 332
Das I., 142
Das M. R., 254, 260
Davidson E. H., 287, 289, 306
Davidson N., 260, 261
Davydov A. S., 176, 183, 184, 188, 192, 219
De Farco J. F., 261
De Groot S. R., 140, 141
Deamer D., 227, 231
Deeb S. S., 306
Degennes P. G., 64, 68,
Dekepper P., 121
Delbruck M., 153
DelGiudice E.. 183
Deshuchamps J., 306
Devault D., 394
Dewel G., 121
Dewette F. W., 20
Dhingra M. M., 385, 394
Dickenson W. J., 6
Dickerson R. E., 241, 253
Dina D., 261

Doglia S., 183
Donoghue D. J., 261
Doolittile R. F., 261
Dooner H. K., 306
Dorfman A., 334, 335
Dorogi P. L., 341
Dose K., 166, 174
Doukas A. G., 394
Dover G. A., 306
Drew H., 253
Driesch H., 279, 334
Drissler F., 183
Drost-Hansen W., 81, 87
Dubas E., 306
Dube S. K., 57, 63
Dubos R. J., 433
Duesberg P. H., 260, 261
Dunia I., 174
Dunyluk R. P., 192
Durston, 316
Dutrieu J., 88
Dutta-Roy B., 61, 76, 77, 184, 192, 193
Dvyatkov N. D., 183
Dwarkin M., 311, 321
Dwek R. A., 401
Dynkin E. B., 111
Dyson F. J., 412, 415, 416
Dzyaloshinskii, I. F., 61

Eawai S., 260
Eberlin J., 343
Ebrey T. G., 386, 394
Echigo A., 88
Edelmann L., 23, 24, 25, 26, 31, 33
Edelstein B B., 313, 315, 319
Edmunds Jr. L. N., 217
Eggars F., 62
Eigen M., 62, 164, 174, 235, 241, 253
El-Sayed M. A., 395
Enenkel A. C., 62
Engineer M. M., 61, 192
Englert F., 416
Engvall E., 335
Enns E. G., 183
Erdos P., 306
Ermentrout G. L., 150
Ernster B., 234
Errede B., 306
Errick J., 334
Errick J. E., 333
Eschrich W., 349, 354
Esser A. H., 433
Eyster J. M., 183

Fallon J. F., 334
Faras A. J., 260
Favera R. D., 261
Feder J., 69
Feigina M. Yu., 394
Feinberg R., 333
Feldman R., 398
Feldman R. A., 261
Fell H. B., 334
Felmister A., 344
Fensom D. M., 346, 354
Field R. J., 138, 141
Fife R. D., 314, 319
Fingh R., 334
Finnegan D. J., 261, 306
Fischinger, P. J., 261
Fogelman-Solief, F., 307
Foidart, J.-M , 334
Folsome C., 234
Forman D., 318, 319
Foster, W.A., 218
Fouvet B., 333
Fox C.F., 192
Fox S.W., 166, 174
Fraenkel A. E., 261
Frank U.F., 142
Frankel J., 285, 289, 310, 319
Franklin W. H., 69
Franks F., 87
Frey-Wyssling A., 36 38, 58, 59
Frisch H. L., 335
Frohlich H., 49, 70, 76, 77, 175, 182, 184,
  185, 192, 194, 213, 217, 286, 289
Funck T.H., 62
Furth J., 260
Furuichi Y., 260

Galanian M. D., 61
Galigher A. E., 217
Gallenshaw J. A., 261
Gallo C. F., 217
Gallo R. C., 261
Gardiner, M., 19
Gardner R.L., 307
Garnier J., 264, 277
Garrod D.R., 318, 319
Gatlin L. L., 264, 277
Gause G. F., 170, 174
Gaze R. M., 311, 319, 321
Gelfand A E., 307
Geller D., 384
Gelman E.P., 261
George C., 20, 111

Georgiew G. P., 261
Gerber G. E., 394
Gibbs J. W., 91, 111, 234
Gierer A., 121, 285, 313, 315, 320
Gilardi R., 394
Gilboa E., 260
Gilula, N. B., 171, 174
Glansdorff, P., 139, 141, 166, 174, 233
Glass L., 307
Glasstone S., 218
Goff S., 260
Goldberg R., 306
Goldbeter, A., 121, 141
Goldstein S., 111
Goles-Chacc E., 307
Gollub J. P., 10
Goodrich K., 111
Goodwin B. C., 278, 279, 284, 288, 289, 290, 316, 320
Gottikh, B. P., 384
Govil G., 396
Graf T., 261
Gray C. P., 394
Gray W., 434
Green D. E., 193, 219
Greer J., 277
Grell E., 62
Grohovsky S. L., 384
Grotthus T. Von, 218
Grunberger D., 384
Gulati J., 30, 33
Gunning B. E. S., 352, 355
Guntaka R. V., 260
Gursky G. V., 383
Gustafson K., 111

Haase A., 260
Hasse R., 140
Haberlandt G., 346, 354
Haken H., 57, 62, 70, 77, 144, 150, 218, 233, 279, 289
Haldane J. B. S., 165
Hall B. D., 333
Ham A. W., 334
Hamanaka T., 394
Hamburger V., 333
Hamkolo B. A., 174
Hammel H. T., 355
Hanafusa H., 260, 261
Hanafusa T., 261
Hanson A. W., 401
Harrison L. G., 315, 320
Harrison R. G., 325

Hartwell L. H., 217
Harvey S. C., 84, 88
Haseltine W. A., 260
Hassell J. R., 335
Hasted J. B. 183
Hayes M. B., 401
Hayman E. G., 335
Hazlewood C. F., 87
Helfrich W., 62
Helms, C., 306
Henderson, R., 394
Henin F., 111
Herliby C., 394
Hershkowitz-Kaufman M., 9
Hess B., 394
Hiernaux J., 116, 121, 141
Hijikuro N., 68
Hinchliffe J. R., 324, 333, 334
Hinton H. E., 87, 88
Ho C. 401
Hodge, A. J., 23, 33
Hoekstra P. J., 84, 87
Hogness D. S., 306
Holland B. W., 193
Holland C. J., 314, 315, 319
Honig B., 394
Horst J., 260
Hosur R. V., 396
Howk, R. S., 260
Hoyle F., 412, 416
Hseih C. L., 394
Hu S., 261
Hubbard R., 394
Hubel D. H., 336, 365, 366, 367, 368
Huggins A. K., 88
Hughes D. S., 20
Hukins D. W. L., 383
Humphreys T., 306
Hunding A., 286, 289
Hunt, 312
Hunter-Szybalska M. E., 217
Hurley J. B., 394
Huxley T. H., 21, 33

Ilyin Y. V. 261
Itakura, K., 253
Ives, D. J. G., 140

Jack A., 383
Jacob H. J., 333
Jacob M., 333
Jacobson M., 311, 320
Jaffe L. F., 287, 289

Jain M. K., 141
Jain V. K., 54, 62
Jakhar R. P. S., 342
Jantsch E., 57, 63
Jardetzky O., 401
Jaskoll T., 334
Jehle H., 62, 193
Jerka-Dziadusz M. 310, 319
Johnson D. R., 324, 333
Johnson S. C., 364, 368
Joho R. H., 260
Jonas V., 261
Jones, A. W. W., 33
Jorquera, B., 334
Julesz, B., 364, 368

Kaczmarek, L. K., 119, 121
Kakudo, M., 394
Kallay, N., 34
Kapur, R. L., 410
Karle, I. L., 394
Karle, J., 394
Katchalsky, A., 141
Kauffman, S. A., 291, 307, 316, 320
Kauzmann, W., 88
Kay, C. M., 62
Kealing, M J., 321
Keilin, D., 78, 85, 87
Kell, D. B., 183
Kelley, R. O., 334
Kenkre, V. M., 192
Kennard, O., 277, 383
Kennedy, D. W., 335
Kerlogue, R. H., 140
Kesting, R. E., 342, 343
Kestner, N. R., 61
Keydar, J., 260
Khorana, H. G., 385, 394
Kieny, M., 333
Kilmartin, J. V., 401
Kimata, K., 335
Kiselev, A. V., 394
Kito, Y., 394
Kittel, C., 68
Kleene, K. C., 306
Klug, A., 383
Knox, J. R., 401
Knox, R. S., 184
Kobayashi, K. K., 68, 69
Koestler, A., 433
Koetzle, T. G., 277
Koga, S., 84, 88

Kolias, N., 61
Kollias, N., 192
Kolpak, F. J., 253, 383
Kondiputi, D. K., 8
Konigsberg, I., 333
Koros, E., 141
Kosher, R. A., 334, 335
Koshland, D. E., 188, 192
Kothari, D. S., 6, 235, 366, 416
Koutroupas, S., 333
Kozloff, E. N., 217
Krauss, M. R., 174
Krishnamurti, D., 69
Krone, W., 306
Kropf, A., 394
Krylov, N. S., 111
Kuffler, S. W., 338
Kuhn, H., 62
Kuhn, H., 235, 253
Kumar, N., 417
Kung, M. C., 394
Kuntz, I. D., 87, 88
Kupper, B., 174

Lacalli, T. C., 315, 320
Lacroix, 288
Lagran, C., 261
Lahiri, A., 77, 192, 193
Lai, M. M. S., 260
Lakshminaranaiah, N., 344
Lama, W. L., 217
Lance-Jones, C., 333
Landau, L., 148, 150
Landmesser, L., 333
Laszlo, E., 433
Lauchli, A., 347, 355
Lavery, R., 369, 383
Lawrence, P. A., 313, 320
Lee, B., 401
Lefever, R., 13, 141
Lehninger, A. L., 231, 354
Lendle, A., 38, 57
Leonard, C. M., 335
Leslie, A. G. W., 253
Levistad, H., 355
Levitt, D., 334
Levitt, J., 82, 87
Levitt, M., 277
Levy, P. M., 68
Lewin, R., 261
Lewis, A., 394, 395
Lewis, C. A., 335
Lewis, H., 287, 290

Lifshitz, E. M., 61, 150
Linden, C. D., 192
Ling, G. N., 21, 22, 23, 24, 25, 26, 30, 33, 34, 87
Lips, W., 354
Livak, K. J., 306
Livshitz, M. A., 192
Lobanov, N. A., 394
Lockhart, C., 110, 111
Lohrmann, 238, 253
Lokeshwar, B. L., 308, 320
Longuet-Higgins, H. C., 47, 61
Lord, E. A., 416
Louvrien, R., 62
Lovelock, E., 225, 233
Lovlin, R. E., 183
Lozier, R. H., 234
Lukasiewicz, J., 433
Lurie, D., 140

Maccabe, J. A., 333
Macdonald, R., 344
Macdougall, M., 334
Mackay, J. P. G., 346, 354
Madhusudana, N. V., 69
Madin, K. A. C., 88
Maharshi Mahesh Yogi, 410
Maisel, J. E., 260
Maizel, J. V., 261
Manavalan, P., 274, 277
Marcus, M. A., 395
Marel, G. Vander, 253, 383
Margulis, L., 170, 174
Marshall, A. G., 140
Maruyama, M., 433
Marwadi, P. R., 342
Mason, W. S., 260
Masurkiewicz, J. E., 335
Mathur, S. C., 61
Mathur, S. S., 342
Matijevic, E., 350, 354
Matthews, B. W., 253
Mauger, A., 333
Maynard Smith, J., 314, 320
Mazur, P., 140
McClean, D. K., 86, 88
McClements, W. L., 261
McClintock, B., 306
McConnell, H. M., 192
McLachlan, A. D., 51, 61
McLean, P. D., 435
McMahon, D., 316, 320
McMillan, W. L., 69

Mears, P., 139, 142
Meinhardt, H., 121, 285, 289, 313, 315, 320
Melander, R., 61
Mellander, W. R., 192
Melnick, N., 334
Memmingsen, E. A., 355
Menon, M. G. K., 5
Meyer, B. S., 140
Meyer, E. F. Jr., 277
Meyer, F., 435
Milani, M., 183
Milburn, J. A., 346, 349, 354, 355
Millang, H. M., 183
Mille, M., 234
Miller, C., 34
Miller, Jr. O. L., 168, 174
Minorski, N., 121
Mishra, B. P., 140
Mishra, R. K., 1, 5, 35, 60, 61, 62, 63, 64, 70, 77, 325, 393, 410
Misner, C. W., 416
Misra, B., 10, 15, 20, 89, 111
Misra, S. N., 141
Mitchell, 191
Mitchell, P. 228, 234, 353, 355
Mitra, S., 61
Mitra, S. K., 69
Mitra, S. W., 260
Mitsui, T., 394
Mitzutani, S., 260
Miyakawa, K., 68
Moeckel, P., 62
Monod, J., 5, 13, 411, 416
Mori, H., 68
Morol, D. S., 69
Morowitz, H. J., 225, 233, 234, 235
Morris, J. G., 183
Morton, R. A., 394
Mukunda, N., 410
Mullins, L. J., 344
Munro, M., 320
Mustfa, M. G., 141

Nagle, J. F., 234
Nakamura, T., 68
Nanjundiah, V., 308, 320
Nanny, D. L., 310, 320
Needham, J., 279, 325, 333
Neet, K. E., 192
Negendank, W., 23, 30, 34
Nelson, O. E., 306
Neubaner, R. L., 261

Neumann, E., 340, 341
Newman, S. A., 84, 87
Newman, S. A., 307, 323, 333, 335
Nicholls, J. G., 338
Nicolis, G., 10, 59, 70, 77, 121, 140, 141, 192, 253
Nicolson, G. L., 344
Nihei, K., 394
Nikara, T., 368
Nilsson, J., 410
Nimtz, C., 183
Northcote, D. H., 347, 355
Northmore, B. P. M., 312, 320
Nosticzius, Z., 141
Noyakovic, L., 64, 68
Noyes, R. M., 138, 141
Nuccitelli, R. 289
Nunomura, K., 88

Ochsenfeld, M M., 24, 30, 33, 34
Oesterhelt, D., 393
Ohlendorf, D. H., 253
Okada, H., 319
Oparin, A. I., 165, 166, 174
Orban, M. 141
Orgel, L. E., 238, 253
Ortoleva, P., 9, 210
Osdoby, P., 334
Osguthrope, D. J., 277
Oskarsson, M., 261
Othmer, H. G., 315, 320
Ottolenghi, M., 394
Ovchinnikov, Y, A., 385, 394

Pachofen, B., 52
Packer, L., 141
Page, K. R., 142
Paigen, K., 306
Pain, R. H., 264, 277
Paleg, L. G., 61
Parikh, J. C., 336
Parks, W. P., 260
Partasarathy, M. V., 348, 355
Pate, E., 315, 320
Pate, J. S., 352, 355
Paton, D. M., 34
Paton, J. D., 343
Pattee, H. H., 434
Pautou, M. P., 333
Peebles, P. T., 260
Pennypacker, J. P., 335
Perle, M. A., 335
Perreault, G. J., 394

Perutz, M. F., 62
Peters, K. S., 394, 395
Pettigrew, J., 365, 367, 368
Phadke R. S., 396
Pilwat, G., 389
Pirschle, K., 35, 57
Pitaevskii, L. P., 61
Pitman, M. G., 355
Pitt, G. A. J., 394
Pohl, D. G., 194
Pohl H. A., 194 217, 287. 289
Ponnuswamy, P. K., 274, 277
Pople, J. A., 401
Popper, K., 20
Porter, S. S., 261
Prasad, K., 141
Pratap, R. V., 336
Pratt, R. M., 334, 335
Prigogine, I., 5, 10, 13, 20, 45, 57, 68, 69, 70, 77, 89, 111, 121, 132, 135, 140 166, 174, 184, 192, 233, 235, 253, 339, 410, 434
Prohofsky, E. W., 183
Pugin, E., 334
Pullman, A., 369, 383, 384
Pullman, B., 369, 383, 384

Quigley, G. J., 253, 383

Racker, E., 234
Rade, W., 62
Rahman, A., 20
Rai, B. P., 141
Ram, J., 141
Rands, F., 260
Rangachari, P. K., 30, 31, 34
Rao, G. S., 57, 60, 61
Raper, K. B., 311, 321
Rapp, P. E., 218
Rapp, U. R., 261
Rastogi, R. P., 122, 128, 137, 140, 141, 142
Ratliff. R. L., 253
Raut, U., 416
Raven, J. A., 349, 355
Reddy, A. K. N., 349, 354
Rees, M. J., 416
Rein, R., 401
Reiner, K. E., 384
Reihhold, L., 352, 354
Reisin, I. L., 34
Reiter, R. S., 334, 335
Rentzepis, P. M., 394, 395

Renyi, A., 306
Reusser, E. J., 141
Rice, S. A., 307
Rich, A., 241, 253, 383, 384
Rich, R., 58
Richards, F. M., 401
Riechert, T., 264, 265, 277
Riggs, T. R., 31, 32, 33, 34
Rius, J. L., 278, 286, 289
Roberts, G. C. K., 401
Robertson, A., 312, 316, 321
Robinson, K. R., 289
Robson, B., 264, 277
Rogers, J. R., 277
Rommens, C., 261
Roper, N. K., 62
Rose, J. K., 260
Rose, S. M., 316, 321
Rosen, D., 84, 88, 183
Rosenfeld, L., 111
Roussel, M., 261
Rowen, R. J., 393
Rowlands, S., 183
Roy, A. K., 61
Rubin, G. M., 306
Rubin, L., 334
Ruoslathti, E., 335
Ruterjans, H., 401
Ruud, B., 355

Saga, Y., 318
Sagan, C., 165, 174
Sander, K., 308, 321
Sanle, S., 261
Santella, R. M., 384
Santry, D. P., 401
Saran, A., 385, 394
Saunders, J. W. Jr., 333, 334
Savage, M., 334, 335
Schadt, M., 62
Scher, C. D., 260
Schlan, J., 260
Schlosber, H., 365, 368
Schmidt, S., 210, 218
Schmidt, F. O., 23, 33
Schneider, T., 69
Schofield, P. K., 218
Scholander, P. F., 346, 347, 355
Schrodinger, E., 7, 231
Schuster, P., 166, 174, 235, 253
Scolmick, E. M., 260
Scott-Russell, J., 224
Searls, R. L., 334

Seeman, P., 344
Segal, G. A., 121, 401
Seitz, P., 87
Selsing, E., 383
Selye, H., 57, 59, 60, 63
Sengel, P., 333
Seshachar, B. R., 410
Sewchand, L. S., 183
Shabd, R., 140, 141, 142
Shakked, Z., 383
Shaller, C., 30, 34
Shapiro, S. L., 394
Shastri, M. L. N., 361, 367
Shatkin, A. J., 260
Shedlovsky, T., 231, 234
Shepherd, J. C. W., 248, 253
Sherlock, R. A., 307
Sherman, F., 306
Sherman, F., 306
Sherman, W. Y., 393, 395
Shimanouchi, T., 277
Shindo, H., 402
Shklovskii, I. S., 165, 174
Shnoll, S. E., 62
Shrivastava, A., 61
Shukla, G. C., 57, 61, 63, 64, 69, 70, 76, 77
Shulman, R. W., 217
Shuze, A. L., 383
Silver, M. H., 339
Sinai, Ya. G., 111
Sinanoglu, O., 47, 61
Singer, M., 333
Singer, S. J., 344
Singh, A. K., 141
Singh, H. J., 141
Singh, J., 140
Singh, K., 140, 141
Singh, S. N., 140
Singh, U. N., 165, 168, 174
Sinha, K. P. 76, 411, 412, 417
Sivaram, C., 416
Slater, E. C., 234
Slavkin, H. C., 334
Smillie, L. F., 62
Smith, F. A., 349, 355
Smith, T., 231
Solursh, M., 334, 335
Spanner, D. C., 346, 355
Spanswick, R. N., 352, 355
Spiegelman, S., 260
Spindel, P., 417
Spoonhower, J. P., 394

Srivastava, M. L., 140
Srivastava, R. C., 140, 141, 342, 343
Stanier, R. Y., 170, 174
Stark, R. J., 334
Steen, T.P., 333
Stehelin, D., 260, 261
Steinberg, M. S., 316, 321
Steinberg, R. A., 38, 58
Steitz, T. A., 383
Stent, G. S., 165, 174
Stern, C., 321
Stoeckenius, W. 234 393, 394
Stoneham, M. E., 192
Strassler, C., 68
Stravenzer, E., 260
Straznicky, C., 312, 321
Stutter, E., 384
Subcasky. W. J., 343
Sudershan, E. C. G., 404, 416
Sueoka, N., 306
Sundaram, K., 262, 277
Sutherland, N. S., 366, 368
Suzuki, H., 394
Swinney, H. L., 10
Szent-Gyorgyi, A., 217
Szybalski, W., 217

Tait, J. J., 87
Takano, S., 253
Takano, T., 253
Takeda, Y., 253
Tanford, C., 401
Technkov, N. A., 261
Teller, E., 47
Temin, H. M., 257, 260
Teorell, T., 141
Terner, J., 394
Thatcher, R. W., 38, 58
Theodosopulu, M., 111
Thom, R., 71, 77, 151, 164
Thomas, H., 69
Thomas, Jr. C. A., 174
Thomas, R., 307
Thornton, C. S , 333
Thoroughgood, P. V., 334
Tickchonebko, V., 258, 261
Tien, H. T., 344
Tigyi, J., 34
Tigyi-Sebes, 24, 25, 34
Todaro, G. J., 261
Tomasek, J. J., 335
Toole, B. P., 334
Trainor, L. E. H., 284, 288, 289

Travnicek, M., 260
Treherne, J. E., 218
Trombitas, K., 24, 25, 34
Troshin, A. S., 344
Tsernoglou, D., 401
Tumanyan, V. G., 383
Turing, A. M., 9, 285, 288, 313, 314, 320, 330, 334
Tyagi, R. S., 57, 61, 62, 63

Unwin, P. N. T., 394
Upadhyaya, B. M., 140
Utsumi, K., 141

Vaadia, Y., 354
Van Beveran, C., 261
Varma, M. K., 137, 141
Varmus, H. E., 260
Venable, J. W., 289
Verghese, C. A., 359, 367
Verma, I. M., 261
Vertel, B. M., 335
Vienken, J., 290
Vincent, A., 343
Viswamitra, M. A., 383
Viswanadhan, V. N., 262
Vogel, O., 313, 321
Vogt, P. K., 260, 261
Von Der Mark, K., 334
Vonder Mark, H., 334
Vreeman, H. J., 344

Waddington, C., 279
Waddington, J. C. B., 62
Wagensberg, J , 140
Waine, J. M., 306
Wald, G., 1, 38, 58, 59, 394, 402
Walgraef, D., 121
Walker, C. C., 307
Walker, L. M., 31, 32, 33, 34
Walker, N. A., 347, 355
Walter, P., 62
Walton, C., 34
Wang, A. H. J., 357, 383
Wang, L. H., 260
Wardlaw, I. F., 348, 355
Warner, A. H., 88
Warner, D. T., 82, 87
Waser, J., 253
Waston, K., 260
Waugh, M., 333
Weatherley, P. E., 346, 354
Webb, S. J., 82, 87, 183, 192, 287, 290

Webster, G. C., 279, 290
Weinberg, S., 10, 411, 416
Weinstein, J. B., 384
Weisbuch, G., 307
Weisel, T. N., 336, 365, 366, 367, 368
Weiss, R. A., 260
Weissman, C., 260
Wetlaufer, D. L., 62
White, T. J., 306
Wickramsinghe, C., 416
Wien, W., 83, 88
Wiener, N., 124
Wigner, E. P., 415, 416
Williams, D., 260
Williams, G. J. B., 277
Williams, R. J. P., 193
Wilson, A. C., 306
Wilson, E. B., 286, 290, 321
Wilson, J. T., 231
Winfree, A. T., 210, 218
Winkler, R., 253
Wintzel, H., 401
Wireman, J. W., 311, 321
Witt, H. T., 234
Wojciechowski, J. A., 417, 434
Wolf, V., 306
Wolpert, L., 121, 147, 150, 285, 287, 290, 309, 312, 313, 321, 335
Wong, A. K. C., 264, 265, 277
Wong-Staal, F., 261

Woude, G. F., 261
Wourms, J. P., 321
Wright, K. L., 192
Wu, H. M., 192
Wyckoff, H. W., 401
Wyhof, J. R., 217

Yadav, S., 343
Yadava, K., 141
Yamada, K. M., 335
Yanagisawa, K., 319
Yato, L. D., 140
Yeoh, G. C. T., 335
Yntema, C. L., 333
Yoon, M. G., 312, 322
Yos, M. T., 62
Yoshimura, M., 415, 416
Yoshizawa, 394
Yosida, K., 111
Young, M. W., 306

Zakrzewska, K., 383
Zalyubovskaya, 183
Zasedatelev, A. S., 383
Zeeman, E. C., 313, 322
Zhdanov, V. M., 258, 261
Ziegler, H., 350, 355
Zimmer, C. H., 383
Zimmerman, U., 287, 290
Zsigmondy, R., 348, 355
Zwilling, E., 333, 334

Subject Index

Accessibility, 370, 371, 381, 382
Acetylcholine Receptor-Channel, 340
Activation Deactivation, 177, 178
Active Transport, 131, 132
Adaptation, 165, 238
Adsorbed State, 23, 25, 26, 27
Adult State, 308
Affinity, 125, 129, 130
Aggregation, 240, 241, 244, 252, 310
Allomorph, 369, 371, 372, 373, 374
Alpha Helical Protein, 219, 220, 221, 223, 224
Ametabolic, 85
Amphiphilic, 229, 230, 232
Antibiotic, 380
Anticodon, 241, 243, 244, 246
Antipsychotic Drug, 342, 343
Apical Ectodermal Ridge, 328, 329, 330
Arrow of Time, 13, 14, 15, 19, 89, 90, 95, 413, 415
Artemia Cyst, 78, 79, 80, 81, 82, 83, 84, 85, 86
Association, 22, 23, 25, 35, 39, 48, 226, 337
Association Induction Hypothesis, 21, 22, 23, 26, 27, 31, 32
Asymmetry, 324, 325
Atmosphere, 225, 226, 227
Attractor, 152, 153, 154, 156, 164, 300, 303
Autocatalytic Process, 8, 113
Autocatalytic Reaction, 135, 136, 140
Autocooperative Adsorption, 27, 29

Bacteriorhodopsin, 385, 386, 392
Bakers Transformation, 103
Baryon, 415
Belousov-Zhabotinskii Reaction, 136, 137, 138
Belousov-Zhabotinskii Reagent, 136
Bifurcations, 71, 73, 112, 115, 116, 117, 118, 119, 121, 148, 149, 152, 154, 286

Bilayer, 342, 343
Bioelectric Singnal, 340
Bioenergetics, 219, 223, 227, 228, 229, 230
Biogenesis, 227, 228, 229
Biological Essentiality, 35, 38
Biosphere, 225, 226, 227
Biphasic Oscillation, 119, 120
Black Body Radiation, 7
Black Lipid Membrane, 342
Blastula, 159, 160
Boolean Network, 299
Bose Condensation, 184, 185
Bose-Einstein-Like Condensation, 52, 213
Boson, 49, 50, 51

Canonical Dynamical Transformation, 7
Carcinogen, 369, 381, 382
Cardinal Adsorbent, 23, 27, 29
Cardinal State, 23
Catalytic Molecule, 166
Catastrophe, 154, 156
Cause, 406, 408, 410
Cellular Differentiation, 153, 159
Cellular Slime-Mold, 310, 311, 318
Cellular Spin Resonance, 213
Chance, 411
Chaos, 6, 7, 9, 120
Chemical Clocks, 6
Chemical Potential, 125, 185, 186, 226
Chemical Wave, 136, 138, 184
Chemiosmosis, 191
Chirality, 240
Choice, 404, 409, 415
Chordrogenesis, 329
Chordroitin Sulfate, 329
Chromophore, 231, 232
Ciliates, 309
Cis-Acting Gene, 292, 293, 296
Cndo-2 Calculation, 398
Codon, 241, 243, 244, 246

Subject Index

Coherence, 7, 404, 406, 407, 409, 410, 411, 415
Coherence Bose-Einstein Like, 54
Coherent, 146, 147, 211
Coherent Behaviour, 6
Coherent Excitation, 175, 176, 177, 185
Coherent Mode, 178, 179
Coherent Oscillations, 287
Coherent Patterns, 6
Collagen, 329
Collective Excitation, 180, 220
Collective Modes, 146, 147
Collision, 16, 17, 19
Colloid, 198
Combinatorial, 302, 303, 304
Complexity, 249
Conceptual Knowledge, 426
Conductivity, 198, 200
Conformation, 240, 241, 245, 263, 341, 369, 380, 388, 389, 390, 391, 393, 400, 401
Connectivity, 120
Connectivity Number, 116
Consciousness, 402, 403, 411, 415, 419, 431
Control Parameter, 146
Convection Patterns 7
Convolution, 240, 241
Cooperative Interaction, 27, 45
Cooperative Phenomenon, 213, 235
Cooperative Transformation, 44
Cooperativity, 184, 186, 191, 246
Correlations, 16, 17, 19, 20, 38, 102, 106, 110, 191, 337
Cosmological Model, 108, 109
Cosmology, 107, 108, 406, 410
Coupled Reactions, 129, 130
Coupling, 128, 129, 130, 131
Creativity, 409, 431
Critical Phenomena, 184
Cryptobiosis, 78, 82, 85
Cryptobiote, 78, 79
Cryptobiotic Cells, 81
Curie's Principle, 128
Cybernetic, 149
Cyclic Group, 288

Darwinian Evolution, 166
Darwins Principle, 146
Degree of Curvature, 361
Deterministic, 91, 103, 165, 227, 233, 300
Dielectric Constant, 82, 84, 189, 195, 199 202, 207, 208, 213, 215, 216, 229, 230

Dielectric Measurement, 83, 86
Dielectric Property, 176, 177, 179, 184, 223
Dielectrophoresis (Dep), 194, 195, 197, 199, 204, 213
Differentiation, 292, 303 304, 327, 329
Dipole Oscillator, 45, 47, 177, 188, 198, 215, 286
Disorder, 15, 16, 146, 227, 230
Dispersion, 293
Disribution Function, 91, 92, 99, 101, 108 110, 337
Dissipation, 7, 124, 132, 133
Dissipative Structure, 166, 236
Distamycin, 380
DNA, 369, 370
DNA-A, 371, 372, 373, 379
DNA-B, 371, 372, 373, 379, 381, 382
DNA-C, 371, 372, 373, 379
DNA-D, 371, 372
DNA-Z 371, 372, 373, 379, 381, 382
Drosophila, 182, 304, 316
Dynamic Equilibrium, 21, 31,
Dynamical Order, 291, 304, 305
Dynamical System, 16, 90, 91, 92, 93, 94, 95, 97, 101, 102, 103, 148, 301, 404
Dynamical System Interactions, 19
Dynamical System Motions, 110
Dynamical System Processes, 16, 19, 100
Dynamical Unfolding, 407

Ecology, 225, 226
Ectoderm, 158, 150, 161, 162, 163, 326, 327
Effect, 405
Electroosmosis 346
Electrical Double Layer, 345, 347, 348, 349, 350
Electro Kinetic Potential, 345
Electrochemical Gradient, 350, 385
Electrochemical Potential 131, 132, 231
Electron Microscopy, 24, 25, 31
Electron Transfer, 223, 224
Electron Transmission, 24
Electrosoliton, 224
Elementary Excitations, 65
Electron Transport, 231
Elongation Factor, 168
Embryo, 309
Embryogenesis, 278, 330
Embryology, 150, 161, 162
Embryonic Development, 323
Embroynic State, 308
Entoderm, 158, 159, 160

Subject Index 447

Entropy, 13, 14, 20 22, 42, 89, 90, 92, 94, 102, 110,
Entropy, 124, 126, 127, 128 129, 132 133, 153, 265, 267
Enzyme, 39 40, 53, 54, 57, 80, 85, 86, 130, 165, 186
Enzyme, 188, 254, 396, 413
Epileptic Seizure, 118, 119 120, 148
Equilibrium, 6, 7, 8, 18, 19, 22, 38, 39, 122, 123, 230
Ergodic, 107
Essential Element, 35, 37, 38
Evolution, 14, 15, 18, 19, 20, 110, 235, 237, 249, 250, 251, 252, 257, 262, 274, 278, 291, 292 299, 322 403, 404, 411, 412, 417, 426, 429, 430, 4?2
Evolution Dynamical, 89, 90, 91, 93, 95, 97, 405
Evolution Thermodynamic, 92, 93
Exciton, 50, 55, 184, 220
Existential System, 419, 420, 421, 422, 430, 431
Expanding Universe, 107
External Pump, 185, 186

Far from Equilibrium Systems, 13, 40, 57, 90, 124, 144, 166, 236, 305
Fibronection, 329, 330, 331, 332
Field-Particle Complimentarity, 279
Fluctuation, 7, 9, 40, 45, 71, 73, 146, 147, 148, 189, 227
Flux, 128, 129, 139, 225
Forcing Structure, 301, 302, 305
Frozen Cell, 31
Functional Organization, 166, 167, 168, 170

Gap Junctions, 171, 172, 173
Gating System, 340
Gene Expression, 291, 292, 330
Genetic Information, 248, 249
Geodesic, 107, 108, 109
Geosphere, 225, 226
Giant Dipole, 185, 187, 189
Gibbs Equation, 125, 127, 135
Global Cycle, 226, 227
Global Potential, 374
Gluon, 414
Glycosaminoglycan, 329
Gravitation, 20
Gravitational Field, 8
Groove-Major, 372, 373 375, 376, 377, 378, 379, 380, 381

Groove-Minor, 372, 373, 375, 376, 377, 378, 379, 380, 381

H+Transport, 345, 349, 350, 351, 352, 353
H—Function, 90, 94, 95, 96, 108
Hadrons, 414, 415
Hamiltonian Distribution, 92
Helical Axis, 371, 372, 373
Helix-Coil Transition, 190
High Energy State, 22
Histidine, 396, 398, 400
Holoblastic, Cleavage, 281, 282, 284, 286, 287, 288
Hyaluromic Acid, 329
Hyaluronate, 331
Hydrodynamics, 7, 8, 9,
Hydrogen Bond 397
Hypercycle, 166, 167, 168, 169, 170
Hysteresis, 197
Hysteresis Loop, 158, 161

Induction, 22
Information, 1, 40, 228, 236, 251, 262, 312, 330, 336, 408, 411, 413
Information Content, 266, 267, 274, 412, 415
Information Density, 265, 267, 274, 275, 276
Information Theory, 264, 276, 411
Informational Molecule, 166, 236
Inharmonic Force, 188
Initial Condition, 91
Instability, 7, 9, 15, 46, 57, 70, 71, 91, 94, 107, 109, 110, 134, 136, 139, 146, 188
Instability Benard, 7, 8
Instability Taylor, 8
Intercalation, 245
Intercellular Communication, 309
Intercellular Junctions, 170
Interfacial Water, 81
Intermolecular Forces, 6
Internal Pump, 40, 57
Irreversibility, 6, 9, 13, 15, 20, 90, 91, 94, 106, 109
Irreversible Processes 6 14, 19, 93, 110, 126, 128
Irreversible Thermodynamics, 122

Jamin's Chains, 346, 347, 348

K+Transport, 349, 350, 351, 352, 353, 354

K—Flows, 94, 102, 103, 104, 105, 107, 108,
K—Layer, 350, 351, 352
K—Partition, 104, 105, 106
Knowledge, 251, 417, 418, 424, 425, 428, 430
Knowledge Construct, 418, 420, 422, 423, 424, 426, 428, 429, 431
Knowledge-Life Relationship, 418

Laser, 145, 146, 147. 175, 170, 186
Laser Microscopic Mass Analysis, 25
Laser-Raman Effect, 179, 180
Learning 236, 237, 239, 249, 251, 252
Lepton, 415
Liapounov Function, 134, 153
Linear Stability Analysis, 114, 137
Liouvillian Distribution, 92
Liouvillian Operator, 96, 109
Liquid Crystal, 54, 64, 65, 66, 67, 68, 190, 325
Liquid Helium, 186
Lithosphere, 225, 226
Living State, 1, 5, 21, 22, 27, 35, 40, 41, 44, 54, 78, 122, 136, 184, 225, 262, 322, 354, 411, 417, 419
Living Systems, 5, 13, 126, 136
Localised Excitation, 186, 221
Long Range Correlation, 6, 184, 185, 189, 192
Long-Range Coherence, 286
Long-Range Order, 279
Lorentz Gas, 104 106
Lotka-Volterra Model, 135
Limit Cycle, 135, 136, 138, 139, 158, 178
Lyotropic Liquid Crystal, 41, 42, 43, 44

Mammalian Smooth Mnscle, 30
Markov Entropy, 265. 269, 274
Markov Process, 90, 92, 93, 94, 95, 96, 110
Markov Process Semi Group, 93, 97, 98, 100, 105
Mass Transfer, 349
Master Equation, 93, 98
Mesectoderm, 159, 160
Membrane, 21, 23, 26, 170, 171, 190, 191, 228, 230, 231, 235, 252, 340, 342, 343, 345, 385
Membrane Liquid, 342, 343
Membrane Oscillation, 136, 139
Membrane Pump, 21, 25
Mesoblast, 328

Mesoderm, 158, 160, 161, 162
Metabolic Network, 228
Metabolic Activity, 85, 123, 133
Metabolic Cooperation, 171, 172
Metabolic Process, 349
Metabolic Transformation, 78
Metastable State, 64, 176, 178, 179, 188, 189, 340
Metastable Ferroelectric State, 176, 177, 186, 187
Metazoan, 291
Micro-Dielectrophoresis (U-DEP), 194, 195, 197, 199, 200, 201, 202, 203, 204
Micro-Dielectrophoresis (U-DEP), 205, 206, 207, 209, 215, 216
Microheterogeneity, 369, 370
Microwave, 180, 181, 182, 287
Mm Wave, 181, 182
Molecular Conformation, 38, 45, 48, 54
Molecular Electrostatic Potential, 370, 371, 372, 382
Morphology, 151, 152, 207, 208, 209, 278, 279, 288
Morphogen, 113, 285, 313, 314, 315, 331
Morphogensis, 9, 151, 152, 156, 185, 278, 279, 281, 285, 286 287, 288, 313, 322
Morphogenetic Field, 112, 279
Multi Layer Polarization, 26, 27, 31
Multicellular System, 113, 170
Multiple Unit System, 112
Multiunit Assembly, 113, 121
Muscle Cells, 23, 24, 25
Muscle Contraction, 219, 221, 222
Muscle Frog, 30
Muscle Uterine, 30
Mutation, 236, 252
Myosin, 222

Netropsin, 380
Neural Induction, 161
Neural Network, 336
Neural Tube, 163
Neuron, 336
NMR, 82, 83, 398, 400, 401
Non Equilibrium, 7, 8, 9
Non Equilibrium Condition, 7
Non Equilibrium State, 134
Non Equilibrium Structure, 8, 9, 136
Non Linearity, 7, 20, 40, 175, 176, 186, 187, 285, 286
Non Commutativity, 406, 407
Non Competitive Inhibition 130

Subject Index 449

Non Equilibrium Theory, 226
Non Linear, 8, 107, 108
Non Linear Differential Functions, 113
Non Linear Systems, 20, 114
Nucleoprotein Complex, 169, 170, 172
Numerical Integration, 114

Oculoegocentre, 362
Oncogene, 257
Onsager Reciprocity, 139
Open System, 122, 123, 124, 125, 126, 136, 186
Order, 15, 16, 22, 146, 147, 148, 154, 227, 230, 287
Order Parameter, 146, 148, 279, 280, 283, 285, 286
Oregonator, 137
Organization, 9, 151, 122
Organization Spatial, 122
Organization Temporal, 122
Origin of Life, 231, 232, 235, 236, 238
Osmotic Activity, 26, 27
Osmotic Equilibrium, 27

Pattern Formation, 113, 327, 278, 311, 312, 318, 329
Pattern Recognition, 148
PCILO, 385, 388
Perception, 402, 404, 409
Perceptual Constancy, 360, 361
Phase Space Trajectory, 91, 92, 104
Phase Transition, 8, 44, 53, 67, 68, 148, 149, 184, 189, 190, 278, 279, 280, 286, 287, 413, 414
Phenomenological Coefficients, 128, 131, 132
Phloem, 348, 352
Photoreaction Cycle, 388, 387, 388, 392
Photosynthesis, 191
PK Value, 396, 397, 398, 400, 401
Poincare Group, 406
Polarisability, 83, 194, 197, 198, 207, 229
Polarisation, 176, 177, 178, 186, 187, 188, 195, 231
Polymorphism, 369
Positional Information, 312, 313, 316, 317, 318
Potential-Line, 370, 371
Potential-Point, 370
Potential-Plane, 370
Potential-Radial, 370
Potential-Surface Envelope, 370, 371, 374, 378, 379, 380

Pre-Biotic, 166, 231, 244, 262
Predator, 154, 157, 158
Prey, 154, 157, 158, 161
Probabilistic Model, 316, 317, 318
Probabilistic Process, 16, 19
Probability, 8, 14, 89, 91, 92, 94, 96, 105
Projection Operator, 93, 96, 99, 100
Proteinoid, 166
Proton Transport, 230
Protoplasmic Streaming, 354

Qβ Replicase, 168
Quantum Theory, 408
Quark, 414

Radio Frequency Oscillation, 194, 197, 203, 206, 210, 211, 214, 216
Raman Study, 388, 393
Random Directed Graph, 296
Randomness, 13, 94, 109
Reaction Diffusion, 112, 114, 116, 152, 184, 286, 287, 314, 331
Reactive Site, 386
Receptor, 165
Recognition, 185, 188, 189
Regeneration, 310
Regulation, 291, 308, 309, 312, 317, 318
Regulatory Architecture, 295, 299
Regulatory System, 292, 299, 303, 305
Relaxation, 198
Relaxation Time, 82, 83
Replicase, 246, 247
Retina, 311, 312
Retrovirus, 257, 258
Reverse Transcriptase, 254, 258
Ribonucleases, 398
Riemann Metric, 153
Robertson Walker Metric, 107

Sap, 345, 346, 347, 348, 352
Scale-Invariance, 311, 312, 314, 315
Scaling Mechanism, 356, 357, 365
Secondary Interaction, 396
Selection, 236, 238, 240, 244, 246, 252, 304
Selection Principle, 110
Selection Attraction, 177, 178, 179
Self Assembly, 278
Self Focusing, 179
Self Organization, 9, 13, 40, 57, 70, 112, 117, 118, 120, 121, 146, 145, 136, 147, 148, 232, 233, 235, 236, 251, 291
Self-Organizing-Systems, 165, 166, 168, 279, 431

Sensation, 402
Sense Knowledge, 426
Sentient Structure, 415, 416
Separability, 405, 406
Sigmoidal Rate Equation, 305
Singular Perturbation Theory, 158
Singularity, 154
Singularity Cusp, 156
Singularity Parabolicumblic, 162
Size-Distance Invariance, 365
Slaving, 146, 147
Solitary Wave, 187, 188
Soliton, 176, 177, 178, 180, 184, 219, 220, 221, 222, 223, 224
Soluble Pathways, 81
Spatial Hypersurface, 107
Spatial Order, 285
Spatial Organization, 145, 279, 281, 286, 287
Spatio Temporal Oscillations, 136, 138
Spatio Temporal Pattern, 118, 147
Spatio-Temporal Order, 166
Spectral Function, 99
Stability, 7, 46, 48, 70, 72, 75, 114, 116, 118, 134, 135, 153
Stability Criterion, 139
Stable Focus, 115
Steady State, 115, 116, 123, 124, 132, 186, 226, 302, 305, 412
Stochastic, 97, 107, 165, 295
Structural Stability, 155
Structural Gene, 293, 296
Structural Organization, 166, 167, 168, 169, 170, 236, 237, 248
Structure, 7, 8, 78, 79, 235
Structure Dissipative, 8, 90, 112, 236
Structure Loose, 45, 47, 48, 49, 52, 188

Superposition, 404, 406, 407
Symmetry Breaking, 9, 15, 16, 18, 19, 70, 76, 91, 92, 93, 94, 95, 98, 100, 105, 109, 110, 156, 157, 279, 285
Symmetry Breaking Time Reversal, 90
Synergetics, 57, 145, 147, 148, 149

Taxonomy, 288
Tectum, 311, 312
Temporal Order, 17, 134
Temporal Organisation, 281
Temporal Oscillation, 136
Temporal Structure, 147
Teorell's Membrane Oscillator, 136, 139
Thermotropic, 44
Time, 89, 98, 101, 107, 109, 184
Time Internal, 89, 94, 98, 99, 100, 101, 107, 108, 109
Trans Cis Isomerization, 386, 387, 392, 393
Trans-Acting Gene, 292, 293, 296
Transcription Translational Complex, 168
Translocation, 345, 346, 348, 349, 386, 387
Transposition, 293, 295, 299

Universe, 411, 412, 415
Unstable Branch, 159

Vertebrate Limb, 323, 324, 325, 326, 327, 329
Vesicle, 229, 230, 231, 232, 233
Vicinal Water, 81, 82
Visuoperceptual Space, 356, 358, 359, 360, 361, 362, 364, 365, 366

Water-Replacement Hypothesis, 82, 83